Recent Development in Novel Green Asphalt Materials for Pavement

Recent Development in Novel Green Asphalt Materials for Pavement

Editors

Qian Chen
Xiaolong Sun
Tao Wang
Guoqiang Sun

MDPI

Basel • Beijing • Wuhan • Barcelona • Belgrade • Novi Sad • Cluj • Manchester

Editors
Qian Chen
Chang'an University
Xi'an
China

Xiaolong Sun
Guangdong University
of Technology
Guangzhou
China

Tao Wang
Beijing Jiaotong University
Beijing
China

Guoqiang Sun
Beijing University
of Technology
Beijing
China

Editorial Office
MDPI AG
Grosspeteranlage 5
4052 Basel, Switzerland

This is a reprint of articles from the Special Issue published online in the open access journal *Coatings* (ISSN 2079-6412) (available at: https://www.mdpi.com/journal/coatings/special_issues/OM5QD3U0L8).

For citation purposes, cite each article independently as indicated on the article page online and as indicated below:

Lastname, A.A.; Lastname, B.B. Article Title. *Journal Name* **Year**, *Volume Number*, Page Range.

ISBN 978-3-7258-1759-7 (Hbk)
ISBN 978-3-7258-1760-3 (PDF)
doi.org/10.3390/books978-3-7258-1760-3

Contents

About the Editors

Qian Chen

Qian Chen is a lecturer of road engineering at the School of Highway, Chang'an University. He was selected to be a part of the Young Key Talents Program of Chang'an Scholars. He is a key member of Key Scientific and Technological Innovation Teams in Shaanxi Province. He has won the first and second prizes of three provincial and ministerial scientific research awards. He has published more than 20 SCI/EI academic papers.

Xiaolong Sun

Xiaolong Sun is an associate professor, science and technology talent of the China Highway Construction Industry Association, part of the outstanding Science and Technology Faculty in Civil Engineering and Architecture in Guangdong province, a Guangdong science and technology commissioner, a Guangdong intellectual property expert, and a visiting scholar at Texas State University Austin, USA. Xiaolong Sun has published over 30 academic papers and granted over 20 patents. Also, he has been the recipient of scientific and technological awards, such as the Guangdong Provincial Technology Invention Award, China Highway Society Science and Technology Award, China Municipal Engineering Association Science and Technology Award, Guangdong Civil Engineering and Architecture Society Science and Technology Award, and Guangdong Mechanical Society Technology Invention Award.

Tao Wang

Tao Wang is an associate professor of Beijing Jiaotong University. He is interested in self-healing asphalt materials; high-viscosity modified asphalt material; bio-asphalt material; phase change asphalt material; recycled asphalt pavement; conductive ultra-thin wearing course; and asphalt aging and anti-aging.

Guoqiang Sun

Guoqiang Sun is a professor within the Beijing University of Technology. He is mainly engaged in teaching and research in the field of smart low-carbon pavement. He teaches several undergraduate/graduate courses, such as "Asphalt and Asphalt Mixtures", "Road Construction Material", and "Pavement Maintenance and Management Technology". He has led 10 projects at various levels. He has published more than 60 publications and applied for and been authorized for more than 10 patents. He has served as a young Editorial Board Member of several Chinese and English journals and he has been a reviewer of more than 20 journals/conferences.

coatings

MDPI

Article

Experimental Study on Mechanism, Aging and Fatigue Performance of Warm Mixing Speed Melting SBS Modified Asphalt Binders

Yazhou Zhuang [1], Jinchao Yue [1], Bo Men [2], Guoqi Tang [3] and Riran Wang [1,*]

1 Yellow River Laboratory, Zhengzhou University, Zhengzhou 450001, China
2 Henan Xuxin Freeway Co., Ltd., Zhumadian 463000, China
3 Guolu Gaoke Engineering Technology Institute Co., Ltd., Beijing 100083, China
* Correspondence: wangrr@zzu.edu.cn

Abstract: In this study, two kinds of quick melting modifier SBS-T and SBS-W, as well as the traditional SBS modifier, were used in the optimization design of asphalt binders. The changes in material structure and fatigue properties of three polymer-modified asphalt after adding 3% Sasobit to warm mix agent were investigated. The feasibility of SBS-T and SBS-W in asphalt binder was discussed from the modification mechanism and fatigue properties. In order to reveal the modification mechanism, the interaction mechanism between the fast-melting SBS modifier and the base asphalt was characterized by Fourier transform infrared spectroscopy (FTIR) and X-ray diffraction (XRD). The temperature sensitivity and viscoelastic properties of SBS-T and SBS-W modified binders were determined by frequency scanning (FS). The fatigue properties of SBS-T and SBS-W modified binders were tested by linear amplitude scanning (LAS). The results of FTIR showed that there was no chemical reaction between the SBS-T and SBS-W and the base asphalt. XRD results showed that SBS-W-modified asphalt has stronger fluidity. The results of FS and LAS showed that the asphalt binder with Sasobit has good stiffness and elastic recovery ability, and the same SBS-T and SBS-W have better temperature sensitivity and deformation resistance. In addition, the fatigue life of asphalt binder under the linear viscoelastic continuous damage theory is increased 3.9 times by SBS-W.

Keywords: warm mixed asphalt; speed melting type; viscoelastic characteristic; performance of fatigue; temperature-reduced production and paving of asphalt mixtures

Citation: Zhuang, Y.; Yue, J.; Men, B.; Tang, G.; Wang, R. Experimental Study on Mechanism, Aging and Fatigue Performance of Warm Mixing Speed Melting SBS Modified Asphalt Binders. *Coatings* **2023**, *13*, 311. https://doi.org/10.3390/coatings13020311

Academic Editor: Valeria Vignali

Received: 15 December 2022
Revised: 19 January 2023
Accepted: 20 January 2023
Published: 30 January 2023

1. Introduction

As a kind of advanced pavement with superior performance, asphalt pavement occupies a high proportion in the world of road engineering, it is the general trend to choose the material with excellent performance for asphalt pavement. At present, the scale of freeway construction is increasing day by day, and the modified asphalt can greatly improve the service life of asphalt pavement, and become the necessary material for high-grade road construction [1]. In recent years, the average annual use of asphalt pavement is more than 6 million tons. Since the late 1990s, because SBS (styrene S-butadiene B-styrene S block copolymer) modified asphalt has better performance, gradually become the most used modified asphalt type, accounting for more than 90%. As the common SBS modifier has a high melting point and is not easily dissolved and dispersed, SBS-modified asphalt is generally produced by the wet process invented abroad more than 40 years ago. However, with the deepening of the application, many defects of technology and management are gradually exposed, and environmental protection and safety problems have attracted widespread attention [2]. Wet SBS modification technology not only has high energy consumption, but also has technical defects such as segregation and thermal decomposition, leading to a series of quality management problems [3,4].

If SBS can be used separately from asphalt, SBS can be directly put into the asphalt mixing plant for use, their storage will not deteriorate, their quality and dosage are transparent, and all problems will be solved. This method is the dry direct injection modification technology. Dry method is not a fresh word in fact, it has been throughout the history of the development of modified asphalt, Sweden as early as the 1960s developed asphalt concrete mixed with rubber particles dry production process, rock asphalt, hard petroleum asphalt, and other dry process has been a large number of applications. Because the processing link in the modified asphalt plant is canceled by the dry process, the energy consumption is much less than that produced by the wet process to produce the same amount of modified asphalt [5]. Therefore, the comprehensive promotion of dry modification technology is not only required by the current situation of the industry, but also in response to the call of the state to create technical support for the industry green highway. Guolu Gaoke Engineering Technology Institute Co., Ltd. has developed the first high-performance, low melting point, micron-level rapid melting SBS modified material (SBS-T) in China: the rapid melting SBS modifier is directly put into the asphalt mixing building for use, so that it can be quickly melted in the normal mix, and directly produce SBS modified asphalt mixture.

Generally, the modified asphalt mixture needs to achieve a better modification effect when the stirring temperature is 175–185 °C. This kind of high-temperature mixing not only seriously increases the degree of asphalt aging, but also leads to a serious increase in energy consumption costs, nitrogen and sulfide and other harmful soot emissions increased significantly [6,7]. In order to solve the problem of high construction temperature of modified asphalt mixture, a variety of asphalt warm mixing technologies emerge at the historic moment, which can reduce the construction temperature by 20–30 °C and still can be normally compacted, which is promoted in the industry as an important technical measure of energy saving and emission reduction. Sasobit, manufactured by Sasol-Wax, is an organic warm thermal additive that has been proven to reduce the viscosity of asphalt adhesives [8–11]. Under the background of national carbon neutrality strategy and green highway development of the Ministry of Transport, Guolu Gaoke Engineering Technology Institute Co., Ltd., (Beijing, China) based on the research and development system of dry SBS modifier, combines small molecule SBS with organic viscosity reduction technology. A warm mix dry SBS modifier (SBS-W) was successfully developed, and the integration of SBS modification and the warm mix was realized for the first time in the industry. When in use, SBS-W can be used as an admixture directly into the asphalt mixing building to produce a modified asphalt mixture. Mixing at 160 °C can be fully melted and achieve the modification effect. While reducing the production temperature and spreading temperature of the modified asphalt mixture to nearly 20 °C, the produced asphalt mixture still reaches a similar level to SBS modified asphalt mixture. There is no need to adjust the rolling process. At the same time, the warm mix dry SBS modification technology inherits a series of advantages in the dry SBS technology system, such as "avoiding the performance attenuation problem of modified asphalt, avoiding the ratio phenomenon of modified indicators, reducing the difficulty of quality supervision of modified asphalt mixture", which meets the functional requirements of warm mix and improves the life of asphalt pavement [12].

Asphalt pavement in service will be due to the cyclic action of vehicle load will reduce the strength. When the vehicle load is cycled to a certain number of times, it will make the asphalt pavement crack and cause fatigue failure, so the asphalt fatigue characteristics play a vital role. Daniel has conducted a large number of tests, which show that the viscoelastic damage characteristics of asphalt mixture do not depend on loading conditions, and each material has its specific Damage Characteristic Curve (DCC), which can be determined by simple monotone loading test [13]. Later, scholars such as Underwood have reasonably Simplified the Viscoelastic Continuum Damage (VECD) model (S-VECD) so that it can better analyze the dynamic fatigue test results of asphalt mixture. Thus the asphalt pavement fatigue performance can be evaluated and predicted more accurately [14]. The whole test process is about 5 min, and the test efficiency is high. It can be used to estimate the fatigue life of asphalt binders at any strain level [15–17].

Hence, this paper has conducted a comprehensive study, in order to select the appropriate modifier to bring excellent performance of asphalt binder to provide guidance. The purpose of this work is to evaluate the modification effects of two fast-melting modifiers (SBS-T and SBS-W) and traditional SBS on asphalt binder, as well as their combined modification effects with Sasobit. To be specific, this work mainly focuses on the following aspects. First, the microscopic characteristics of SBS/SBS-T/SBS-W modifiers and modified binders were characterized by XRD and FTIR infrared spectroscopy to determine the chemical composition and microstructure of the modifiers and further reveal the modification mechanism. Then, the viscoelastic parameters of different kinds of modified asphalt binder were analyzed by FS frequency scanning test, WLF equation and CAM model. Finally, the LAS fatigue acceleration test was carried out by DSR, and the fatigue damage resistance of various modified asphalt in base asphalt binder was studied based on VECD theoretical model.

2. Theoretical Background

2.1. Linear Viscoelastic Theory

Asphalt is a characteristic viscoelastic material. In the range of linear viscoelastic (LVE), its rheological properties are meaningfully impacted by loading conditions, such as temperature and frequency. Whereas, the rheological properties of asphalt binder under disparate loading conditions are often prolonged and labor-intensive. Therefore, the researchers developed a predictive model to measure the rheological arguments of asphalt at random temperatures and frequency, and then used the constitutive equation to calculate the rheological properties.

These models need to apply the time-temperature superposition principle (TTSP). TTSP, known as the time-temperature zoom and frequency-temperature duplication principle, is a very resultful instrument to evaluate the rheological properties of asphalt. TTSP method was used to select the referable temperature T_r, and a continuous curve was obtained by translation fitting. The obtained data were the sweep test data at diverse temperatures in the orientation of the referable temperature. The displacement argument is requested to institute the main curve at det temperature, which is called the temperature displacement factor φ_T. In this paper, the classical WLF nonlinear equation is used to fit the temperature displacement factor φ_T in TTSP.

$$\lg\varphi_T = \frac{-D_1(T - T_r)}{D_2 + (T - T_r)} \quad (1)$$

where T is the test temperature; T_r stands for the referential temperature (25 °C) of the master curves; and the matching arguments of the WLF are denoted as D_1 and D_2.

$$|G^*| = G_g\left[1 + \left(\frac{\omega_c}{\omega_r}\right)^v\right]^{-\frac{m}{v}} \quad (2)$$

where ω_c, m, v are the viscoelastic parameters obtained by the fitting; G_g is the glassy modulus of asphalt materials, taken as 1 GPa in this study; ω_r is the reduced angular frequency, which can be calculated by Equation (3):

$$\omega_r = \varphi_T \times \omega \quad (3)$$

where ω is the test angular frequency.

2.2. S-VECD Model

Many researchers have conducted a lot of experimental studies on the S-VECD theory, and the research consequence shows that the S-VECD theory model can precisely analyze and predict the fatigue damage nature of asphalt materials [18]. Schapery's work potential theory (WPT), which is based on the principle of non reversing thermodynamics, is the theoretical basis for the establishment of the S-VECD model. In this theory the expression of material damage rate is given by Equation (4):

$$\frac{dS}{dt} = \left(-\frac{\partial W^R}{\partial S}\right)^{\alpha} \tag{4}$$

where α is the characteristic constant introduced by Schapery into the theory of the work potential for elastic media materials and later extended to viscoelastic materials. The value of α is also related to the loading mode; Lee and Kim et al. [19] concluded that $\alpha = 1 + 1/m$ in the strain control pattern and $\alpha = 1/m$ in the stress control mode. S is defined as the damage variable to characterize the damage inside the material; t is the test time; the supposititious strain energy density is denoted as W^R, and it can be calculated as Equation (5):

$$W^R = \frac{1}{2} \cdot DMR \cdot C(S) \cdot \left(\gamma_p^R\right)^2 \tag{5}$$

where the dynamic modulus ratio (DMR) is the ratio of the tested initial modulus $|G^*|_{initial}$ to the dynamic state modulus $|G^*|_{LVE}$. This argument was imported to remove the adaptability. The range of DMR is generally from 0.9 to 1.1.

C is the pseudo modulus, then, the relationship between t, C and the damage variable S can be fitted by using Equation (6):

$$C(S) = \frac{\tau_p}{\gamma_p^R \times DMR} \tag{6}$$

where τ_p is the peak shear stress; the peak pseudo strain ${\gamma_p}^R$ during this loading cycle can be calculated by Equation (7):

$$\gamma_p^R = \frac{1}{G_R} \cdot |G^*|_{LVE} \cdot \gamma_p \tag{7}$$

where γ_p is the peak shear strain, and G_R is the reference shear modulus, which is taken as 1 MPa in this study to simplify the calculation process and make the imaginary strain dimensionless.

In combination with Equations (4)–(8), the accumulative variation in the damage variate S over time t is acquired as follows:

$$S(t) = \sum_i^N \left[\frac{DMR}{2}\left(\gamma_p^R\right)^2 (C_{i-1} - C_i)\right]^{\frac{\alpha}{1+\alpha}} \left[t_{Ri} - t_{Ri-1}\right]^{\frac{1}{1+\alpha}} \tag{8}$$

$$t_{Ri} = \frac{t_i}{\varphi_T} \tag{9}$$

where i is the loading period, and t_R is the reduced time.

Based on the experimentally measured data, the values of C and S can be calculated, and their relationship between them can be fitted according to Equation (10):

$$C(t) = 1 - C_1 \cdot S(t)^{C_2} \tag{10}$$

where C_1, C_2 are the best-fitting parameter.

2.3. Failure Definition and Criterion for Asphalt Binder

Pseudo strain energy (PSE) theory eliminates the effect of the viscoelasticity of asphalt materials and quantifies the damaging effect caused by the load separately. The test data should be analyzed in pseudo strain coordinates, and Figure 1 shows the definition of the imaginary strain energy in the LAS test. Taking the "undamaged line" as a reference, at the beginning of the test, the mechanical response is the same as that of the "undamaged line", reflecting that the material has not been damaged. With further loading, the strain enlarges step by step and the practical answer of the asphalt tends to deviate from the "undamaged line", demonstrating that the asphalt is getting destroyed and that damage accumulates [20].

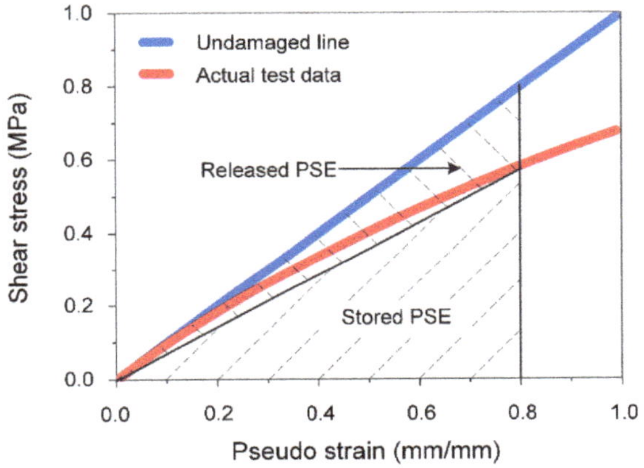

Figure 1. Definition of pseudo strain energy.

Then the stored PSE can be calculated by using Equation (11) for that point, denoted as $W^R{}_s$. The total PSE ($W^R{}_t$) applied to the asphalt from outside during the loading process can be expressed as the area of the triangle enclosed by the trunnion axis, calculated by using Equation (12). Therefore, as an approximation, Equation (13) shows that the PSE released by the material because of $W^R{}_r$ can be calculated by subtracting the hoarded PSE ($W^R{}_s$) from the $W^R{}_t$. $W^R{}_s$ and $W^R{}_r$ can be represented by the two shaded areas in Figure 1.

$$W_s^R = \frac{1}{2} \times \tau_p \times \frac{\gamma_p^R}{DMR} = \frac{1}{2} \times C \times \left(\gamma_p^R\right)^2 \tag{11}$$

$$W_t^R = \frac{1}{2} \times \tau_{undamage} \times \gamma_p^R = \frac{1}{2} \times \left(\gamma_p^R\right)^2 \tag{12}$$

$$W_r^R = W_t^R - W_s^R = \frac{1}{2} \times (1-C)\left(\gamma_p^R\right)^2 \tag{13}$$

Therefore, with the use of Equations (11)–(13), the values of $W^R{}_s$ and $W^R{}_t$ in the LAS test can be counted for every datum point and the results are plotted in Figure 2. At the beginning of the test, the asphalt can basically store all the PSE input from the load, and $W^R{}_r$ is nearly zero. Throughout the experiment, the amplitude of the loading gradually increases, $W^R{}_s$ increase, and $W^R{}_r$ also increases. The release of energy indicates that the material is gradually being damaged. A sharp peak point in $W^R{}_s$ follows, and after reaching the maximum, it begins to decline, manifesting that the asphalt gradually loses its ability to store energy and release increasing amounts of energy. The maximum value of $W^R{}_s$ is the maximum amount of external energy that the material can store, which characterizes the fatigue performance of the material itself and has a good agreement compared to the phase angle response, so it is considered that fatigue failure occurs in asphalt binder when $W^R{}_s$ reaches a maximum. Then the average $W^R{}_r$ for each cycle, $\overline{W_r^R}$, can be calculated from the beginning of the test to the fatigue failure of the material. G^R, the PSE release rate during this period, can be calculated by Equation (14):

$$G^R = \frac{\overline{W_r^R}}{N_f} = \frac{\frac{A}{N_f}}{N_f} = \frac{A}{\left(N_f\right)^2} \tag{14}$$

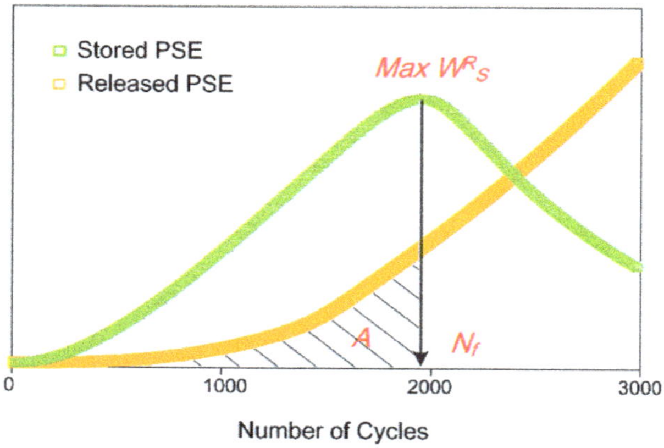

Figure 2. PSE-based failure definition.

As displayed in Figure 2, the shadow area A is the area of the $W^R{}_r$ curve before fatigue failure. The calculated G^R and its corresponding N_f are plotted on double logarithmic coordinates, and the relationship between G^R and N_f can be well-fitted by a power function [21]:

$$G^R = aN_f{}^b \tag{15}$$

where a and b are the best-fitting parameters.

$$N_f = \left[\frac{K}{a} \times (\gamma)^{2+2\alpha\left(\frac{C_2}{Q}\right)}\right]^{\frac{1}{b+1-\frac{C_2}{Q}}} \tag{16}$$

where K and Q can be counted by Equations (17)–(19).

$$K = \frac{1}{2} \times C_1 \times (|G^*|_{LVE})^2 \times P^{\left(-\frac{C_2}{Q}\right)} \times \frac{1}{\left(\frac{C_2}{Q}\right)+1} \tag{17}$$

$$Q = 1 - \alpha \times C_2 + \alpha \tag{18}$$

$$P = \frac{f \times 2^\alpha}{Q(C_1 \times C_2)^\alpha (|G^*|_{LVE})^{2\alpha}} \tag{19}$$

P is the calculated parameter. After fitting the test data, the fatigue life of asphalt materials under different strain levels can be predicted.

3. Materials and Test Methods

3.1. Materials

The matrix asphalt with physical properties as shown in Table 1 was selected. Three polymers, SBS, SBS-T and SBS-W, were selected as polymer modifiers, and Sasobit as warm mix agent. The Sasobit modifier, as well as its parameters, was provided by Henan Lupeng Transportation Technology (Zhengzhou, China). The technical information is shown in Table 2. Both SBS, SBS-T, and SBS-W, as well as their parameters, were provided by Beijing Guolu Hi-tech Co., Ltd. (Beijing, China). Table 3 lists the information of all modifiers.

Table 1. Technical information of base asphalt.

Parameters	Units	Values	Technical Requirement
Penetration test	0.1 mm	70	60–80
Softening Point test	°C	51.8	>46
Rotational viscosity (135 °C)	Pa·s	0.63	-
Ductility test	cm	>100	>100

Table 2. Technical information of modifiers.

Modifiers	Parameters	Units	Values
SBS	Oil content	%	0.70
	S/B ratio	-	30/70
	Total ash	%	0.20
	Tensile strength	MPa	18.0
	Voliatility	%	1.00
	Elongation	%	700
Sasobit®	Viscosity at 135 °C	Pa·s	5.47×10^{-3}
	Viscosity at 150 °C	Pa·s	3.26×10^{-3}
	Flashing point	°C	290
	Melting point	°C	100
	Penetration at 25 °C	0.1 mm	1
	Penetration at 60 °C	0.1 mm	8

Table 3. Technical information of fast melting modifiers.

Modifiers	Parameters	Units	Values
SBS-W	Appearance	Particle	-
	Individual weight	g	0.20
	Total ash	%	18.0
	Dry mix dispersibility	-	No particle residue
SBS-T	Appearance	Green particle	-
	Individual weight	g	0.25
	Total ash	%	0.42
	Dry mix dispersibility	-	No particle residue

3.2. Preparation of Sample

Three kinds of polymer-modified asphalt (PMA) were prepared with 5% SBS, 5% SBS-T, and 6% SBS-W as additives. 3% Sasobit, which acted as a warm mixing agent, was added to three polymer-modified asphalt. The primary steps of preparing modified asphalt are as follows: the matrix asphalt is placed in the oven at 120 °C until it is entirely melted, and then SBS or SBS-T or SBS-W modifier is added into the matrix asphalt, and the shear rate is 5000 r/min with a high-speed shear machine for 30 min at 170 °C. Then, after the preliminary of the PMA was completed, a portion of the PMA was heated in an oven for 30 min at 140 °C, in which 3% Sasobit warm mix was added and mixed for 30 min on a high-speed shearing machine at 5000 r/min.

3.3. Microscopic Characteristics

The differences and similarities of the internal crystal phases of SBS, SBS-T, and SBS-W polymer-modified asphalt and the changes of the internal crystal phases of the three polymer-modified asphalt before and after the addition of Sasobit were compared. The sharp X-ray diffractometer (Empyrean) produced by Panaco in Almelo, the Netherlands was used. The target material of the optical tube was Cu-Kα, the tube voltage was 45 KV, and the tube current was 40 mA. The preparation process of the XRD test sample is as follows: the PMA sample is heated to liquid condition, a little amount of asphalt is dropped into the center of the slide, and heat preservation is made to flatten it to form a smooth surface asphalt film. In this study, the scanning range of diffraction angle (2θ) is 10~90°, the scanning rate is 0.05°/s, and the wavelength is 0.154 nm. A software called MDI Jade 6.5 (V6.5, 2019, Materials Data, Livermore, CA, USA) was used to analyze the test data. XRD tests are conducted in accordance with AASHTO TP101.

Fourier transform infrared spectroscopy (FTIR, Nicolet Continuum, Thermo Fisher Scientific, Waltham, MA, USA) was used to study the variations of their chemical compositions. The preparation process of the infrared spectrum sample is as follows: the modified asphalt sample is dissolved with CS2 organic solution, an amount of the liquor is dropped on the KBr slide, the liquor is completely air seasoning to form asphalt film, and therewith put into the FTIR sample chamber for testing. The frequency of spectral scanning used in this study is 64 times, and the scanning range is 4000–400 cm^{-1}. FTIR tests are conducted in accordance with AASHTO TP101.

3.4. Test Methods

The Discovery HR-1 Dynamic Shear Rheometer (DSR) of the TA instruments (New Castle, DE, USA) was used for all tests. The experiment was focused in the range of medium temperature (10–40 °C). To ensure the precision of the data, every test was repeated three times, and the average value was taken to record the data. Tests were conducted according to AASHTO TP101.

3.5. Aging Methods

In this study, asphalt was first subjected to short-term laboratory aging using a rotating film oven test (RTFOT, ASTM D2872). Then the asphalt residue after the RTFOT test was subjected to PAV and UV aging. The residue of RTFOT was placed on a stainless steel plate according to the asphalt film thickness specified in the test, and then placed in a pressure vessel with 2.1 MPa air pressure, and aged for 20 h at the selected aging temperature of 100 °C. Finally, vacuum degassed the residue. That's how the PAV aging test went.

The specific operation process of the UV aging test is to take a 20 g modified asphalt sample after short-term aging and pour it into a stainless steel asphalt aging dish with a diameter of 140 mm and a depth of 10 mm. In order to ensure the effect of ultraviolet aging, it is automatically leveled and covered with an aging dish in the oven, and the asphalt film with a thickness of about 1 mm can be formed, so that ultraviolet light can be uniformly irradiated on the asphalt film. Then the aging tray containing the modified asphalt sample is placed on the sample rack inside the UV aging chamber. The test temperature was set at 45 ± 0.5 °C, the rotation speed was set at 5 r/min, and the radiation intensity measured by the UV radiation instrument was set at 140 w/m^2. The UV irradiation time is 7 days. The aging methods were carried out according to AASHTO TP101 [22].

3.6. Frequency Scan (FS) Test

FS test can describe the viscoelastic characteristic of asphalt binder at any loading temperature and frequency. A strain control mode and 0.1% strain amplitude were applied at 15 °C, 25 °C, and 35 °C, and the frequency range was selected from 0.1 rad/s to 100 rad/s.

3.7. Linear Amplitude Scanning (LAS) Test

LAS test is based on the VECD accelerated fatigue test theory and has been included in the AASHTO TP101 asphalt anti-fatigue evaluation test specification. In this study, the concept of constant strain amplitude rate (CSR) was adjusted by varying the scanning time in addition to standard LAS tests on raw, PAV aged, and UV-aged asphalt. The scanning times were 500 s and 1000 s, respectively. The test temperature was 25 °C, and the loading frequency was 10 Hz. The CSR of the standard LAS test is equal to 30% of the total strain amplitude in a test time of 300 s divided by 0.01%/s. Then, the CSR values tested in this study are respectively 0.1%/s and 0.06%/s.

Through the above tests, the modification mechanism and fatigue properties of modified asphalt are analyzed, and the following conclusions are drawn.

4. Results and Discussion

4.1. XRD Analysis

Although asphalt itself is not crystal, but is a kind of thick cyclic aromatic hydrocarbon, its feature construction also shows the diffraction pattern of layered crystals like graphite in XRD. There are mainly four diffraction peaks: γ peak, (002) peak, (100) peak and (110) peak. The peaks of the first three peaks are roughly located at $2\theta = 20°$, $25°$ and $44°$. The γ peak comes from the diffraction phenomenon of saturated hydrocarbon in the asphalt component, and the (002) peak comes from the diffraction phenomenon of aromatic hydrocarbon in the thick ring. The crystallinity of asphalt is positively correlated with the peak value. The XRD patterns of the six compound-modified asphalt are shown in Figure 3.

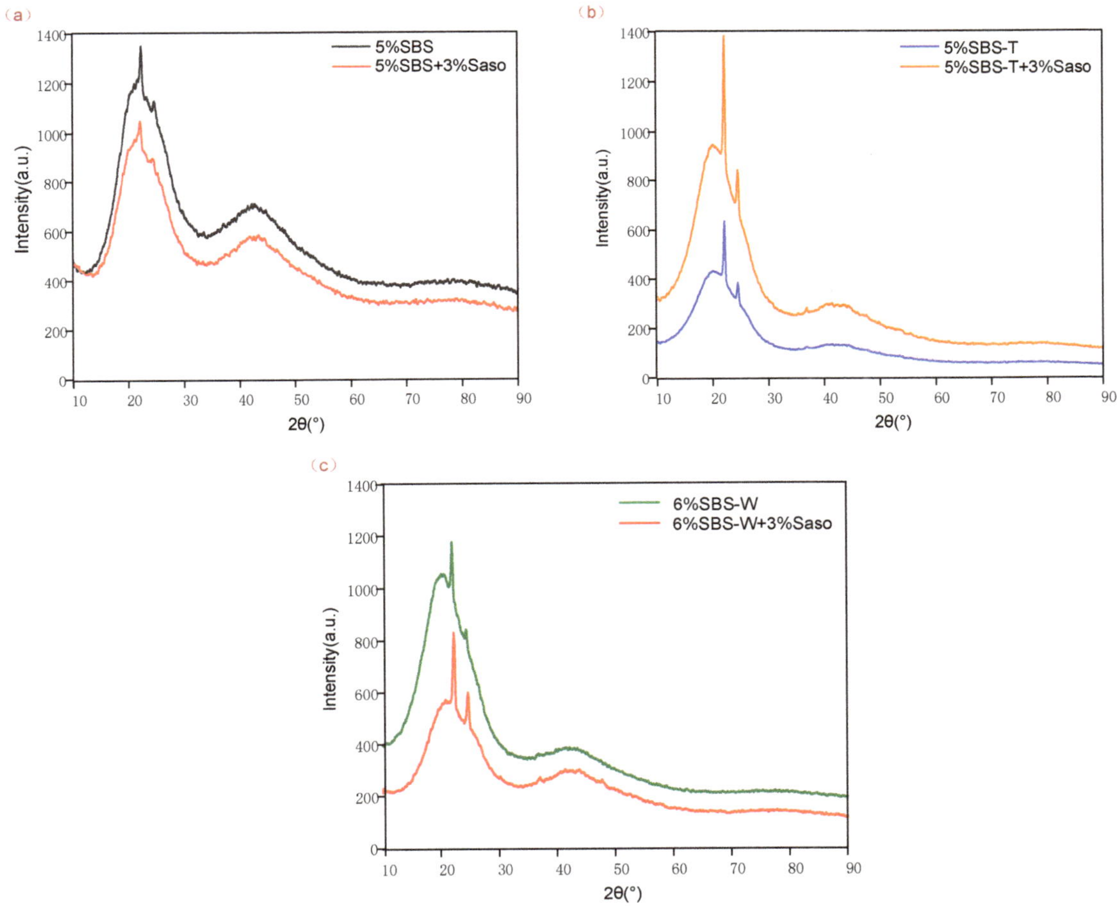

Figure 3. XRD pattern of modified asphalt. (**a**) SBS/Saso; (**b**) SBS-T/Saso; (**c**) SBS-W//Saso.

According to Figure 3a–c, the six modified asphalt all contain substances with the obvious crystal structures. After the addition of Sasobit, no new diffraction peaks appear. All the six modified asphalt samples have γ diffraction peaks near 2θ = 22° and (002) peaks near 2θ = 24°. Before and after SBSMA added 3%Sasobit, the (002) peak was not obvious, indicating that SBS has a strong absorbing ability of aromatic content in asphalt material, and the addition of Sasobit will not have a remarkable effect on it. Nevertheless, after the addition of Sasobit, the γ diffraction peak and (002) peak become more obvious, which indicates that Sasobit will weaken the absorption capacity of saturated and aromatic hydrocarbons in the fast-melting SBS-modified asphalt. MDI Jade 6.5 software was used to analyze the XRD data, and the results of the peak search report were listed in Table 4.

Table 4. XRD peak search report of modified asphalt.

Modified Asphalt Pitch	Distance between Layers (Å)	2θ (°)	Peak Height	Area of Peak	FWHM
5%SBS	3.9685 3.5725	22.384 24.901	255 120	4194 2323	0.380 0.388
5%SBS+3%Saso	4.0023 3.6249	22.174 24.537	140 113	4824 1886	0.385 0.186
5%SBS-T	4.0639 3.5847	21.852 24.817	106 176	1087 4453	0.114 0.282
5%SBS-T+3%Saso	4.0167 3.6244	22.112 24.542	1611 530	39,164 12,856	0.268 0.232
6%SBS-W	4.0237 3.6436	22.073 24.409	652 215	10,219 2969	0.281 0.262
6%SBS-W+3%Saso	3.9816 3.6052	22.310 24.674	765 373	20,496 11,126	0.299 0.333

The XRD pattern shows that the crystal structure of PMA remains intact after the addition of Sasobit. In asphalt binder, the content of saturated fraction and aromatic fraction mainly affect the fluidity of asphalt, and the content of aromatic fraction will make asphalt binder show more viscosity. Therefore, compared with polymer-modified bitumen, the modified asphalt with Sasobit and SBS-W modified asphalt show stronger fluidity on the macro level.

4.2. FTIR Analysis

In order to deeply understand the modification mechanism of the fast-melting SBS modifier on the polymer-modified asphalt binder, this study adopted the Fourier infrared spectrometer model Nicolet Continuum manufactured by Thermo Fisher Science and Technology of the Waltham, MA, USA to analyze the chemical characteristic groups of different types of SBS modified asphalt. In this paper, six kinds of modified asphalt were studied by infrared spectroscopy, and the action mechanism of speed melting SBS modifier polymer modified asphalt was preliminarily determined by analyzing the changes of chemical characteristic groups of asphalt modified by a wet and dry polymer. Figure 4a–c shows the FTIR spectra of SBS-modified asphalt, SBS T modified asphalt, SBS W modified asphalt and the modified asphalt binder with Sasobit added, respectively. The characteristic peak positions of the modified asphalt and the infrared absorption of relevant functional groups are listed in Table 5.

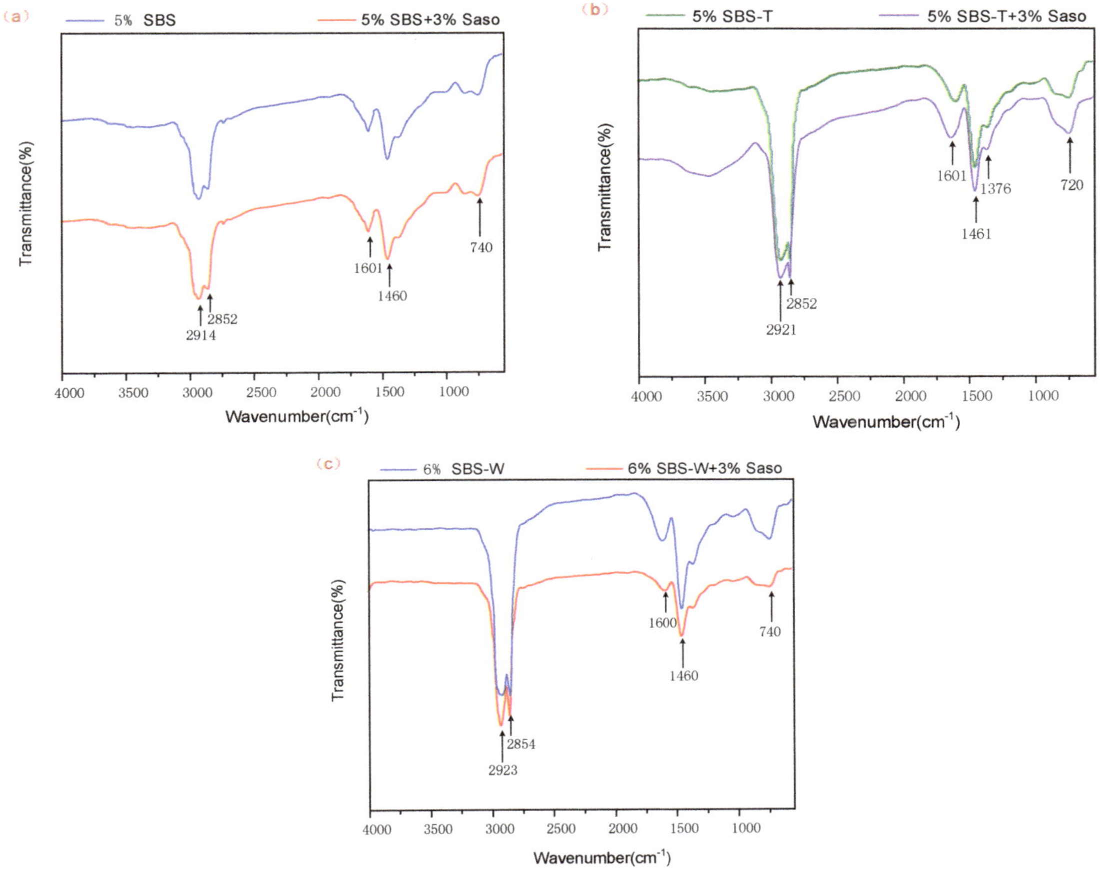

Figure 4. Fourier infrared spectroscopy of modified asphalt. (**a**) SBS/Saso; (**b**) SBS-T/Saso; (**c**) SBS-W//Saso.

Table 5. Characteristic peak information of functional groups of modified asphalt by infrared spectrum.

Characteristic Peak Position	Types of Functional Groups
2921, 2923, 2914 cm^{-1}	Alkanes and cycloalkanes C–H
2854 cm^{-1}	–CH2– Vibration of stretch
1600 cm^{-1}	C=C (Benzene ring skeleton vibration) and C=O
1460–1430 cm^{-1}	–CH3, –CH2–, –CHIn-plane stretching vibration
1376 cm^{-1}	–CH3 Shear vibration absorption peak
800–650 cm^{-1}	Benzene ring substituent C–H

On account of the data information in the chart, the following conclusions can be drawn:

(1) All six asphalt cements have sharp vibration characteristic peaks at about 2925–2850 cm^{-1}, which represents the oscillate assimilate peak of methylene CH2–. It can be observed that the position of characteristic peaks remained unchanged after the addition of Sasobit to the three polymer-modified asphalt, indicating that no complex chemical reaction occurred between Sasobit and the polymer-modified asphalt.

(2) The characteristic peaks of the infrared spectra of SBSMA, SBS-TMA, and SBS-WMA are somewhat different, indicating that SBS and fast-melting SBS additives have different reactions with the base asphalt. SBS-TMA has a weak peak at 1376 cm^{-1}, which is caused by the shear vibration of the methyl group in the SBS-T copolymer molecular chain. Because the proportion of SBS-T copolymer in asphalt is relatively small, so the characteristic peak here is not obvious.
(3) The three polymer-modified asphalts have obvious characteristic peaks at 738 cm^{-1}, and the characteristic peaks at 738 cm^{-1} are the benzene ring =CH. Due to the small proportion of the characteristic peak in the modified asphalt, the characteristic peak is weak.
(4) After the addition of 3%Sasobit, the absorption peaks of the three polymer-modified asphalt were increased in the range of 2930–2800 cm^{-1}, because the content of saturated hydrocarbons was relatively high. This phenomenon may occur due to Sasobit's reduced capacity of the polymer additive to absorb saturated HC in the base asphalt.

4.3. G^* and δ Master Curves

Table 6 shows the logarithmic value of the shift factor when 25 °C is the reference temperature. The constructed G^* main curve is shown in Figure 5. As can be seen from Figure 5 as a whole, for the three SBS composite modified binders, the higher the content, the higher the dynamic modulus response. At the same loading frequency, the addition of Sasobit significantly improves the dynamic modulus, especially in the low/high-temperature range. This indicates that the addition of Sasobit can improve the stiffness and high-temperature deformation resistance of polymer-modified asphalt. Under the original condition, the dynamic modulus of the bituminous binder with the mixture ratio of SBS-WMA is the highest when Sasobit is not added, which indicates that the high temperature rutting resistance and low-temperature resistance of the composite modified bituminous binder are the best. The dynamic modulus of SBSMA at high frequency/low temperature is higher than that of SBS-TMA, indicating that SBSMA has brilliant properties at low temperatures. The dynamic modulus of SBS-TMA at low frequency/high temperature is higher than that of SBSMA, indicating that SBS-TMA has better anti-rut performance at high temperature [23]. By comparing the data in Figure 5a,c, it can be found that G^* of PAV aging binder is higher than that of UV aging binder. The fitting parameters m, D1, D2, and so on were obtained from the G^* main curve of the modified binder, and the specific values were shown in Table 7.

Table 6. Fitting parameters of WLF equation.

Asphalt Condition	WLF Fitting Parameters	5%SBS	5%SBS+3%Saso	5%SBS-T	5%SBS-T+3%Saso	6%SBS-W	6%SBS-W+3%Saso
Origin	D_1	28.132	32.455	36.78	37.496	27.188	33.727
	D_2	231.466	251.461	289.73	346.297	264.120	301.786
PAV	D_1	28.893	37.471	33.707	35.114	28.018	34.452
	D_2	258.890	290.873	298.747	289.434	250.125	281.156
UV	D_1	28.073	29.698	27.756	33.718	23.267	29.405
	D_2	230.688	330.426	252.773	298.458	213.281	237.465

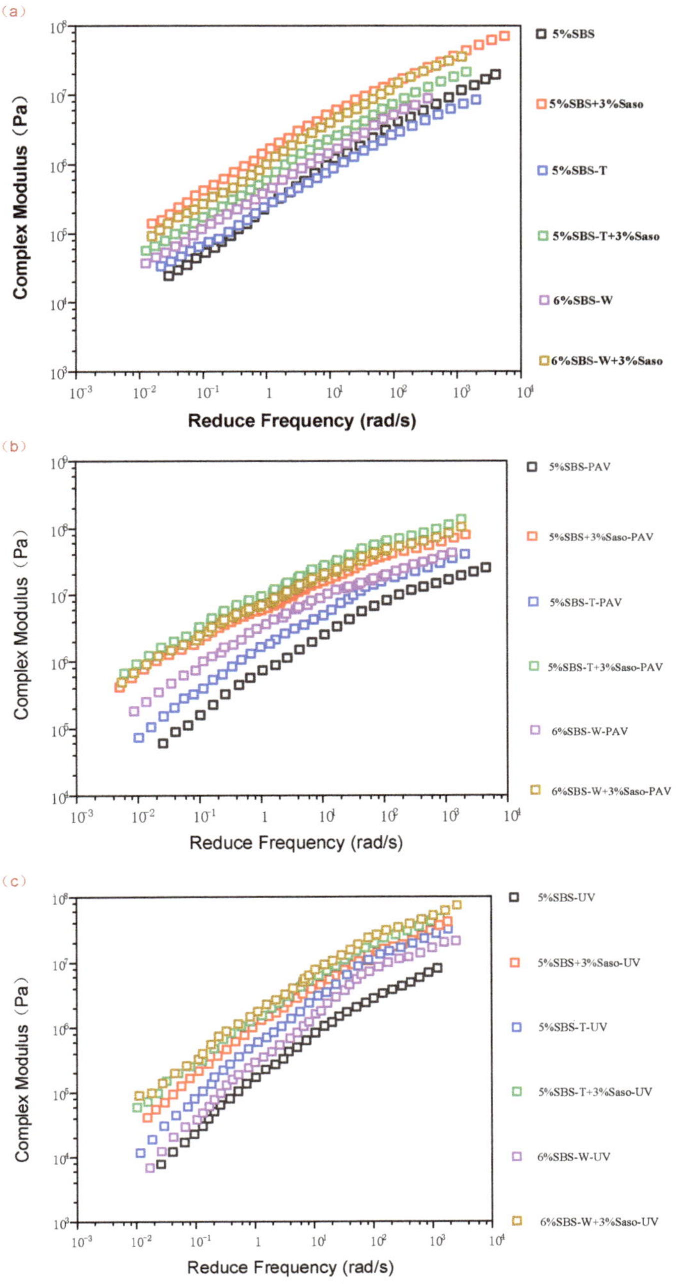

Figure 5. Main curve of dynamic modulus of modified asphalt. (**a**) SBS/Saso; (**b**) SBS-T/Saso; (**c**) SBS-W//Saso.

Table 7. CAM equation fitting parameters of modified asphalt.

Asphalt Condition	CAM Fitting Parameters	5%SBS	5%SBS+3%Saso	5%SBS-T	5%SBS-T+3%Saso	6%SBS-W	6%SBS-W+3%Saso
Origin	ω_c	0.306	0.425	0.673	0.266	0.010	0.006
	m	1.721	1.697	1.469	1.371	1.289	1.157
	v	0.147	0.144	0.149	0.147	0.153	0.161
PAV	ω_c	0.790	0.740	0.415	0.448	0.480	0.340
	m	1.161	1.135	1.138	1.112	1.270	1.167
	v	0.120	0.118	0.134	0.131	0.142	0.144
UV	ω_c	0.312	0.747	0.729	0.689	1.100	0.138
	m	1.710	1.552	1.549	1.445	1.283	1.226
	v	0.142	0.142	0.143	0.143	0.154	0.156

4.4. Stress-Strain Curve

Figure 6 describes the stress-strain response of three composites modified asphalt binders under LAS fatigue tests at two aging levels. The shear stress of all the modified binders showed obvious peaks of different width amplitudes. It has been shown that the wider the peak stress width is, the greater the deformation the asphalt binder can bear. Usually, this change in width is most likely due to a change in the asphalt composition to resist deformation [24,25].

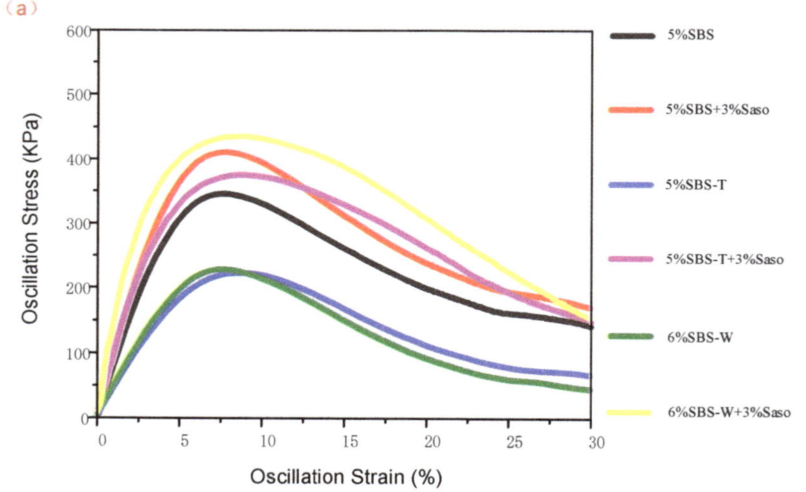

Figure 6. *Cont.*

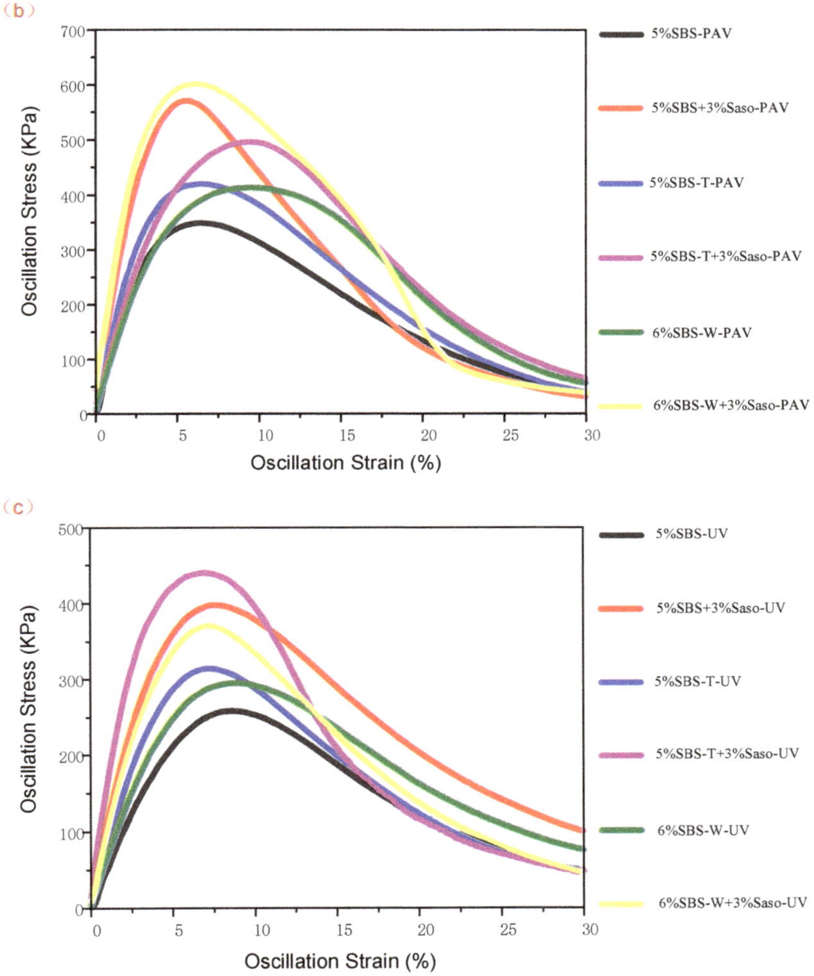

Figure 6. Stress-strain curves of modified asphalt. (**a**) SBS/Saso; (**b**) SBS-T/Saso; (**c**) SBS-W//Saso.

The curves in the figure show that the peak stress and peak width of polymer-modified asphalt after adding Sasobit are significantly increased, which indicates that Sasobit can improve the stiffness and deformation resistance of asphalt binder. Figure 6b,c both show that the strength of UV- or PAV-SBS MA is lower than that of SBS-T or SBS-W modifiers. These results indicate that compared with SBSMA, the speed-melting SBS-modified asphalt has better strength after aging, that is, better fatigue resistance.

4.5. Damage Characteristic Curve

On the basis of PSE theory, the variation trend of the virtual modulus C and damage variable S of the modified asphalt material in the process of fatigue test can be calculated according to the formula, that is, the damage characteristic curve (DCC) of the material.

It can be clearly seen from Figure 7 that with the progress of the experiment, the asphalt binder gradually develops from the undamaged state of C = 1 to the damaged state, and the value of C gradually decreases while the value of S gradually increases. This change can quantify the damage evolution process of the material. The shape of the DCC curve of asphalt binder before and after adding Sasobit is different, indicating that Sasobit changes the injure evolution process of asphalt. As can be seen from Figure 7a, the DCC curves of the three types of modified asphalt with Sasobit are all above the three types of polymer modified asphalt, indicating that at the same damage degree, the integrity of the warm mixed polymer modified asphalt is better and the ability to resist deformation is stronger. When fatigue failure is reached, the values of the virtual modulus C_f and damage variable S_f of each material are different. Compared with the three polymer-modified asphalt cement, the Sasobit-added asphalt binder has a smaller C_f value and a larger S_f value. Since the virtual modulus C represents the integrity of the material and the damage variable S represents the damage degree of the material, this situation indicates that the material with Sasobit has a larger damage degree when it reaches fatigue failure, and the material has a strong anti-fatigue ability.

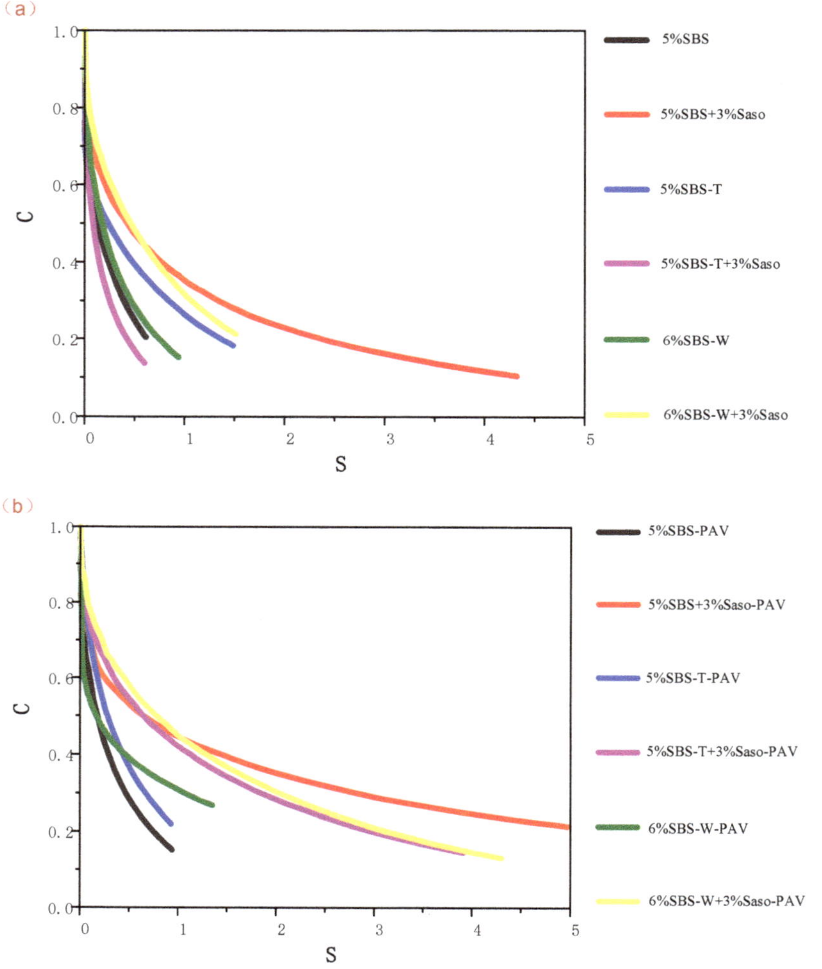

Figure 7. *Cont.*

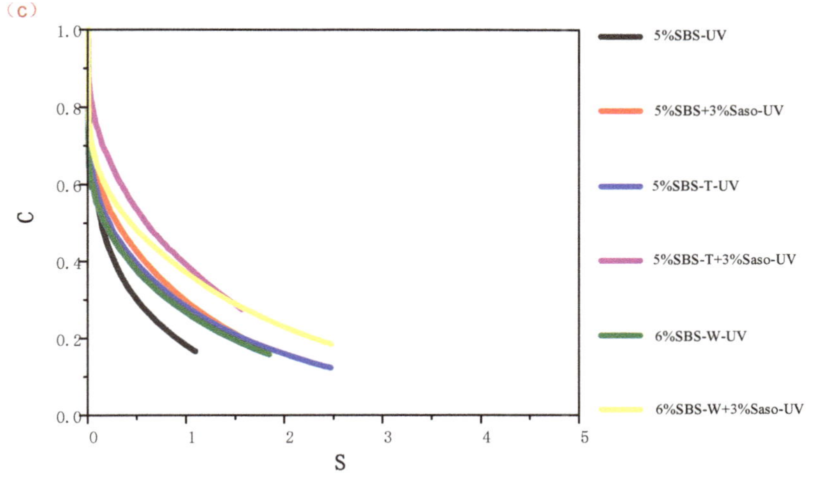

Figure 7. DCC curves of modified asphalt binder: (**a**) SBS/Saso; (**b**) SBS-T/Saso; (**c**) SBS-W//Saso.

The results show that the speed-melting SBS modified asphalt has better fatigue properties than SBS MA at the same strain level, especially under the condition of uv aging. This indicates that compared with the traditional SBS modifiers, the oxidation degree of SBS T and SBS W modifiers is lower in the UV aging process.

As can be seen from Table 8, when the damage variable S is the same, the lower the values of C1 and C2 are, the higher the comparative virtual modulus C is. According to the data, C1 and C2 values decreased after Sasobit was added into the three polymer modified asphalt, indicating that Sasobit can improve the fatigue resistance of polymer modified asphalt, which is consistent with the results shown in the DCC curve image.

Table 8. DCC curve fitting parameters.

Asphalt Condition	Fitting Parameters	5%SBS	5%SBS+3%Saso	5%SBS-T	5%SBS-T+3%Saso	6%SBS-W	6%SBS-W+3%Saso
Origin	C_1	0.628	0.625	0.737	0.731	0.878	0.724
	C_2	0.267	0.237	0.221	0.216	0.236	0.230
PAV	C_1	0.768	0.598	0.842	0.821	0.572	0.555
	C_2	0.350	0.160	0.286	0.267	0.260	0.255
UV	C_1	0.699	0.679	0.603	0.598	0.737	0.642
	C_2	0.232	0.215	0.330	0.208	0.194	0.183

4.6. Fatigue Performance Analysis of Asphalt

According to the values of a and b in Table 9, the fatigue lives of different types of modified asphalt cements at 3%, 6%, and 9% strain levels were calculated and plotted in Figure 8a–c.

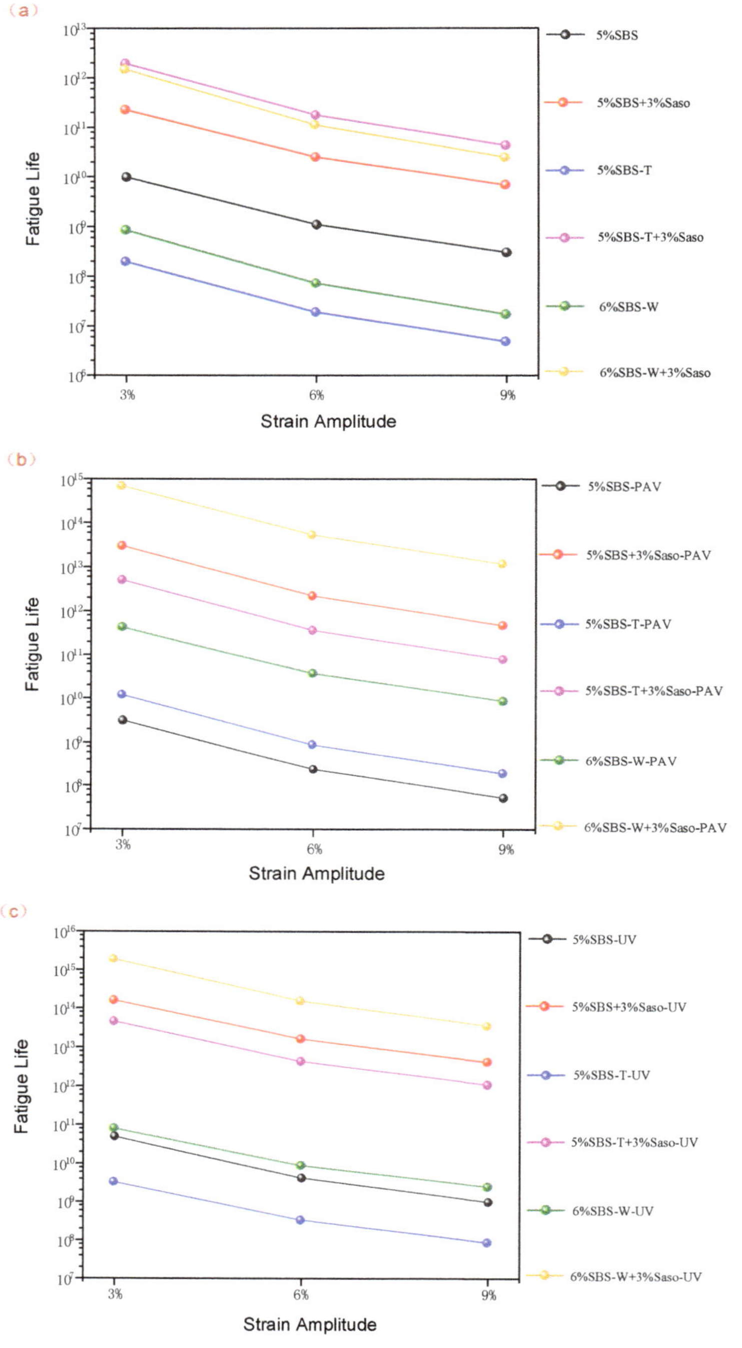

Figure 8. Fatigue life of modified asphalt binder under different strain levels. (**a**) SBS/Saso; (**b**) SBS-T/Saso; (**c**) SBS-W//Saso.

Table 9. Fatigue equation parameters of the binders.

Asphalt Condition	Fitting Parameters	5%SBS	5%SBS+3%Saso	5%SBS-T	5%SBS-T+3%Saso	6%SBS-W	6%SBS-W+3%Saso
Origin	a	2.99×10^8	3.03×10^7	3.54×10^{10}	1.87×10^7	3.63×10^9	5.92×10^9
	b	−1.841	−2.087	−1.646	−3.644	−2.048	−3.096
PAV	a	9.97×10^{11}	6.32×10^{10}	4.62×10^{10}	3.09×10^{10}	3.48×10^{10}	2.42×10^7
	b	−5.114	−4.077	−1.553	−1.963	−2.046	−3.383
UV	a	2.36×10^{10}	2.71×10^{10}	3.45×10^{10}	1.94×10^9	3.13×10^{10}	1.66×10^9
	b	−1.745	−1.537	−2.059	−1.547	−1.497	−1.547

As can be seen from Figure 8, the fatigue life of the modified asphalt binder after Sasobit is added significantly increases, indicating that Sasobit improves the fatigue resistance of the polymer-modified asphalt. As shown in Figure 8a, the fatigue life of SS-T modified asphalt and SS-W modified asphalt with Sasobit is larger than that of SBSMA under non-aging conditions, indicating that the anti-fatigue performance is better than that of SBSMA. Under the conditions of PAV aging and UV aging, the fatigue life of SBS-W-modified asphalt is the maximum. It can be seen that the speed melting SBS modified asphalt has better fatigue damage resistance.

5. Conclusions

The purpose of this work is to evaluate the modification effects of two fast-melting modifiers (SBS-T and SBS-W) and traditional SBS on asphalt binder, as well as their combined modification effects with Sasobit. First, the microscopic characteristics of SBS/SBS-T/SBS-W modifiers and modified binders were characterized by XRD and FTIR infrared spectroscopy to determine the chemical composition, microstructure, and defect level of the modifiers and further reveal the modification mechanism. Then, the viscoelastic parameters of different kinds of modified asphalt binder were analyzed by FS frequency scanning test, WLF equation and CAM model. Finally, the LAS fatigue acceleration test was carried out by DSR, and the fatigue damage resistance of various modified asphalt in base asphalt binder was studied based on VECD theoretical model. The main findings are as follows:

(1) According to the XRD diffraction peak data, the peak height of the fast-melting modified binders (SBS-T and SBS-W) is higher, their crystallization properties are better, and their crystal phase content is higher than that of the control SBS MA. With the addition of Sasobit, the crystal structure in PMA remains intact. FTIR showed that no chemical changes occurred after Sasobit was added to the asphalt binder.

(2) Compared with SBSMA, the speed-melting modifiers (SBS-T and SBS-W) have higher complex modulus, better fatigue resistance, and lower temperature sensitivity. SBS-W improves stiffness and stability by three times. The addition of Sasobit can also improve the stiffness and stability of polymer-modified asphalt.

(3) The LAS test based on the linear VECD theoretical model shows that the modified asphalt with 6%SBS-W+3%Sasobit mixture ratio has the longest fatigue life, indicating that it has the best anti-fatigue failure ability. The fatigue life is increased 3.9 times by SBS-W. In addition, Sasobit can prolong the fatigue life of asphalt binders.

6. Recommendations

In future research, we will increase the scale to study the fatigue properties of the fast-melting modified asphalt mixture. Sasobit is relatively expensive, and future projects will use more SBS-W, an inexpensive warm mix modifier.

Author Contributions: Y.Z.: Methodology, Investigation, Writing—original draft. J.Y.: Data curation. B.M.: Data curation. G.T.: Resources. R.W.: Validation. All authors have read and agreed to the published version of the manuscript.

Funding: Funding support from the Project of the Henan Provincial Department of Transportation (Grant No. 2021-2-13) and the Key Science and Technology Project of Henan Province (Grant No. 212102310937) were greatly appreciated.

Institutional Review Board Statement: Not applicable.

Informed Consent Statement: Not applicable.

Data Availability Statement: Not applicable.

Conflicts of Interest: The authors declare no conflict of interest.

References

1. Sukhija, M.; Saboo, N. A comprehensive review of warm mix asphalt mixtures-laboratory to field. *Constr. Build. Mater.* **2021**, *274*, 121781. [CrossRef]
2. Zhu, J.; Birgisson, B.; Kringos, N. Polymer modification of asphalt: Advances and challenges. *Eur. Poly. J.* **2014**, *54*, 18–38. [CrossRef]
3. Yu, H.; Chen, Y.; Wu, Q.; Zhang, L.; Zhang, Z.; Zhang, J.; Miljković, M.; Oeser, M. Decision support for selecting optimal method of recycling waste tire rubber into wax-based warm mix asphalt based on fuzzy comprehensive evaluation. *J. Clean. Prod.* **2020**, *265*, 121781. [CrossRef]
4. Cao, R.; Leng, Z.; Yu, H.; Hsu, S.C. Comparative life cycle assessment of warm mix technologies in asphalt rubber pavements with uncertainty analysis. *Resour. Conserv. Recycl.* **2019**, *147*, 137–144. [CrossRef]
5. Mao, Y.; Liu, Y.; Hao, P.W.; Wang, H.N. Process Parameters of Rubber Powder Particles Modified Asphalt. *Adv. Mater. Res.* **2014**, *900*, 491–498. [CrossRef]
6. Ali, B. A review of the warm mix asphalt (WMA) technologies: Effects on thermo-mechanical and rheological properties. *J. Clean. Prod.* **2020**, *259*, 120817.
7. Abd, D.M.; Al-Khalid, H.; Akhtar, R. An investigation into the impact of warm mix asphalt additives on asphalt mixture phases through a nano-mechanical approach. *Constr. Build. Mater.* **2018**, *189*, 296–306. [CrossRef]
8. Xiao, F.; Punith, V.S.; Amirkhanian, S.N. Effects of non-foaming WMA additives on asphalt binders at high performance temperatures. *Fuel* **2012**, *94*, 144–155. [CrossRef]
9. Ameri, M.; Afshin, A.; Shiraz, M.E.; Yazdipanah, F. Effect of wax-based warm mix additives on fatigue and rutting performance of crumb rubber modified asphalt. *Constr. Build. Mater.* **2010**, *262*, 120882. [CrossRef]
10. Yu, H.; Zhu, Z.; Leng, Z.; Wu, C.; Zhang, Z.; Wang, D.; Oeser, M. Effect of mixing sequence on asphalt mixtures containing waste tire rubber and warm mix surfactants. *J. Clean. Prod.* **2020**, *246*, 119008. [CrossRef]
11. Yang, X.; You, Z.; Hasan, M.R.M.; Diab, A.; Shao, H.; Chen, S.; Ge, D. Environmental and mechanical performance of crumb rubber modified warm mix asphalt using evotherm. *J. Clean. Prod.* **2017**, *159*, 346–358. [CrossRef]
12. Yue, M.; Yue, J.; Wang, R.; Xiong, Y. Evaluating the fatigue characteristics and healing potential of asphalt binder modified with Sasobit and polymers using linear amplitude sweep test. *Constr. Build. Mater.* **2021**, *289*, 123054. [CrossRef]
13. Wang, R.; Xiong, Y.; Yue, M.; Hao, M.; Yue, J. Investigating the effectiveness of carbon nanomaterials on asphalt binders from hot storage stability, thermodynamics, and mechanism perspectives. *J. Clean Prod.* **2020**, *276*, 124180. [CrossRef]
14. Bahia, H.U.; Zhai, H.; Bonnetti, K.; Kose, S. Non-linear viscoelastic and fatigue properties of asphalt binders. *J. Assoc. Asph. Paving Technol.* **1999**, *68*, 1–34.
15. Bahia, H.U.; Zhai, H.; Zeng, M.; Hu, Y.; Turner, P. Development of binder specification parameters based on characterization of damage behavior. *J. Assoc. Asph. Paving Technol.* **2002**, *70*, 442–470.
16. Zhou, F.; Mogawer, W.; Li, H.; Andriescu, A.; Copeland, A. Evaluation of fatigue tests for characterizing asphalt binders. *J. Mater. Civ. Eng.* **2013**, *25*, 610–617. [CrossRef]
17. Wang, C.; Castorena, C.; Zhang, J.; Richard Kim, Y. Unified failure criterion for asphalt binder under cyclic fatigue loading. *Road Mater. Pave. Des.* **2015**, *16*, 125–148. [CrossRef]
18. Underwood, B.S.; Baek, C.; Kim, Y.R. Simplified viscoelastic continuum damage model as platform for asphalt t concrete fatigue analysis. *Transp. Res. Rec.* **2012**, *2296*, 36–45. [CrossRef]
19. Lee, H.J.; Kim, Y.R. Viscoelastic constitutive model for asphalt concrete under cyclic loading. *J. Eng. Mech.* **1998**, *124*, 32–40. [CrossRef]
20. Wang, R.; Yue, M.; Xiong, Y.; Yue, J. Experimental study on mechanism, aging, rheology and fatigue performance of carbon nanomaterial/SBS-modified asphalt binders. *Constr. Build. Mater.* **2021**, *268*, 121189. [CrossRef]
21. Wang, C.; Wang, H.; Zhao, L.; Cao, D. Experimental study on rheological characteristics and performance of high modulus asphalt binder with different modifiers. *Constr. Build. Mater.* **2017**, *155*, 26–36. [CrossRef]

22. AASHTO. Standard method of test for estimating fatigue resistance of asphalt binders using the linear Amplitude Sweep. In *AASHTO TP 101*; AASHTO: Washington, DC, USA, 2016.
23. Underwood, B.S. A continuum damage model for asphalt cement and asphalt mastic fatigue. *Int. J. Fatigue* **2016**, *82*, 387–401. [CrossRef]
24. Sabouri, M.; Mirzaiyan, D.; Moniri, A. Effectiveness of Linear Amplitude Sweep (LAS) Asphalt Binder Test in Predicting Asphalt Mixtures Fatigue Performance. *Constr. Build. Mater.* **2018**, *171*, 281–290. [CrossRef]
25. Ameri, M.; Seif, M.R.; Abbasi, M.; Khiavi, A.K. Viscoelastic fatigue resistance of asphalt binders modified with crumb rubber and styrene butadiene polymer. *Pet. Sci. Technol.* **2017**, *35*, 30–36. [CrossRef]

coatings

MDPI

Article

Sprayed-Polyurea-Modified Asphalt: Optimal Preparation Parameters, Rheological Properties and Thermal Properties

Qinyuan Peng [1], Xiaolong Sun [1,2,*], Zhisheng Liu [3], Jiao Jin [4], Huayang Yu [5] and Yingmei Yin [1]

1 School of Civil and Transportation Engineering, Guangdong University of Technology, Guangzhou 510006, China
2 Highway Engineering Key Laboratory of Sichuan Province, Southwest Jiaotong University, Chengdu 611756, China
3 The Key Laboratory of Road and Traffic Engineering, Ministry of Education, Tongji University, Shanghai 201804, China
4 School of Traffic and Transportation Engineering, Changsha University of Science and Technology, Changsha 410114, China
5 School of Civil Engineering and Transportation, South China University of Technology, Guangzhou 510641, China
* Correspondence: xls1998@gdut.edu.cn

Abstract: For promoting modifying application of sprayed polyurea (SPUA) in asphalt pavement materials, the effects of sprayed polyurea materials on high-temperature and fatigue performance of asphalt binders were investigated from different aspects. First, the optimal preparation parameters of sprayed-polyurea-modified asphalt binders (SPMAs) were determined by designing an orthogonal test. Then, the high-temperature and fatigue properties of sprayed-polyurea-modified asphalt binders with different contents were characterized by rheological testing methods, including Brookfield rotary viscosity (RV) test, performance grading (PG) test, multiple stress creep recover (MSCR) test, linear amplitude sweep (LAS) test and time sweeping (TS) test. Finally, the thermal properties of the asphalt binders were analyzed by differential scanning calorimetry (DSC) test. The results showed that the optimum preparation parameters were determined by the extreme difference analysis method and analysis of variance (ANOVA) method, and the shearing time was 40 min, the shearing rate was 6000 rpm and the shearing temperature was 150 °C. Sprayed polyurea positively affected high-temperature performance of asphalt binders and could improve fatigue resistance of asphalt binders. Moreover, the Brookfield rotary viscosity test, multiple stress creep recover test and linear amplitude sweep test had high sensitivity to the performance of sprayed-polyurea-modified asphalt binder, which could help to distinguish the effect of sprayed polyurea dosing on performance of asphalt binders accurately. The differential scanning calorimetry test showed that sprayed polyurea was beneficial to high-temperature stability of asphalt binders, which explains the reason why sprayed-polyurea-modified asphalt binders have excellent high-temperature performance from a microscopic perspective.

Keywords: modified asphalt binder; sprayed polyurea resin; high-temperature property; fatigue property; thermal property; correlation

Citation: Peng, Q.; Sun, X.; Liu, Z.; Jin, J.; Yu, H.; Yin, Y. Sprayed-Polyurea-Modified Asphalt: Optimal Preparation Parameters, Rheological Properties and Thermal Properties. *Coatings* **2023**, *13*, 544. https://doi.org/10.3390/coatings13030544

Academic Editor: Valeria Vignali

Received: 3 February 2023
Revised: 23 February 2023
Accepted: 25 February 2023
Published: 2 March 2023

1. Introduction

Compared with cement concrete pavement, asphalt pavement has advantages of short maintenance cycle, easy maintenance and good driving performance and has gradually replaced cement concrete pavement and become the main application of pavement structure [1–4]. However, it has been found in practical applications and research that common asphalt materials are prone to deformation under prolonged or excessive loading and performance degradation under light, high and low temperatures and rain, resulting in shortened lifespan of asphalt pavements [1]. The available research results show that polymer modification technology is an effective way to improve performance of asphalt binders [2]. However, while polymers improve performance of asphalt binders, they also add other new problems. For example, the excellent high- and low-temperature performance of SBS-modified asphalt binder is favored by researchers, but its high cost and poor compatibility limit application in low- and medium-grade pavements [5,6]. Ethylene–vinyl acetate copolymer, polyethylene and other modified asphalt both have good high-temperature performance and economy, but low-temperature performance is difficult to meet expectations [7–10]. For this reason, researchers are still searching for modifiers with good modification effects [11,12]. This paper investigated the effect of a high-performance resin material, SPUA, on high-temperature performance and fatigue performance of asphalt binders. SPUA is a kind of polymer material containing urea bonds, carbamate and other functional groups [13]. After curing, SPUA has extremely high tensile strength, impact strength, waterproofing, anti-corrosion and wear resistance and other excellent properties. It has been widely used in many fields, such as the military industry, coating, waterproofing, anticorrosion and engineering construction [14–16]. However, in the asphalt road field, fewer researchers are studying the preparation method and modification effect of SPMA. It is well known if the modification effect is excellent determines whether the modified material can be widely used in practical engineering. Therefore, it is necessary to conduct research on the road performance aspects of SPMA, evaluate the applicability of SPUA to asphalt binders and provide a relevant basis for research, application and promotion of SPUA in asphalt pavements.

High-temperature performance and fatigue performance of asphalt binders have been the focus of researchers in response to the progressively harsher high-temperature environment and increasing traffic volume [17–21]. Many complete and mature test methods and evaluation systems have been developed for high-temperature performance of asphalt binders. Among them, due to their higher sensitivity, accuracy and precision, rheological-based testing methods have gradually become mainstream, for example, the PG test, MSCR test, zero shear viscosity test (ZSV), elastic viscosity index, frequency sweeping, etc. [22,23]. For fatigue resistance performance of asphalt binders, the main indicators include fatigue factor ($|G^*|\sin\delta$), damage factor ($\tan\delta$) and fatigue life based on the TS test and LAS test. However, $|G^*|\cdot\sin\delta$ and $\tan\delta$ are fatigue performance parameters obtained in low loading times and linear viscoelastic range, but these two indicators lack relevance with actual fatigue performance of asphalt pavements [24–26].

In summary, researchers have proposed many relevant indicators for high-temperature performance and fatigue resistance of asphalt binders. However, due to the different loading methods and calculation principles between the tests, there are different degrees of difference among the indicators, which affects evaluation of real performance of asphalt binders [17,26,27]. Therefore, to accurately evaluate degree of influence of SPUA on asphalt binders, it is necessary to analyze applicability of different performance indicators to SPMA.

The main objective of this study focused on the optimal preparation process parameters, high-temperature performance and fatigue performance evaluation of SPMA. First, the optimum preparation process parameters of SPMA were determined based on orthogonal test combined with range analysis and variance analysis. On this basis, the high-temperature performance and fatigue resistance of SPMA with different contents were studied. Among them, the high-temperature performance characterization methods include PG test and MSCR test, and the fatigue resistance test includes LAS test and TS test. Finally, the applicability of different indicators to the performance evaluation of SPMA was investigated by correlation analysis.

2. Materials, Preparation and Test Methods

2.1. Materials

2.1.1. Asphalt Binder

The base asphalt binder used in this paper was 70# base asphalt binder (Kunlun brand, Petrochina Fuel Oil Co., Ltd., Zhuhai, China), and its main performance indicators are listed in Table 1.

Table 1. Properties of base asphalt binder.

Properties		Unit	Test Result	Specification
Performance grade		°C	64	ASTM D7643-10
Softening point		°C	46.25	ASTM D36
Ductility (10 °C)		cm	60	ASTM D113
Penetration (25 °C)		0.1 mm	65	ASTM D5
Dynamic viscosity (60 °C)		Pa·s	218	ASTM D2170
After TRTOF Aging	Mass variation	%	−0.114	ASTM D2872
	Penetration ratio	%	≥81	ASTM D5
	Ductility (10 °C)	cm	≥6	ASTM D113

2.1.2. SPUA Base Material

The modifier used in this paper is sprayed polyurea (SPUA), a high-performance resin material containing many strong polar functional groups: urea bonds and carbamate [10]. Figure 1 shows the molecular formulas of urea bond and carbamate. The technical indicators of SPUA are shown in Table 2. In this study, we expect to process the SPUA base material to a suitable size by grinding it and applying it in asphalt binders. However, during the grinding process, it was found that the excellent stress absorption properties of SPUA at normal temperature greatly deteriorated the efficiency of the grinding machine. Therefore, this study took advantage of the high brittleness of the polymer at the glass transition temperature to grind the SPUA base material in a low-temperature environment. Figure 2 showed the particle size distribution of the SPUA modifier after grinding. The solid red line shows the volume proportion of SPUA particles. The red dotted lines and orange dots indicate the frequency with which the SPUA particles are smaller than that size. After the grinding process, 90% of the SPUA modifier particles were below 122.5 μm and the maximum particle size was 74.5 μm. This proved that the low-temperature grinding technology has high processing efficiency. Finally, SPUA particles obtained by low-temperature grinding and processing were used as asphalt binder modifiers in this study, and selected quality of asphalt binder was 6%, 9% and 12% as the modifier content.

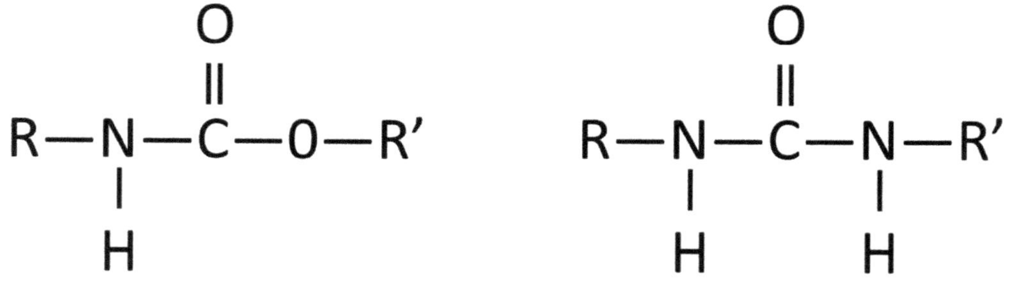

(a) Molecular structure of the urea bond (b) Molecular structure of the carbamate

Figure 1. Molecular structure of the urea bond and carbamate.

Table 2. Technical properties of SPUA.

Solid Content/%	Viscosity/cps	Tensile Strength/MPa	Elongation at Break/%
81–85	≤800	28	375

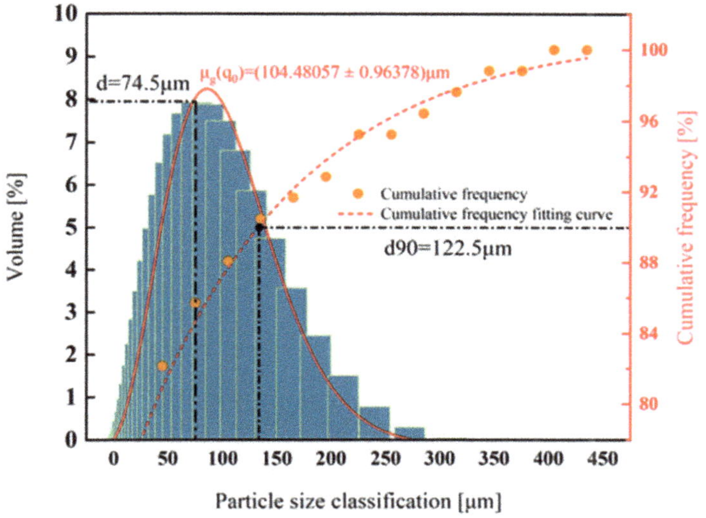

Figure 2. Particle size distribution of the SPUA modifier.

2.2. Test Method

The main research routes are shown in Figure 3. First, the optimum shearing time, rate and temperature of the SPMA were determined by an orthogonal test. On this basis, the high-temperature performance and fatigue performance of SPMA with different contents were studied. Among them, the high-temperature performance included the RV test, PG test and MSCR test. The fatigue performance tests included the LAS test and the TS test. Finally, applicability of various performance indexes to the SPMA was evaluated.

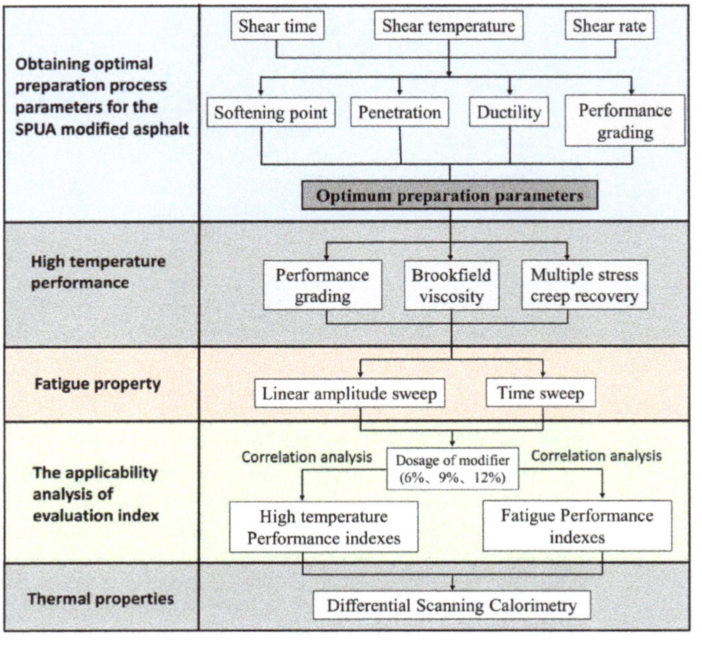

Figure 3. Research routes.

2.2.1. Orthogonal Experimental Design

Previous research results have indicated that shear time, rate and temperature are important factors affecting the cementing properties of modified asphalt [26–28]. Proper shear time and temperature are conducive to swelling of polymer modifiers in asphalt binders and promote interaction between asphalt binders and modifiers. The appropriate shear rate has a positive effect on the dispersion and uniformity of the modifier in the asphalt binder.

Therefore, in this study, the $L_9(3^4)$ orthogonal test table was made with shear time, rate and temperature as factors. However, the interaction between factors was not considered in this study, and the fourth column of the orthogonal test table was used as a blank column [28]. The content of SPUA in orthogonal test is 9%. This is because when the content of SPUA changes from 9% to 6% or 12%, only the content change is 3%, which reduces the influence of the content change on the test results.

In the early preparation, it was found that the base asphalt binder used in this study was sensitive to temperature: when the temperature was low (120 °C), the base asphalt binder could be completely melted, exhibiting good fluidity; when the temperature was close to 160 °C, there was an obvious smoke phenomenon of asphalt binder, accompanied by deterioration of ductility. This was caused by the high content of light components in the base asphalt binder used. Given this, to avoid excessive influence of process parameters on the original properties of base asphalt binder and to corporately characterize the real effect of SPUA modifier on the properties of base asphalt binder, the level of the orthogonal test was selected within a reasonable range in this study [29,30]. The specific process parameters and orthogonal test tables are listed in Tables 3 and 4, respectively.

Table 3. The factor and levels table of the orthogonal experiment.

	Factors		
	Shear time (A)	Shear rate (B)	Shear temperature (C)
Levels	3000 rpm	10 min	120 °C
	4000 rpm	25 min	135 °C
	5000 rpm	40 min	150 °C

Table 4. Orthogonal test table.

Test Number	Shear Time (A)	Shear Rate (B)	Shear Temperature (C)	Test Program
1	120 °C	10 min	5000 rpm	$A_1B_1C_1$
2	120 °C	25 min	6000 rpm	$A_1B_2C_2$
3	120 °C	40 min	7000 rpm	$A_1B_3C_3$
4	135 °C	10 min	6000 rpm	$A_2B_1C_2$
5	135 °C	25 min	7000 rpm	$A_2B_2C_3$
6	135 °C	40 min	5000 rpm	$A_2B_3C_2$
7	150 °C	10 min	7000 rpm	$A_3B_1C_3$
8	150 °C	25 min	5000 rpm	$A_3B_2C_1$
9	150 °C	40 min	6000 rpm	$A_3B_3C_2$

2.2.2. Modified Asphalt Binder Preparation Process

First, the base asphalt binder was put into the oven at 135 °C for 1.5 h until it was completely melted. The corresponding mass of SPUA was divided into three equal parts and added to the base asphalt binder in turn. To ensure the initial mixing of SPUA and the base asphalt binder, a glass rod should be used to stir manually for 1 min after each SPUA addition. Then, the asphalt binder was placed in an oil bath at the corresponding shear temperatures (120 °C, 135 °C and 150 °C) and held for 30 min. The asphalt binder was then prepared using a high-speed shear according to the process parameters determined in the orthogonal test table. The sheared asphalt binder was put into a 135 °C oven for 1 h to provide conditions for the swelling of SPUA. Finally, the glass rod was used to slowly stir the finished asphalt binder to eliminate the tiny air bubbles. The modified asphalt binder preparation process is shown in Figure 4.

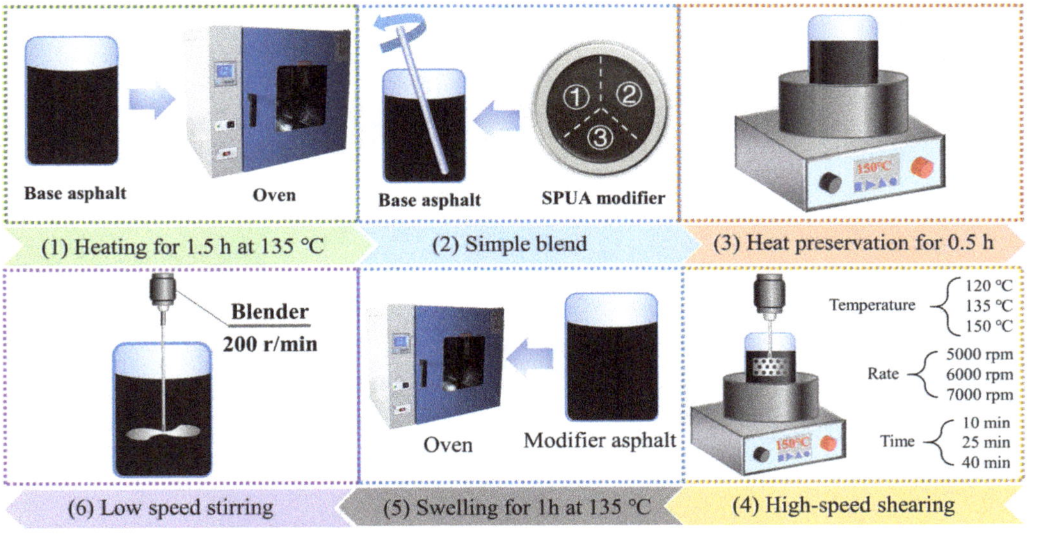

Figure 4. The modified asphalt preparation process.

2.2.3. Basic Performance Evaluation

The softening point, penetration (25 °C) and ductility (10 °C) of the modified asphalt were measured according to ASTM D113, ASTM D5 and ASTM D36, respectively. The testing instruments of softening point, penetration, and ductility were the WSY-025F asphalt softening point tester, WSY-026C automatic asphalt penetration tester, and LYY-10A asphalt elongation tester produced by Wuxi Petroleum Company of China.

2.2.4. High-Temperature Rheological Properties

a. Brookfield viscosity test

Brookfield viscosity was performed test according to ASTM D4402. The test temperatures were selected as 90 °C, 115 °C, 135 °C, 155 °C and 175 °C. Brookfield viscosity instrument was NDJ-1F produced by Changji Company in Shang-hai, China.

b. Performance grade test

In this study, the Smartpave102 dynamic shear rheometer from Anton Paar (Graz, Austria) was used. The performance grade (PG) test was performed according to ASTM D7643-10. The test was performed with a 25 mm diameter rotor, and the gap between parallel plates was controlled at 1 mm. The loading strain and frequency were 12% and 10 rad/s. The test temperature range was 46–76 °C, and the temperature rise interval was 6 °C.

c. Multiple stress creep recovery test

The multiple stress creep recovery (MSCR) test was performed according to ASTM-D7404-10a. The test temperature was chosen to be 64 °C. The rotor of the rheometer was 25 mm with a clearance control of 1 mm. First, the asphalt binder was loaded 20 times in a "load-recovery" cycle under a 0.1 kPa load. In this case, the first 10 cycles of loading were to equilibrate the specimens. Then, the asphalt binder was cyclically loaded 10 times under a 3.2 kPa load. In each cycle of loading, the asphalt binder specimens were loaded for 1 s and unloaded for 9 s. The specific calculation equations are as follows:

$$R = \frac{\gamma - p - \gamma_{nr}}{\gamma - p - \gamma_0} \tag{1}$$

$$J_{nr} = \frac{\gamma_{-r} - \gamma_0}{\tau} \tag{2}$$

$$J_{nr-diff} = \frac{[J_{nr}(3.2\ \text{kPa}) - J_{nr}(0-1\ \text{kPa})]}{J_{nr}0.1\ \text{kPa}} \times 100\% \tag{3}$$

where γ_p was the peak strain within each cycle (%), γ_{nr} was the residual strain within each cycle (%) and γ_0 was the initial strain within each cycle (%); τ was the load (kPa).

2.2.5. Fatigue Performance

The fatigue test temperatures in this study were all 25 °C. The rotor diameter was chosen to be 8 mm and the clearance was controlled at 2 mm.

a. Linear amplitude sweep test

The linear amplitude sweep (LAS) test was conducted in accordance with AASHTO TP 101. The non-destructive characteristic value α of the asphalt binder was first determined by frequency scanning. The strain of the amplitude sweep increased linearly from 0% to 30% within 5 min. The calculation method is as follows:

First, the cumulative strength of the damage $D(t)$ was calculated.

$$D(t) \cong \sum_{i=1}^{N} \left[\pi\gamma_0^2(C_{i-1} - C_i)\right]^{\frac{\alpha}{1+\alpha}} \cdot (t_i - t_{i-1})^{\frac{1}{1+\alpha}} \tag{4}$$

where C_i was the integrity parameter, calculated as $C(t) = \frac{|G^*|(t)}{|G^*|_{initial}}$, $|G^*|$ was complex shear modulus; γ_0^2 was the strain value (%) of the measured data points and t was the test time.

Then, $C(t)$ and $D(t)$ were fitted according to Equation (5) and the curve fitting coefficients C_1 and C_2 were obtained.

$$C(t) = C_0 - C_1(D)^{C_2} \tag{5}$$

where C_0 was the initial value of C, C_0 = 1; C_1 and C_2 were the damage fitting parameters.

D_f was then defined as the damage value of the asphalt material at the peak shear stress corresponding to the failure. the formula for D_f was given in Equation (6).

$$D_f = \left(\frac{C_0 - C\ at\ peak\ stress}{C_1}\right)^{\frac{1}{C_2}} \tag{6}$$

Finally, the fatigue life N_f of the asphalt material was calculated by Equation (7).

$$N_f = A(\gamma_{max})^{-B} \tag{7}$$

where N_f was the fatigue life; A and B were parameters; γ_{max} was the maximum allowable strain, including 2.5% and 5%. Equations for A and B were shown in Equations (8) and (9).

$$A = \frac{f\left(D_f\right)^k}{k(\pi C_1 C_2)^{\alpha}} \tag{8}$$

$$B = 2\alpha \tag{9}$$

where f was the loading frequency (10 Hz); $k = 1 + (1 - C_2)\alpha$. C_1 a–d C_2 were damage curve fitting parameters.

b. Time sweeping test

The time sweeping (TS) was in strain control mode with a strain of 5%. This was determined from strain scans to ensure that the asphalt binders were in a linear viscoelastic range during the test. The *DER* is calculated as follows:

$$W_i = \int \delta(t)\frac{d\varepsilon(t)}{d(t)}dt = \pi\delta_i\varepsilon_i sin(\delta_i) \quad (10)$$

$$W_c = \sum_{i=1}^{n} W_i \quad (11)$$

$$DER_i = \frac{\sum_{i=1}^{n} W_i}{W_n} = \frac{W_c}{W_n} \quad (12)$$

where W_i was the dissipation energy at the cycle i; W_c was the cumulative dissipation energy; DER_i was the cumulative dissipation energy ratio during the cycle i; $\sigma(t)$ was tle stress, t was the time; $\varepsilon(t)$ was the strain; δ_i was the phase angle at the cycle i; W_n was thI dissipation energy at the cycle n.

2.2.6. Differential Scanning Calorimetry Test

DSC3 differential scanning calorimeter produced by Mettler Toledo, Greifensee, Switzerland, was adopted in this section. The samples were about 8–9 mg, and the samples were made of aluminum. The test temperature was −20–200 °C, and the heating rate was 10 °C/min.

3. Results and Discussion

3.1. Orthogonal Test Analysis

The results of orthogonal tests were often analyzed by the extreme difference analysis method and ANOVA, which were used to determine the optimal production conditions. Among them, the extreme difference analysis method can determine the degree of influence of the factor on the performance index based on the R-index of the factor. The ANOVA method can determine whether a factor is significant for a performance indicator based on the *p*-value. The results of the orthogonal test are shown in Figure 5.

3.1.1. Extreme Difference Analysis Method

The results of the extreme difference analysis for the failure temperature, softening point, penetration and ductility tests are listed in Tables 5–8, respectively. From the R in Tables 5–7, the optimal process parameters combination corresponding to failure temperature, softening point and pinning degree was $A_3B_3C_3$. The influence degree of the factors on the performance indexes in descending order were C (shear temperature) > A (shear time) > B (shear rate). This indicated that shear temperature was the most important factor affecting the failure temperature, softening point and penetration of the SPMA. Failure temperature and softening point were the indexes of high-temperature performance of asphalt binder [18]. The penetration represented the consistency of the asphalt, which indirectly reflected the high-temperature performance [31]. Therefore, this may be the reason why the three indicators had the same optimal process mix.

It was noteworthy that all the above three performance indexes reached their optimal values at level 3, which pinpointed that increasing the shear time, rate and temperature were all beneficial to improve the high-temperature performance and consistency of the SPMA. According to the available research results, this was mainly due to the following reasons: the increase in shear time, temperature and rate accelerated the swelling of the modifier, promoted the combination of the modifier with the asphalt mastic and improved the dispersion of the modified asphalt [26]. Second, the increased shear time, temperature and rate promoted volatilization and transformation of light components within the asphalt, accelerating the oxidation and aging of the asphalt components [32].

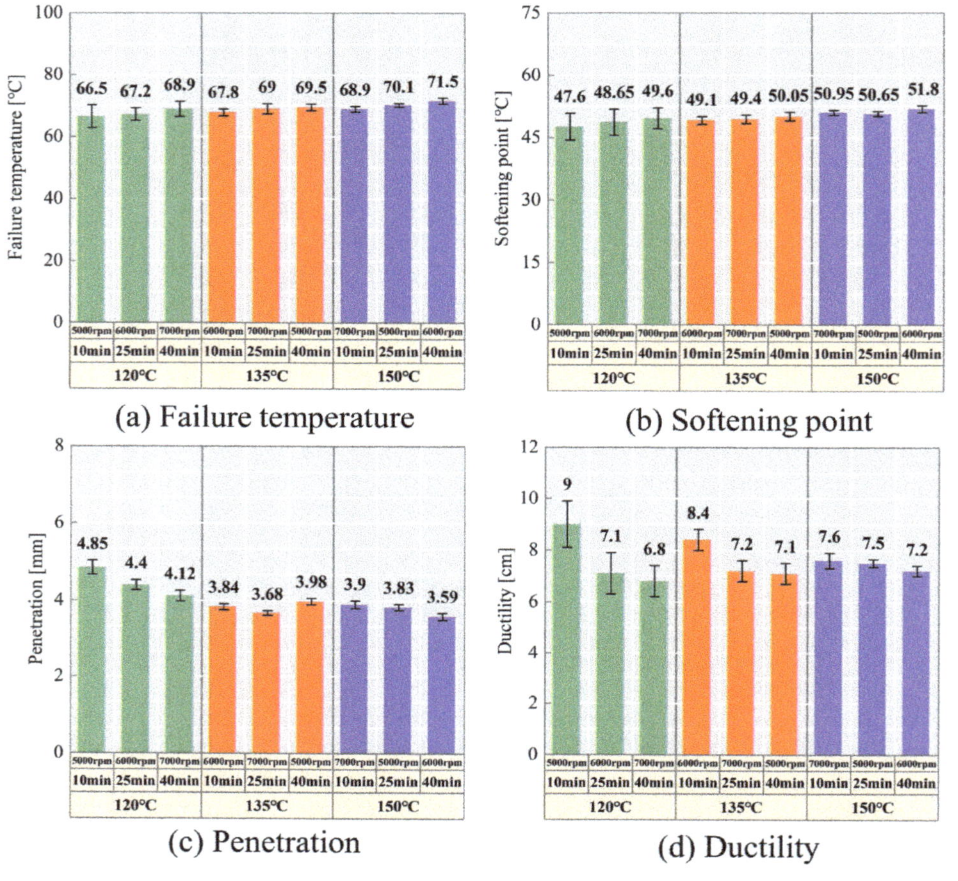

Figure 5. Orthogonal experimental results.

Table 5. Failure temperature range analysis results.

Test Number	Shear Time	Shear Rate	Shear Temperature	Blank Group	Failure Temperature
1	1	1	1	1	66.5
2	2	2	1	2	67.2
3	3	3	1	3	68.9
4	1	2	2	3	67.8
5	2	3	2	1	69
6	3	1	2	2	69.5
7	1	3	3	2	68.9
8	2	1	3	3	70.1
9	3	2	3	1	71.5
K1	203.20	206.10	202.60	207.00	-
K2	206.30	206.50	206.30	205.60	-
K3	209.90	206.80	210.50	206.80	-
k1	67.73	68.70	67.53	69.00	-
k2	68.77	68.83	68.77	68.53	-
k3	69.97	68.93	70.17	68.93	-
R	2.23	0.23	2.63	0.47	-
$R_C > R_A > R_B$, optimal process: $A_3B_3C_3$					

Table 6. Softening point range analysis results.

Test Number	Shear Time	Shear Rate	Shear Temperature	Blank Group	Softening Point
1	1	1	1	1	47.6
2	2	2	1	2	48.65
3	3	3	1	3	49.6
4	1	2	2	3	49.1
5	2	3	2	1	49.4
6	3	1	2	2	50.05
7	1	3	3	2	50.95
8	2	1	3	3	50.65
9	3	2	3	1	51.8
K1	147.65	148.30	145.85	148.80	-
K2	148.70	149.55	148.55	149.65	-
K3	151.45	149.95	153.40	149.35	-
k1	49.22	49.43	48.62	49.60	-
k2	49.57	49.85	49.52	49.88	-
k3	50.48	49.98	51.13	49.78	-
R	1.27	0.55	2.52	0.28	-
$R_C > R_A > R_B$, optimal process: $A_3B_3C_3$					

Table 7. Penetration range analysis results.

Test Number	Shear Time	Shear Rate	Shear Temperature	Blank Group	Penetration
1	1	1	1	1	4.9
2	2	2	1	2	4.4
3	3	3	1	3	4.1
4	1	2	2	3	3.8
5	2	3	2	1	3.7
6	3	1	2	2	4.0
7	1	3	3	2	3.9
8	2	1	3	3	3.8
9	3	2	3	1	3.6
K1	12.59	12.66	13.37	12.12	-
K2	11.91	11.83	11.50	12.28	-
K3	11.69	11.70	11.32	11.79	-
k1	4.20	4.22	4.46	4.04	-
k2	3.97	3.94	3.83	4.09	-
k3	3.90	3.90	3.77	3.93	-
R	0.30	0.32	0.68	0.16	-
$R_C > R_B > R_A$, optimal process: $A_3B_3C_3$					

From Table 8, the factors affecting the index of ductility were A (shear time) > B (shear rate) > C (shear temperature) in order, and the optimal preparation parameters were $A_1B_1C_1$. It can be found that the ductility decreased with an increase in the three factors. The reason for this was mainly due to the transformation of the components and the aging of the asphalt to improve the stiffness of the asphalt binder.

The extreme difference analysis method can reflect the influence of different factors on performance, while it was still unclear whether the influence of different levels on the performance index was significant. Therefore, the following will combine the results of the ANOVA and further optimize the preparation parameters in terms of energy consumption and preparation efficiency.

Table 8. Ductility range analysis results.

Test Number	Shear Time	Shear Rate	Shear Temperature	Blank Group	Ductility
1	1	1	1	1	9.0
2	2	2	1	2	7.1
3	3	3	1	3	6.8
4	1	2	2	3	8.4
5	2	3	2	1	7.2
6	3	1	2	2	7.1
7	1	3	3	2	7.6
8	2	1	3	3	7.5
9	3	2	3	1	7.2
K1	25.00	23.60	22.90	23.40	-
K2	21.80	22.70	22.70	21.80	-
K3	21.10	21.60	22.30	22.70	-
k1	8.33	7.87	7.63	7.80	-
k2	7.27	7.57	7.57	7.27	-
k3	7.03	7.20	7.43	7.57	-
R	1.30	0.67	0.20	0.53	-
$R_A > R_B > R_C$, optimal process: $A_1B_1C_1$					

3.1.2. Analysis of Variance

From Tables 9–12, shear temperature had a significant effect on the failure temperature, softening point and penetration. Shear time had a significant effect on the failure temperature and softening point, and the shear rate had no significant effect on any of the four indices. Therefore, the shear rate can be reduced appropriately.

Table 9. Failure temperature variance analysis results.

Test Number	Shear Time	Shear Rate	Shear Temperature	Blank Group
Sj	7.50	0.08	10.42	0.38
DOF	2.00	2.00	2.00	2.00
MSE	3.75	0.04	5.21	0.19
F	19.61	0.22	27.25	1.00
P	0.05	0.82	0.04	0.50
Sig	√	×	√	

Note: √ indicated that the selected factors had a significant impact on the index and × indicated that the selected factors had no significant effect on the index.

Table 10. Softening point variance results.

Test Number	Shear Time	Shear Rate	Shear Temperature	Blank Group
Sj	2.57	0.49	9.76	0.12
DOF	2.00	2.00	2.00	2.00
MSE	1.28	0.25	4.88	0.06
F	20.72	3.99	78.76	1.00
P	0.05	0.20	0.01	0.50
Sig	√	×	√	-

Note: √ indicated that the selected factors had a significant impact on the index and × indicated that the selected factors had no significant effect on the index.

Table 11. Penetration variance results.

Test Number	Shear Time	Shear Rate	Shear Temperature	Blank Group
Sj	0.15	0.18	0.86	0.04
DOF	2.00	2.00	2.00	2.00
MSE	0.07	0.09	0.43	0.02
F	3.53	4.34	20.64	1.00
P	0.22	0.19	0.05	0.50
Sig	×	×	√	-

Note: √ indicated that the selected factors had a significant impact on the index and × indicated that the selected factors had no significant effect on the index.

Table 12. Ductility variance results.

Test Number	Shear Time	Shear Rate	Shear Temperature	Blank Group
Sj	2.88	0.67	0.06	0.43
DOF	2.00	2.00	2.00	2.00
MSE	1.44	0.33	0.03	0.21
F	6.72	1.56	0.15	1.00
P	0.13	0.39	0.87	0.50
Sig	×	×	×	-

Note: × indicated that the selected factors had no significant effect on the index.

In addition, by comparing the k values in the results of the extreme difference analysis, it can be found that, for the failure temperature, softening point and penetration, the difference between k3 at the optimal level and k2 was not significant. For the ductility, the difference between k1 at the optimal level and k2 was not significant. Therefore, considering the high- and low-temperature performance and energy consumption, $A_3B_2C_3$ (shear time, rate and temperature of 40 min, 6000 rpm and 150 °C, respectively) was selected as the best process parameter for preparing the SPMA.

3.2. *High-Temperature Rheological Properties*

3.2.1. Brookfield Viscosity

(1) Apparent viscosity

The SPMA with different dosages were prepared using optimal process parameters. Figure 6 manifested the viscosity–temperature curves of the asphalt binder, showing the effect of modifier and temperature on the apparent viscosity of the asphalt binder. Enhancement in temperature intensified the thermal movement of molecules, manifested as a gradual decrease in the viscosity–temperature curve of asphalt binder, easier to overcome the frictional resistance between molecules and produce flow deformation [33]. To ensure that the asphalt binder had good constructability, the specification requires the apparent viscosity of less than 3 Pa·s at 135 °C. Figure 5 displayed that the viscosity of the different asphalt binders was already less than 3 Pa·s at 115 °C. This demonstrated that the modified asphalt can be applied at lower temperatures, which reduced energy consumption and waste gas emissions [34].

In addition, the modifier enhanced the viscosity of the asphalt binder. With the boost in dosage, the improvement in viscosity was more obvious. For example, at 135 °C, the viscosity of 6%, 9% and 12% modified asphalt binders increased by 10.8%, 28.5% and 52.2%, respectively, compared to the base asphalt binder. This revealed that addition of modifiers impeded the flow of asphalt binder and had a positive effect on the high-temperature performance [35,36].

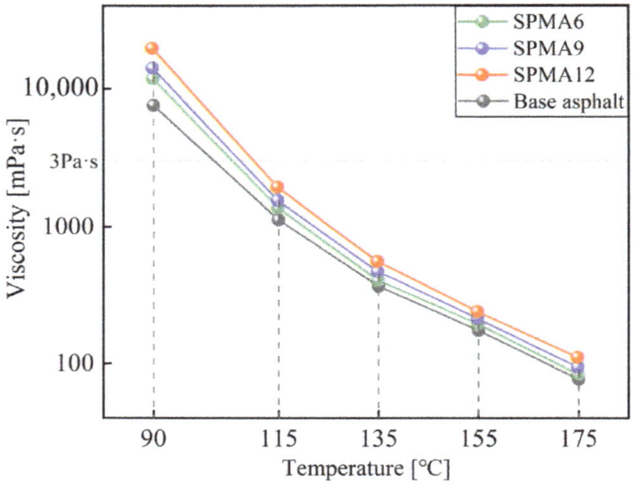

Figure 6. Apparent viscosity of asphalt binders.

(2) Activation energy

In addition, the temperature sensitivity of the asphalt binder was calculated by Arrhenius Equation (13) [37]. The specific calculation formula is as follows:

$$\eta = A \cdot e^{\frac{E_\alpha}{RT}} \quad (13)$$

where η was the apparent viscosity of asphalt binder, A was the fitting parameter, E_α (kJ/mol) was the activation energy of asphalt mastic, R was the gas constant, R = 8.314 J/(mol·K) and K was the absolute temperature.

After processing, the Arrhenius Equation (13) can be expressed by Equation (14).

$$ln\eta = lnA + \frac{E_\alpha}{R} \times \frac{1}{T} \quad (14)$$

The activation energy of the asphalt binders is exhibited in Figure 7. The specific calculation results are shown in Table 13. The R^2 was greater than 0.97, which expressed a good fitting result. With the addition of 6%, 9% and 12% modifiers, the activation energy of the asphalt binders grew by 7.4%, 8.8% and 13.0% compared to the base asphalt binder. As a result, the asphalt binder required more external energy when it produced flow deformation. This implied that mounting the modifier content improved the temperature sensitivity of the asphalt binder [34,38]. It may be related to the following reasons: first, the polymer modifier will absorb the light component of the asphalt binder in the process of swelling, which indirectly enhanced the content of heavy components in the asphalt system [39]. Second, when the polymer modifier was mixed with the asphalt binder, the stiffness and internal friction resistance will be increased, which will hinder the flow of the asphalt binder [37,40]. Under the synergistic action of the above two reasons, the activation energy of asphalt binder improved with the addition of modifier content.

Table 13. The fitting result of activation energy.

Asphalt Type	E_α (Slope, kJ/mol)	A (Intercept)	R^2
Base asphalt	21.5	−7.4	0.98
SPMA6	23.1	−8.1	0.98
SPMA9	23.4	−8.2	0.97
SPMA12	24.3	−8.5	0.97

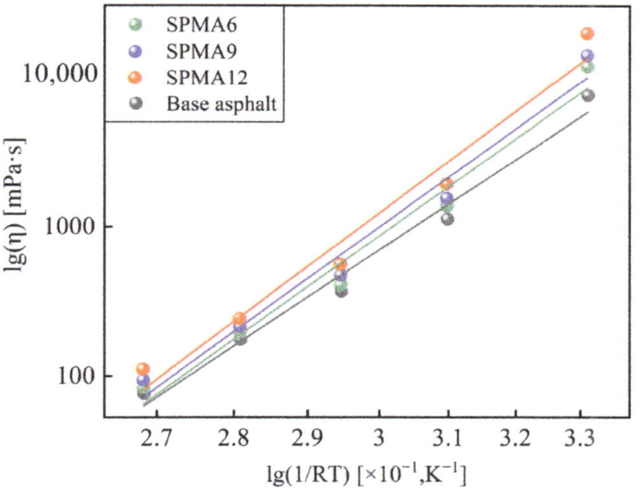

Figure 7. Activation energy of asphalt binders.

3.2.2. PG Test

(1) Rutting factor

Figure 8a shows the $G^*/\sin\delta$ of asphalt binder as a function of temperature. It was obvious that the enhancement in temperature reduced the $G^*/\sin\delta$, revealing a negative effect on the deformation resistance at high temperatures. Meanwhile, in agreement with the prediction, the addition of modifiers improved the $G^*/\sin\delta$. Moreover, the high-temperature failure temperature of the asphalt binder was calculated according to the specification. In the specification, the high-temperature failure temperature was defined as the temperature at the $G^*/\sin\delta$ of 1 kPa, and the calculation results are shown in Figure 8b. It can be found that the failure temperature gained by 5.89 °C, 6.23 °C and 9.8 °C respectively with the addition of modifier dosage. This meant that the modifier improved the high-temperature deformation resistance of the asphalt binders, which was equivalent to expanding the ambient temperature range that the asphalt binder can withstand during the service life. This was due to the absorption of light components and polymer network formation by the modifier [17].

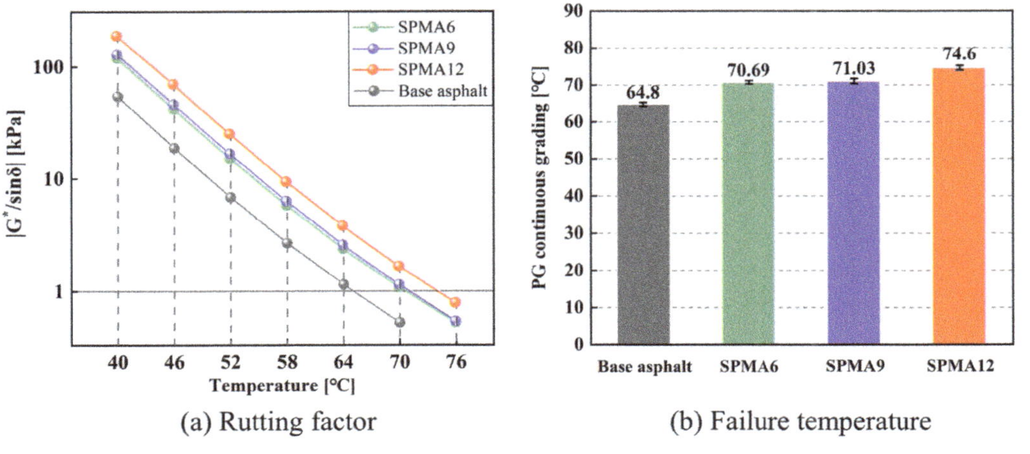

Figure 8. Rutting factor of asphalt binders.

It was noteworthy that the failure temperatures of the modified asphalt with 6%, 9% and 12% dosage were increased by 9.1%, 9.6% and 15.1% relative to the base asphalt binder. It can be found that the $G^*/\sin\delta$ of the modified asphalt binder did not produce a significant increase when increasing from 6% to 9%, and the same phenomenon was reflected in the $G^*/\sin\delta$ curve. This may be related to the sensitivity of the test method to the viscoelastic properties of the asphalt binder.

In order to investigate the applicability of the $G^*/\sin\delta$ of the PG test to the characterization of the high-temperature properties, two improved rutting factors: $G^*/(\sin\delta)^9$ and $G^*/[1 - 1/(\tan\delta\sin\delta)]$ were also used in this paper to analyze the test results. The specific results are manifested in Figure 9. The results demonstrated that the two improved rutting factors can obviously reflect the performance difference between modified asphalt and base asphalt but still cannot effectively reflect the high-temperature performance of modified asphalt at 6% and 9% dosage.

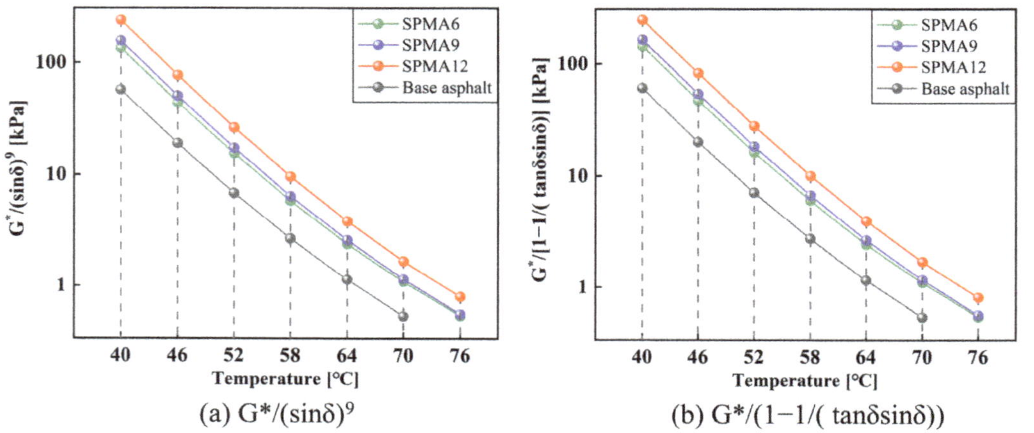

Figure 9. Improved rutting factor.

(2) Phase angle

Figure 10 reveals the results of the phase angle for the PG test. As the temperature increased, the phase angle showed a gradual increase. It meant that the improvement in temperature caused the asphalt binder to gradually lose its elastic properties and transformed into a fully cohesive material. Compared with the base asphalt, the SPUA can effectively reduce the phase angle, and the reduction was enhanced with an increase in the dosage. For example, at 40 °C, the phase angles of the modified asphalt with 6%, 9% and 12% dosage decreased by 3.5°, 6.1° and 7.6° compared to the base asphalt. This signaled that the modifier played a positive role in the elastic recovery performance of the asphalt binder, enabling it to recover quickly after deformation. This was because the modifier changes the colloidal structure of the asphalt binder and improves the internal elasticity of the components.

(3) Temperature susceptibility

In addition, The $G^*/\sin\delta$-T curves in logarithmic coordinates were used to characterize the temperature sensitivity:

$$ln|G^*| = AlnT + B \tag{15}$$

Parameter A was the temperature sensitivity index; its larger absolute value represented higher temperature sensitivity of asphalt binder. The fitting results are shown in Figure 11 and Table 14. The temperature sensitivity of asphalt binder was boosted with the addition of modifier dosage in the range of 40–70 °C, which signified that, the greater the dosage of SPMA, the more obvious the magnitude of high-temperature performance deterioration in the process of temperature rise. This was because of the incompatibility between the modifier and the asphalt binder, and the strength of the bonding interface was susceptible to weakening due to temperature effects, which reduced the contribution of the bonding properties to the strength of the asphalt binder. The higher the modifier dose, the more the bonding interface inside the asphalt binder and the greater the strength weakening at elevated temperatures, which explained the increase in temperature sensitivity of asphalt binder with additional modifier dosage from 40–70 °C.

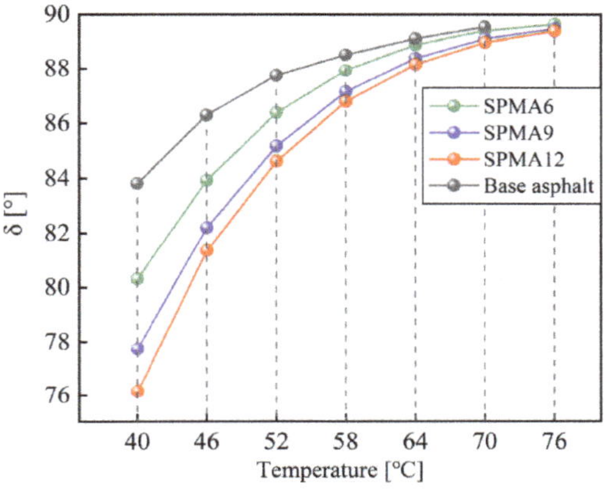

Figure 10. The phase angle of asphalt binders.

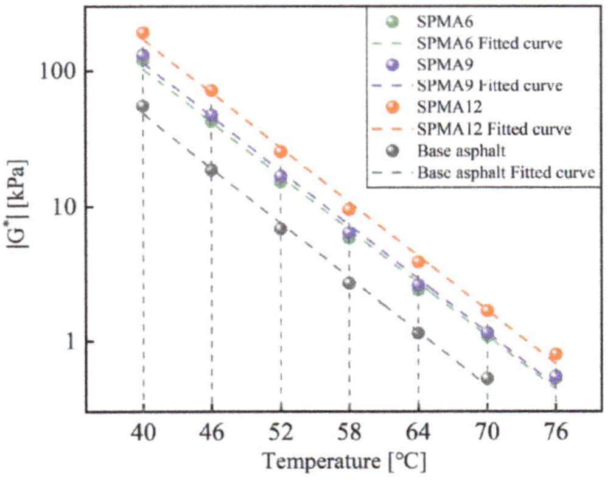

Figure 11. Temperature sensibility.

Table 14. The fitting result of temperature sensibility.

Asphalt Type	∣A∣ (Slope)	B (Intercept)	R^2
Base asphalt	8.32	18.10	0.99
SPMA6	8.53	18.79	0.99
SPMA9	8.61	18.97	0.99
SPMA12	8.64	19.20	0.99

Notably, this was not consistent with the temperature-sensitive results for Brookfield viscosity. This was probably due to the different contributions of internal friction resistance, bonding properties, etc., to temperature sensitivity in different temperature ranges. In the range of 40–70 °C, the internal friction resistance and the bonding properties of the modifier and the asphalt binder together played a positive role in the temperature sensitivity.

However, as the temperature enhanced to within the Brookfield viscosity test temperature range, the asphalt binder took on a liquid state. At this time, the bonding effect between the asphalt binder and the modifier disappears and the internal friction resistance into the impact for the main factors of temperature sensitivity. The internal friction resistance was positively correlated with the modifier dosage, so the temperature sensitivity of the asphalt binder improves with increasing admixture at the Brookfield viscosity test temperature.

3.2.3. MSCR Test

In response to the inability of the PG test to effectively distinguish the difference in high-temperature performance between 6% and 9% modified asphalt binders, this study used the MSCR test for asphalt binders. Figure 12 shows the stress–strain relationship of the asphalt binder during the loading process. As can be seen from the figure, the strain of asphalt binders in the cyclic loading process constantly made the accumulation, resulting in irrecoverable deformation. Compared with 0.1 kPa load, the asphalt binder under 3.2 kPa load had higher cumulative deformation. This explained the greater susceptibility to rutting on asphalt roads with high traffic volumes and overloaded vehicles [18].

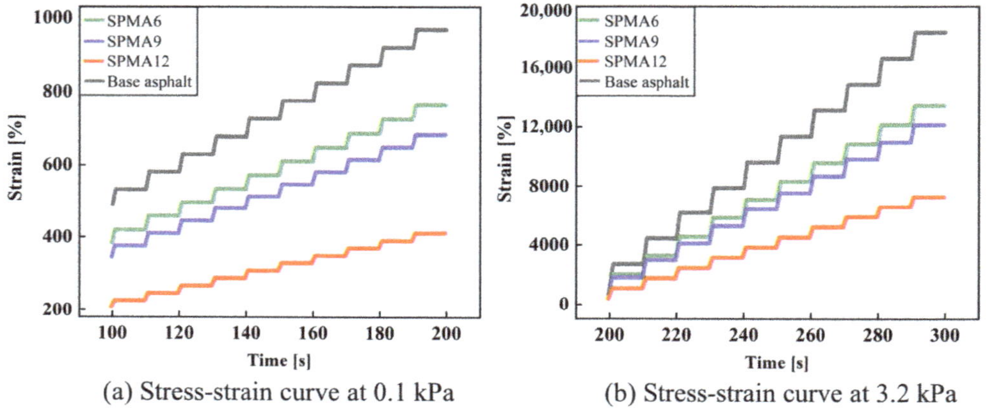

Figure 12. MSCR test results of asphalt binders.

Meanwhile, the strain value of asphalt binder decreased with an increase in modifier dosage. This indicated that the addition of the modifier reduced the deformation of asphalt binders, playing a positive effect on the anti-deformation performance. It had a positive effect on the deformation resistance, and the high-temperature performance can be further improved by increasing the modifier content.

Interestingly, at a load of 0.1 kPa, there was a significant difference between the 6% and 9% dosage of modified asphalt, implying a higher sensitivity of the MSCR test to the viscoelastic properties of the asphalt binders.

In addition, two other viscoelastic indices of asphalt binders can be obtained from the MSCR test: the recovery rate (R) and the irrecoverable creep flexibility (J_{nr}), where the R reflects the deformation recovery performance of the asphalt binder after unloading and J_{nr} reflects the level of permanent deformation of the asphalt binder under the load. As shown in Figure 13, the R of the base asphalt was only 0.02 at 0.1 kPa load, which almost completely lost the elastic recovery performance, and, after adding the modifier, the R of the asphalt binder improved significantly and gradually enhanced with an increase in modifier dosage The R of SPMA improved by 0.85, 1.59 and 1.72 over the base asphalt binder with the addition of 6%, 9% and 12% doping modifiers, respectively. At 3.2 kPa load, with an increase in modifier dosage, the trend of R of asphalt binder was the same as that at 0.1 kPa load. The R of asphalt binders increased by 0.048, 0.058 and 0.148 at 6%, 9% and 12% modifier dosage, respectively. This illustrated that increasing the amount of modifier was conducive to improvement in elastic recovery properties of asphalt binder. This was because the modifier improved the elastic component of the asphalt binders.

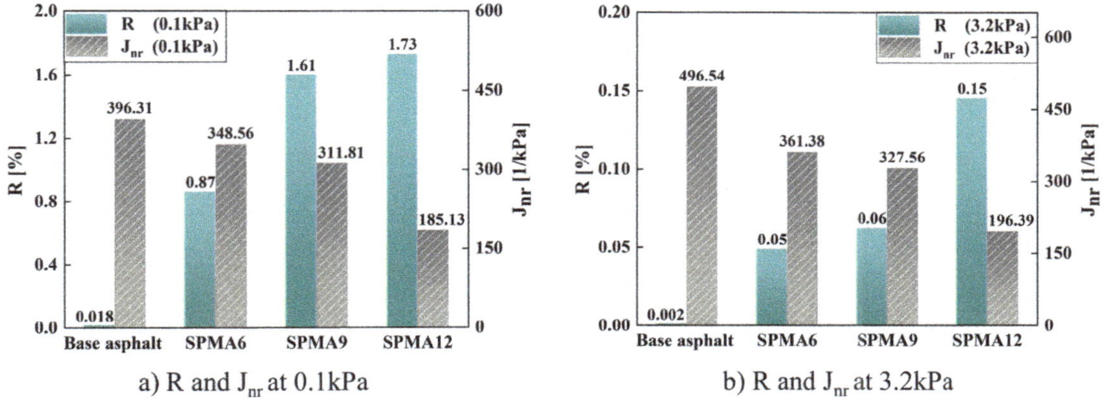

Figure 13. Calculating values of R and J_{nr}.

From Figure 13, the J_{nr} of asphalt binder decreases with the addition of modifier. At 0.1 kPa load, J_{nr} of asphalt binders with 6%, 9% and 12% modifier content was reduced by 12.5%, 2.3% and 53.3% compared with base asphalt binder, respectively. At 3.2 kPa load, J_{nr} of asphalt binders with 6%, 9% and 12% modifier content decreased by 27.2%, 34.0% and 60.4% compared with that of base asphalt binder, respectively. This meant that the modified asphalt with larger dosage had higher deformation resistance.

In addition, $J_{nr\text{-}diff}$ was used to characterize the stress sensitivity of the asphalt binder in the MSCR test. Figure 14 demonstrates the relationship between $J_{nr\text{-}diff}$ values and modifier dosage. As can be seen from the figure, the addition of the modifier reduced the stress sensitivity of the base asphalt binder. However, the stress sensitivity of the modified asphalt gradually increased with an increase in the modifier. It should be noted that, although the modified asphalt binders with high admixture were more sensitive to stress, this did not mean that the high admixture modifier was detrimental to the high-temperature performance of the asphalt binders. This was because the index only reflected the sensitivity of the asphalt binder to the applied stress.

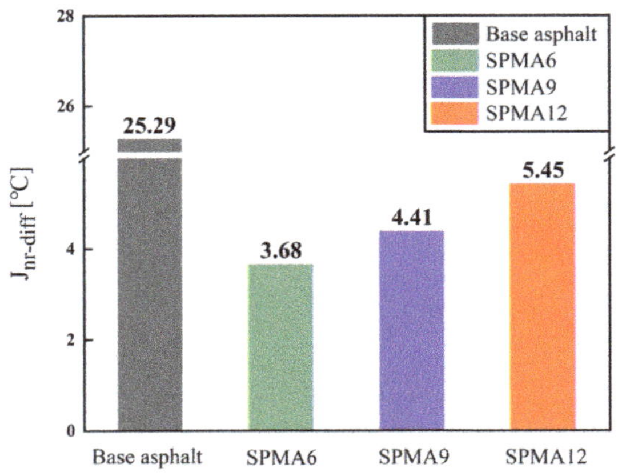

Figure 14. Calculating values of $J_{nr\text{-}diff}$.

3.3. Fatigue Performance Characterization

3.3.1. Fatigue Performance Analysis Based on LAS Test

Figure 15 describes the stress–strain curves of the asphalt binder in the LAS test. With an increase in strain, the stress rose rapidly at first and then decreased rapidly after reaching the maximum value. The relevant literature referred to the strain corresponding to the peak stress in the curve as the yield strain and defines it as the breaking point of the asphalt binder [41]. The yield strains of the base asphalt binder and the modified asphalt with different admixtures were 7.6%, 7.99%, 8.28% and 9.26%, respectively. This indicated that the modifier advanced the hardness of the asphalt binder [42].

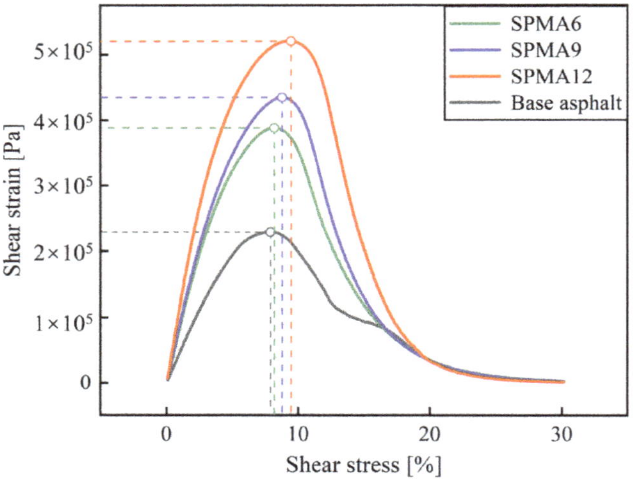

Figure 15. Stress–strain curve of asphalt binders.

Figure 16 displays the fatigue damage characteristic curves obtained by fitting the S-VECD model. The vertical coordinate of S-VECD was the integrity parameter (C) and the horizontal coordinate was the cumulative damage intensity. The high C signified that the asphalt mastic had excellent fatigue resistance [43]. It can be found that, when the damage intensity was small, the integrity parameters of different asphalt binders were the same. As the damage intensity gradually escalated, the integrity of the base asphalt decreased rapidly and the difference with the SPMA gradually became distinct. This suggested that both the base asphalt binder and the modified asphalt binder had essentially the same fatigue resistance in the early stages of loading. However, the fatigue resistance of the base asphalt binder decayed rapidly as the damage intensity gradually grew, demonstrating a higher strain sensitivity of the base asphalt binder.

To quantitatively analyze the effect of modifiers on the fatigue resistance of asphalt binders, the fatigue performance parameters A and B and the fatigue life of asphalt binders were calculated. Parameter A represented the integrity of the asphalt binder, and, the larger parameter A, the better the fatigue resistance of the asphalt binder. Parameter B represented the strain sensitivity of the asphalt binder in the fatigue test; the larger parameter B, the stronger the strain sensitivity of the asphalt binder [44]. As shown in Figure 17a, the integrity of the asphalt binder boosted with an increase in modifier dosage: 6%, 9% and 12% modifier dosage of asphalt binder increased the parameter A by 99.1%, 135.3% and 539.6%, respectively, compared to the base asphalt binder, while the B parameter illustrated that the modifier was detrimental to the stress sensitivity of the asphalt binder under cyclic loading. This may predict a more substantial decay in the fatigue life of the SPMA with enhancing strain.

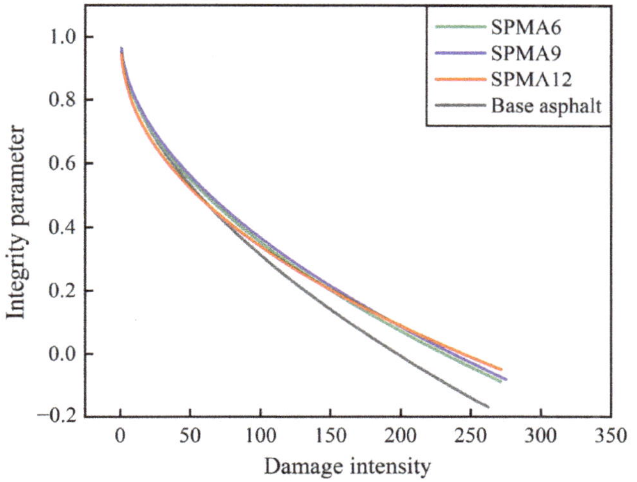

Figure 16. C–D curve of asphalt binders.

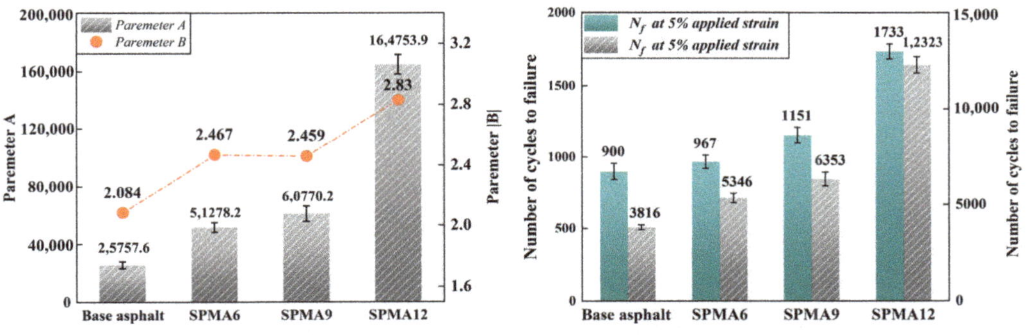

(a) Parameter A&B (b) Number of cycles to failure

Figure 17. LAS test result of asphalt binders.

The fatigue life of the asphalt binder is shown in Figure 17b. When the strain was 2.5%, the fatigue life enhanced by 40.1%, 66.5% and 223.9%, respectively, with an increase in modifier content. When the strain is 5%, the fatigue life improved by 7.4%, 27.8% and 92.6%, respectively, with the increase in modifier content. This indicated that increasing the modifier content can significantly improve the fatigue properties of asphalt binder [45,46].

It can also be noted that, compared to the fatigue life at 2.5% applied strain, the fatigue life of the base asphalt binder and modified asphalt binder at 5% applied strain was reduced by 76.4%, 81.9%, 81.8% and 85.9%, respectively. This meant that the fatigue life of the modified asphalt binders was more sensitive to stress, and this sensitivity was more pronounced in the high admixture of the modified asphalt binder, which was consistent with the results predicted by parameter B. Similarly, an increase in sensitivity did not imply an improvement in fatigue life or decay. In conclusion, the fatigue life of asphalt binders can be effectively improved by increasing the content of modifiers.

3.3.2. Fatigue Performance Analysis Based on Time Scanning Test

Figure 18 displays the relationship between loading times and normalized modulus ($|G^*|/|G^*|$initial) of the asphalt binder. The $|G^*|/|G^*|$initial of the asphalt binder gradually decreased as the number of loadings increased. This was due to the accumulation of fatigue damage in the asphalt binder during cyclic loading, which caused its integrity to be destroyed and the modulus to decay rapidly [46]. The fatigue life Np50 of the base asphalt binder and modified asphalt binders were 1302, 1398, 1500 and 2202, respectively. Compared to the base asphalt binder, the Np50 of the modified asphalt binder with 6%, 9% and 12% dosage increased by 7.3%, 15.2% and 69.1%, respectively. This was as expected; the modifier enhanced the $|G^*|/|G^*|$initial of the asphalt binders and improved the fatigue performance. This was consistent with the LAS test results. The difference was that the $|G^{*`}|/|G^*|$initial curves of the modified asphalt binder at 6% and 9% admixture are similar. This may be related to the loading methods of the TS test and LAS tests. A 0% to 30% strain, amplitude sweep was used in the LAS test to accelerate the fatigue damage, where some of the strains exceeded the linear viscoelastic range of the asphalt binders. In contrast, the TS test used 5% of the strain in the linear viscoelastic energy range. This may be the reason why the LAS test was more sensitive to viscoelastic energy and can effectively distinguish the fatigue performance of different asphalt binders.

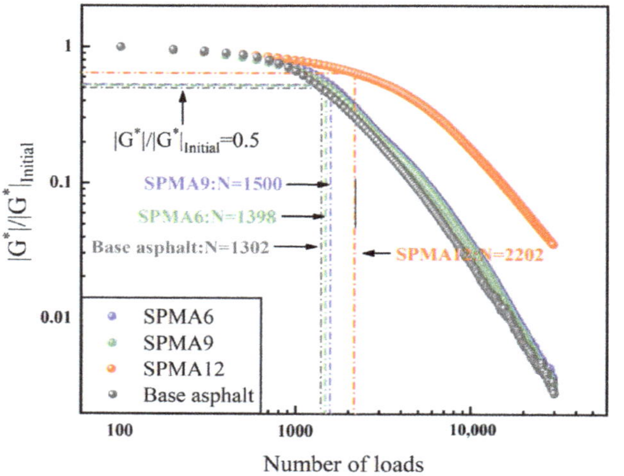

Figure 18. Normalized modulus curve of asphalt binders.

Meanwhile, the fatigue performance of the asphalt binder was analyzed from the perspective of energy dissipation. Figure 19 demonstrates the DER of asphalt binders versus the number of cyclic loading. At the early stage of the experiment, DER and N followed the lossless state curve with a slope of 1. As fatigue damage gradually accumulates, the DER gradually deviated from the lossless state curve. The number of loads corresponding to a 20% deviation of the DER-N curve from the nondestructive state curve was used as the fatigue life Np20 of the asphalt binder [17,25]. As shown in Figure 19, the modifiers had a positive effect on the fatigue life of the asphalt binder. The fatigue life Np20 of the base and modified asphalt binder were 601, 702, 799 and 1379, respectively. Compared to the base asphalt binder, the Np50 of the modified asphalt binder with 6%, 9% and 12% dosage increased by 16.8%, 32.9% and 129.5%, respectively. However, the DER-N curves of the modified asphalt binders at 6% and 9% admixture were similar.

It is worth noting that there was a significant difference between the fatigue life Nf obtained using the LAS test, Np50 obtained using the normalized modulus calculation and Np20 obtained using the dissipative energy calculation. This was because the selection of applied loads, frequencies and fatigue failure determination criteria in the TS test was artificially determined and highly empirical [25]. The fatigue life of the same asphalt binder can range from a few thousand to tens of thousands depending on the fatigue failure judgment criteria of the TS test. In contrast, the fatigue failure of asphalt binders should be a material's characteristic and should have obvious material dependence. Therefore, from the sensitivity of the test to the asphalt binder and the accuracy and completeness of the evaluation system, the LAS test was more suitable as a test method for the fatigue performance of asphalt binders.

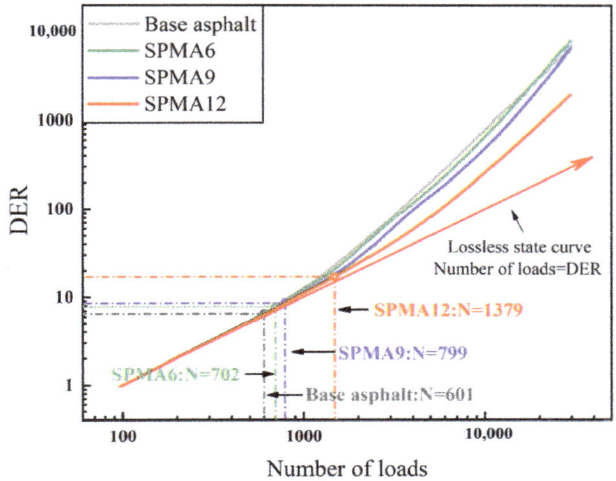

Figure 19. DER curve of asphalt binders.

3.4. Applicability of High-Temperature and Fatigue Indicators to the Performance Evaluation of Asphalt Binders

In summary, the SPUA had an excellent performance on asphalt binders for both high-temperature performance and fatigue performance improvement. However, it can be seen from Figures 20 and 21 that some of the tests were unable to effectively differentiate the viscoelastic properties of the modified asphalt binders in some of the admixture ranges (e.g., 6% and 9%), and this will be an obstacle to application and promotion of SPUA modifiers in practical engineering. Therefore, it was necessary to investigate the applicability of different tests and their corresponding metrics.

In this study, the correlation between the amount of modifier dosing and the performance index was used to reflect the applicability of the evaluation index. The greater the correlation, the better the applicability of the index. Among them, the high-temperature performance indexes included failure temperature, softening point, R, J_{nr} and Brookfield viscosity at 135 °C. The specific calculation results are listed in Table 15. The correlation coefficients revealed a good linear relationship between the SPUA modifier admixture and each high-temperature performance index. Among them, the correlation coefficients of apparent viscosity at 135 °C, J_{nr} at two loading levels and R at 3.2 kPa were greater than 0.9, while the correlations of softening point, failure temperature and R at 0.1 kPa were relatively low. This expressed that the apparent viscosity and MSCR tests are more sensitive to the viscoelastic properties of asphalt binders and can evaluate the effect of modifier admixture on the high-temperature properties of asphalt binders more precisely. Therefore, it was recommended to use a Brookfield viscosity test (such as 135 °C) and MSCR test (R and J_{nr}) as the evaluation index of high-temperature performance of the SPMA.

Similarly, correlations between several fatigue performance indices of modified asphalt binders and modifier admixtures were calculated, as shown in Table 16. It can be found that all four fatigue lives showed good linear relationships with the modifier doping. The correlation between the fatigue life and modifier dose obtained from the LAS test was greater than 0.85, while the correlation between the fatigue life N_{p50} and N_{p20} obtained from the TS test was 0.84 and 0.85, respectively. This indicated that the LAS test was more suitable as a means to evaluate the fatigue performance of asphalt binder characterization.

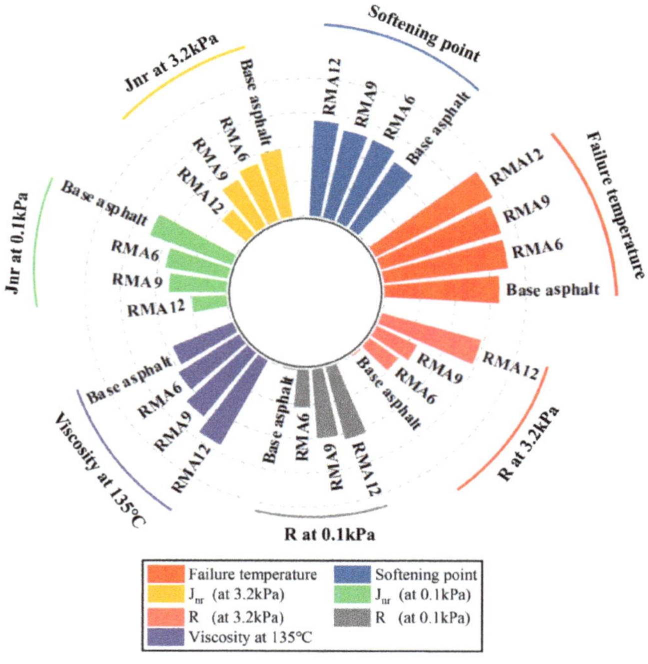

Figure 20. High-temperature performance index of asphalt binders.

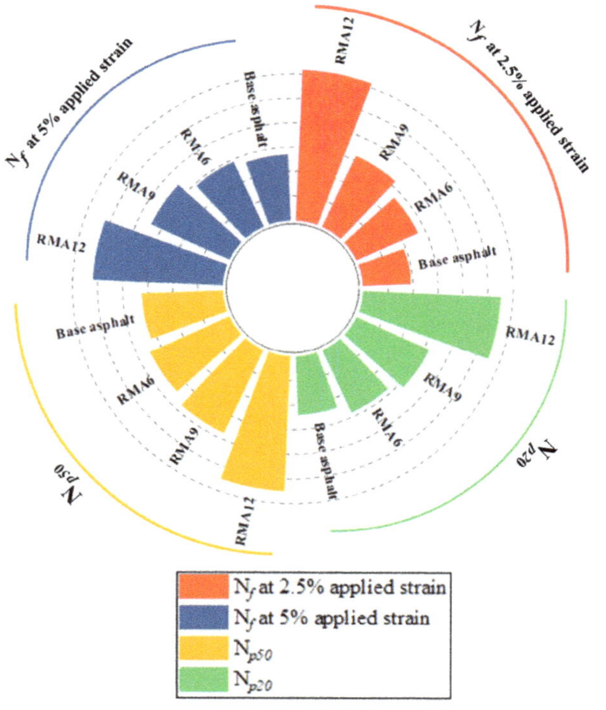

Figure 21. Fatigue Performance index of asphalt binders.

Table 15. Standard deviation of high-temperature index.

High-Temperature Index	Slope	Intercept	Correlation Coefficient
Failure temperature	65.2	66.242	0.81
Softening point	61.7	47.62	0.83
Viscosity at 135 °C	2533.3	252.33	0.99
J_{nr} (at 0.1 kPa)	−2725.0	527.1	0.91
J_{nr} (at 3.2 kPa)	−2750.0	542.6	0.90
R (at 0.1 kPa)	13.9	0.14	0.83
R (at 3.2 kPa)	0.8	−0.005	0.93

Table 16. Standard deviation of high-temperature index.

Fatigue Index	Slope	Intercept	Correlation Coefficient
N_f at 2.5% applied stress	12,767	135	0.91
N_f at 5% applied stress	116,283	−2458.2	0.86
N_{p50}	13,400	494	0.84
N_{p20}	11,283	−55.9	0.85

3.5. Thermal Properties Analysis

Figures 22 and 23 are the result of DSC test. It can be seen from the figure that the overall trend of DSC curves of base asphalt binder and three kinds of modified asphalt was basically the same. The DSC curves and thermodynamic parameters between base asphalt binder and modified asphalt binders had obvious changes. Compared with the base asphalt binder, the DSC curves of the SPMA6 moved up significantly, and the peak temperature rose from 26.3 °C of the base asphalt binder to 38.5 °C, the peak range from 13.2–40.4 °C to 29.3–48.6 °C and the enthalpy change decreased from 10.1 J/g of the base asphalt binder to 2.18 J/g. When the content was 9%, the DSC curve moved up again to a certain extent, but its thermodynamic parameters were almost the same as those of 6% modified asphalt binder. When the modifier content reached 12%, the DSC curve and thermodynamic parameters changed obviously again, the DSC curve moved up, the peak temperature of the endothermic peak further increased and the enthalpy change decreased.

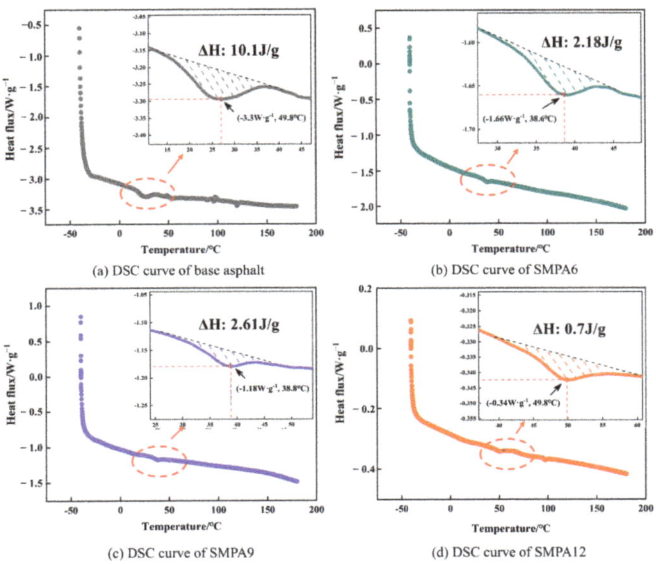

Figure 22. Thermodynamic parameters of asphalt binders.

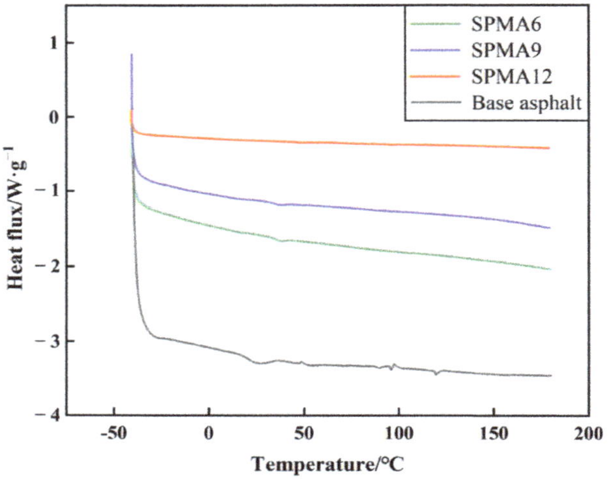

Figure 23. DSC curve of asphalt binders.

This showed that the modifier had a significant effect on the thermal stability of asphalt binders [36]. When the modifier was added, the thermal stability of the corresponding modified asphalt binder was significantly improved. This meant that higher temperature was needed in the process of aggregation transformation in SPUA modified asphalt binder during heating; that is, more energy was needed to complete the transformation of the asphalt phase. The macroscopic performance was improvement in temperature stability of modified asphalt. This explained why SPUA-modified asphalt binder had higher high-temperature performance.

4. Conclusions

The purpose of this study was to evaluate the effect of SPUA as the modifier on asphalt binders. First, the optimum process parameters of the modified asphalt binder were determined based on the orthogonal test. On this basis, by changing the amount of modifier in the asphalt binder, the high-temperature and fatigue properties of the asphalt binder were tested and the applicability of each index to the modified asphalt binder was analyzed. According to the limited experimental results obtained in this study, the following conclusions can be drawn:

(1) Combined with extreme difference analysis and ANOVA, the best preparation process parameters of modified asphalt binder under the condition of 9% content were determined as follows: shear time, rate and temperature of 40 min, 6000 rpm and 150 °C, respectively.
(2) The SPUA enhanced the apparent viscosity and improved the high-temperature shear resistance while reducing the temperature sensitivity in the test temperature range.
(3) Addition of SPUA admixture had a positive effect on high-temperature deformation resistance and elastic recovery performance of asphalt binders but hurt temperature sensitivity. There was higher sensitivity of the MSCR test to the viscoelastic properties of asphalt binders compared to the PG test.
(4) Increasing the modifier admixture could promote the modifier's effect on fatigue resistance of asphalt binders, but this was detrimental to stress sensitivity of asphalt binders.
(5) The performance indexes obtained from the MSCR test, Brookfield viscosity and LAS test were more correlated with modifier dosage. Therefore, the MSCR test and Brookfield viscosity test were recommended to evaluate high-temperature performance of asphalt binders and the LAS test to evaluate fatigue resistance of asphalt binders.

(6) The SPUA modifier had a positive effect on thermal stability of asphalt binders, and the effect increased with an increase in dosage. Higher thermal stability corresponded to better high-temperature performance.
(7) Excellent high-temperature performance and fatigue resistance of SPMA showed that SPUA material has great potential and application value in asphalt pavement.

Author Contributions: Conceptualization, X.S. Methodology, X.S.; Software, Z.L. and H.Y.; Validation, Z.L. and Y.Y.; Data curation, J.J.; Writing—original draft, Q.P.; Writing—review & editing, Q.P.; Visualization, Q.P. All authors have read and agreed to the published version of the manuscript.

Funding: This paper describes research activities mainly requested and supported by the open research fund of Highway Engineering Key Laboratory of Sichuan Province, Southwest Jiaotong University under grant numbers HEKLSP2022-08, by supported by the Fundamental Research Funds for the Central Universities under grant number 300102212516, supported by Open Fund of National Engineering Research Center of Highway Maintenance Technology (Changsha University of Science & Technology) kfj220104, supported by Special Fund for Science and Technology Innovation Strategy of Guangdong Province under grant numbers pdjh2021a0149 and pdjh2022b0161. The sponsorship and interest are gratefully acknowledged, supported by the Key Laboratory of Transport Industry of Road Structure and Material (Research In statute of Highway, Ministry of Transport), Beijing, PRC), supported by Guangxi Key Laboratory of Road Structure and Materials under grant numbers 2021gxjgclkf001.

Institutional Review Board Statement: Not applicable.

Informed Consent Statement: Not applicable.

Data Availability Statement: Not applicable.

Conflicts of Interest: The authors declare no conflict of interest.

References

1. Sun, X.L.; Qin, X.; Liu, Z.S.; Yin, Y.M.; Zou, C.; Jiang, S. New preparation method of bitumen samples for UV aging behavior investigation. *Constr. Build. Mater.* **2020**, *233*, 117278. [CrossRef]
2. Liu, S.; Jin, J.; Yu, H.; Gao, Y.; Du, Y.; Sun, X.; Qian, G. Performance enhancement of modified asphalt via coal gangue with microstructure control. *Constr. Build. Mater.* **2023**, *367*, 130287. [CrossRef]
3. Wang, C.; Li, Y.; Wen, P.; Zeng, W.; Wang, X. A comprehensive review on mechanical properties of green controlled low strength materials. *Constr. Build. Mater.* **2022**, *363*, 129611. [CrossRef]
4. Yang, X.; Tang, H.; Cai, X.; Wu, K.; Huang, W.; Zhang, Q.; Li, H. Evaluating reclaimed asphalt mixture homogeneity using force chain transferring stress efficiency. *Constr. Build. Mater.* **2023**, *365*, 130050. [CrossRef]
5. Yao, X.G.; Li, C.; Xu, T. Multi-scale studies on interfacial system compatibility between asphalt and SBS modifier using molecular dynamics simulations and experimental methods. *Constr. Build. Mater.* **2022**, *346*, 128502. [CrossRef]
6. Liang, M.; Liang, P.; Fan, W.Y.; Qian, C.D.; Xin, X.; Shi, J.T.; Nan, G. Thermo-rheological behavior and compatibility of modified asphalt with various styrene–butadiene structures in SBS copolymers. *Mater. Des.* **2015**, *88*, 177–185. [CrossRef]
7. Siddig, E.; Feng, C.P.; Ming, L.Y. Effects of ethylene vinyl acetate and nanoclay additions on high-temperature performance of asphalt binders. *Constr. Build. Mater.* **2018**, *169*, 276–282. [CrossRef]
8. Brasileiro, L.; Moreno-Navarro, F.; Martínez, R.; Sol-Sánchez, M.; Matos, J.; Rubio-Gámez, M. Study of the feasability of producing modified asphalt bitumens using flakes made from recycled polymers. *Constr. Build. Mater.* **2019**, *208*, 269–282. [CrossRef]
9. Celauro, C.; Bosurgi, G.; Sollazzo, G.; Ranieri, M. Laboratory and in-situ tests for estimating improvements in asphalt concrete with the addition of an LDPE and EVA polymeric compound. *Constr. Build. Mater.* **2019**, *196*, 714–726. [CrossRef]
10. Chen, Q.; Wang, C.; Li, Y.; Feng, L.; Huang, S. Performance development of polyurethane elastomer composites in different construction and curing environments. *Constr. Build. Mater.* **2023**, *365*, 130047. [CrossRef]
11. Sun, X.; Ou, Z.; Xu, Q.; Qin, X.; Guo, Y.; Lin, J.; Yuan, J. Feasibility analysis of resource application of waste incineration fly ash in asphalt pavement materials. *Environ. Sci. Pollut. Res.* **2023**, *30*, 5242–5257. [CrossRef] [PubMed]
12. Sun, X.; Peng, Q.; Zhu, Y.; Jin, J.; Xu, J.; Yin, Y.; Ng, A.H.M. Modification and enhancing effect of SPUA material on asphalt binder: A study of viscoelastic properties and microstructure characterization. *Case Stud. Constr. Mater.* **2023**, *18*, e01781. [CrossRef]
13. Zhu, H.J.; Wang, X.; Wang, Y.T.; Ji, C.; Wu, G.; Zhang, L.; Han, Z.Y. Damage behavior and assessment of polyurea sprayed reinforced clay brick masonry walls subjected to close-in blast loads. *Int. J. Impact Eng.* **2022**, *167*, 104283. [CrossRef]
14. Liu, Z.C.; Wu, J.; Yu, J.; Xu, S.L. Damage assessment of normal reinforced concrete panels strengthened with polyurea after explosion. *Case Stud. Constr. Mater.* **2022**, *17*, e01695. [CrossRef]

15. Liu, Q.; Guo, B.Q.; Chen, P.W.; Su, J.J.; Arab, A.; Ding, G.; Yan, G.H.; Jiang, H.Y.; Guo, F. Investigating ballistic resistance of CFRP/polyurea composite plates subjected to ballistic impact. *Thin Walled Struct.* **2021**, *166*, 108111. [CrossRef]
16. Ameri, M.; Seif, M.; Abbasi, M.; Molayem, M.; KhavandiKhiavi, A. Fatigue performance evaluation of modified asphalt binder using of dissipated energy approach. *Constr. Build. Mater.* **2017**, *136*, 184–191. [CrossRef]
17. Wen, Y.K.; Guo, N.S.; Wang, L.; Jin, X.; Li, W.; Wen, H.F. Assessment of various fatigue life indicators and fatigue properties of rock asphalt composite. *Constr. Build. Mater.* **2021**, *289*, 123147. [CrossRef]
18. Gao, J.F.; Wang, H.N.; Liu, C.C.; Ge, D.D.; You, Z.P.; Yu, M. High-temperature rheological behavior and fatigue performance of lignin modified asphalt binder. *Constr. Build. Mater.* **2020**, *230*, 117063. [CrossRef]
19. Sun, X.; Zhang, Y.; Peng, Q.; Yuan, J.; Cang, Z.; Lv, J. Study on adaptability of rheological index of nano-PUA-modified asphalt based on geometric parameters of parallel plate. *Nanotechnol. Rev.* **2021**, *10*, 1801–1811. [CrossRef]
20. Du, Y.; Xu, L.; Deng, H.B.; Deng, D.; Ma, C.; Liu, W.D. Characterization of thermal, high-temperature rheological and fatigue properties of asphalt mastic containing fly ash cenosphere. *Constr. Build. Mater.* **2020**, *233*, 117345. [CrossRef]
21. Sun, X.L.; Qin, X.; Liu, Z.S.; Yin, Y.; Jiang, S.; Wang, X. Applying feasibility analysis and catalytic purifying potential of novel modifying agent used in asphalt pavement materials. *Constr. Build. Mater.* **2020**, *245*, 118467. [CrossRef]
22. Li, B.; Liu, P.; Zhao, Y.; Li, X.; Cao, G. Effect of graphene oxide in different phases on the high temperature rheological properties of asphalt based on grey relational and principal component analysis. *Constr. Build. Mater.* **2023**, *362*, 129714. [CrossRef]
23. Li, Y.; Hao, P.; Zhao, C.; Ling, J.; Wu, T.; Li, D.; Liu, J.; Sun, B. Anti-rutting performance evaluation of modified asphalt binders: A review. *J. Traffic Transp. Eng.* **2021**, *8*, 339–355. [CrossRef]
24. Cao, W.; Wang, C. Fatigue performance characterization and prediction of asphalt binders using the linear amplitude sweep based viscoelastic continuum damage approach. *Int. J. Fatigue* **2019**, *119*, 112–125. [CrossRef]
25. Wang, C.; Zhang, H.; Castorena, C.; Zhang, J.; Kim, Y. Identifying fatigue failure in asphalt binder time sweep tests. *Constr. Build. Mater.* **2016**, *121*, 535–546. [CrossRef]
26. Kim, H.; Mazumder, M.; Lee, S.-J.; Lee, M.-S. Characterization of recycled crumb rubber modified binders containing wax warm additives. *J. Traffic Transp. Eng.* **2018**, *5*, 197–206. [CrossRef]
27. Zhang, J.; Huang, W.; Zhang, Y.; Yan, C.; Lv, Q.; Guan, W. Evaluation of the terminal blend crumb rubber/SBS composite modified asphalt. *Constr. Build. Mater.* **2021**, *278*, 122377. [CrossRef]
28. Hong, Z.; Yan, K.; Wang, M.; You, L.; Ge, D. Low-density polyethylene/ethylene–vinyl acetate compound modified asphalt: Optimal preparation process and high-temperature rheological properties. *Constr. Build. Mater.* **2022**, *314*, 125688. [CrossRef]
29. Xu, C.; Zhang, Z.; Zhao, F.; Liu, F.; Wang, J. Improving the performance of RET modified asphalt with the addition of polyurethane prepolymer (PUP). *Constr. Build. Mater.* **2019**, *206*, 560–575. [CrossRef]
30. Lei, L.; Li, D.; Chen, Y.; Tian, Y.; Pei, J. Dynamic chemistry based self-healing of asphalt modified by diselenide-crosslinked polyurethane elastomer. *Constr. Build. Mater.* **2021**, *293*, 123480.
31. Wang, Z.; Ye, F. Experimental investigation on aging characteristics of asphalt based on rheological properties. *Constr. Build. Mater.* **2020**, *231*, 117158. [CrossRef]
32. Dong, R.; Zhao, M.; Xia, W.; Yi, X.; Dai, P.; Tang, N. Chemical and microscopic investigation of co-pyrolysis of crumb tire rubber with waste cooking oil at mild temperature. *Waste Manag.* **2018**, *79*, 516–525. [CrossRef] [PubMed]
33. Jiang, X.; Li, P.; Ding, Z.; Yang, L.; Zhao, J. Investigations on viscosity and flow behavior of polyphosphoric acid (PPA) modified asphalt at high temperatures. *Constr. Build. Mater.* **2019**, *228*, 116610. [CrossRef]
34. Neda, K.; Ataollah, S.; Mousavi, R. Rheological behavior of asphalt binders and fatigue resistance of SMA mixtures modified with nano-silica containing RAP materials under the effect of mixture conditioning. *Constr. Build. Mater.* **2021**, *303*, 124433.
35. Yongjun, M.; Chaoliang, G.; Qixiong, Z.; Yue, Q.; Weikang, K.; Liupeng, F. Study on multiple Damage-Healing properties and mechanism of laboratory simulated recycled asphalt binders. *Constr. Build. Mater.* **2022**, *346*, 128468.
36. Xiaolong, S.; Qin, X.; Guotao, F.; Yongqiang, Z.; Zhengbing, Y.; Qian, C.; Junshen, Y. Effect Investigation of Ultraviolet Ageing on the Rheological Properties, Micro-Structure, and Chemical Composition of Asphalt Binder Modified by Modifying Polymer. *Adv. Mater. Sci. Eng.* **2022**, *2022*, 7190428.
37. Bo, L.; Yufan, C.; Fang, L.; Jianlong, Z. Evaluation of rheological and aging behavior of modified asphalt based on activation energy of viscous flow. *Constr. Build. Mater.* **2022**, *321*, 126347.
38. Hazim, A.-S.; Haryati, Y.; Mohd, S.; Khleel, S.; Putra, J.; Norazah, B.; Radhi, R.; Abdul, H.N. Effects of maltene on the attributes of reclaimed asphalt pavement: Performance optimization. *Constr. Build. Mater.* **2021**, *302*, 124210.
39. Ming, L.; Zhengmei, Q.; Xuehao, L.; Cong, Q.; Ning, G.; Zhaoxin, L.; Linping, S.; Zhanyong, Y.; Jizhe, Z. The Effects of Activation Treatments for Crumb Rubber on the Compatibility and Mechanical Performance of Modified Asphalt Binder and Mixture by the Dry Method. *Front. Mater.* **2022**, *9*, 845718.
40. Li, P.; Jiang, X.; Ding, Z.; Zhao, J.; Shen, M. Analysis of viscosity and composition properties for crumb rubber modified asphalt. *Constr. Build. Mater.* **2018**, *169*, 638–647. [CrossRef]
41. Xu, X.; Guo, H.; Wang, X.; Zhang, M.; Wang, Z.; Yang, B. Physical properties and anti-aging characteristics of asphalt modified with nano-zinc oxide powder. *Constr. Build. Mater.* **2019**, *224* (Suppl. 1), 732–742. [CrossRef]
42. Ashish, P.K.; Singh, D.; Bohm, S. Evaluation of rutting, fatigue and moisture damage performance of nanoclay modified asphalt binder. *Constr. Build. Mater.* **2016**, *113*, 341–350. [CrossRef]

43. Han, L.; Qingshan, X.; Haibo, D.; Hong, Z.; Yanjun, Q. Rheological properties of model wax doped asphalt binders. *Constr. Build. Mater.* **2022**, *350*, 128865.
44. Jiang, J.; Zhao, Y.; Lu, G.; Dai, Y.; Ni, F.; Dong, Q. Effect of binder film distribution on the fatigue characteristics of asphalt Binder/Filler composite based on image analysis method. *Constr. Build. Mater.* **2020**, *260*, 119876. [CrossRef]
45. Daryoosh, D.; Mahdi, H.; Saqib, G.; Shane, U.B. Combined effect of waste polymer and rejuvenator on performance properties of reclaimed asphalt binder. *Constr. Build. Mater.* **2021**, *268*, 121059.
46. Kumar, D.A.; Dharamveer, S. Evaluation of fatigue performance of asphalt mastics composed of nano hydrated lime filler. *Constr. Build. Mater.* **2020**, *269*, 121322.

coatings

MDPI

Article

Mechanism of Sodium Dodecyl Diphenyl Ether Disulfonate Filled Hydrotalcite Inhibiting the Photo-Degradation of Polyvinyl Chloride under Different Ranges of Ultraviolet Wavelength Irradiation

Enguo Zhou [1], Yuan Liu [1,2], Huajin Yuan [1,3], Xiaoling Cheng [1,2,*], Yuanhong Zhong [1,2,*], Jiebing He [3] and Xi Lu [3]

1 Key Laboratory of Clean Chemistry Technology of Guangdong Regular Higher Education Institutions, Guangdong Engineering Technology Research Center of Modern Fine Chemical Engineering, School of Chemical Engineering and Light Industry, Guangdong University of Technology, Guangzhou 510006, China
2 Jieyang Branch of Chemistry and Chemical Engineering Guangdong Laboratory (Rongjiang Laboratory), Jieyang 515200, China
3 GCH Technology Co., Ltd., Guangzhou 510540, China
* Correspondence: ggcxl@gdut.edu.cn (X.C.); zhongyuanhong@gdut.edu.cn (Y.Z.)

Abstract: This content introduces a novel Ultraviolet (UV)-shielding material, Zn_2Al-MADS-LDH (MADS-LDH), which was synthesized through co-precipitation method to insert sodium dodecyl diphenyl ether disulfonate (MADS) into the interlayer of Zn_2Al-LDH layered double hydroxide (LDH), to improve the photoaging resistance of polyvinyl chloride (PVC). The characterization results indicated that MADS-LDH had a host-guest interaction between the LDH host layer and MADS guest anion, and it exhibited superior UV absorption capabilities than Zn_2Al-CO_3-LDH (CO_3-LDH) and a broader absorption spectrum compared to MADS. A series of LDHs/PVC film composite materials containing LDHs nanosheets were prepared by incorporating the prepared LDHs into a PVC matrix via a solvent casting method. As expected, the MADS-LDH/PVC film composite materials exhibited enhanced photoaging resistance. The results of photoaging tests indicated that MADS-LDH inhibits the rate of carbonyl generation during photoaging of MADS-LDH/PVC film composite materials, resulting in a decrease in the carbonyl index (ΔCl) and relative degradation rate (RDR) compared to pristine PVC film and CO_3-LDH/PVC film composite materials. Furthermore, the study evaluated the influence of different UV light wavelength ranges, such as UVB (280~315 nm), UVC (200~280 nm), and UV (200~400 nm), on the aging performance of PVC film and LDHs/PVC film composite materials. The results demonstrated that UV had the highest aging effect on PVC composite films, followed by UVC and UVB. Therefore, the MADS-LDH is a highly efficient and promising UV-shielding material with excellent potential for wide applications in the field of PVC.

Keywords: hydrotalcite; polyvinyl chloride; intercalation; UV-shielding materials; film composite materials

Citation: Zhou, E.; Liu, Y.; Yuan, H.; Cheng, X.; Zhong, Y.; He, J.; Lu, X. Mechanism of Sodium Dodecyl Diphenyl Ether Disulfonate Filled Hydrotalcite Inhibiting the Photo-Degradation of Polyvinyl Chloride under Different Ranges of Ultraviolet Wavelength Irradiation. *Coatings* **2023**, *13*, 985. https://doi.org/10.3390/coatings13060985

Academic Editor: Dimitrios Tasis

Received: 30 March 2023
Revised: 17 May 2023
Accepted: 19 May 2023
Published: 25 May 2023

1. Introduction

Polyvinyl chloride (PVC) is a highly versatile and commonly used polymer materials in modern industries [1,2]. However, as a polymer material, PVC is susceptible to degradation and aging under the influence of external factors such as light, water, heat, and oxygen. This degradation can lead to discoloration, brittleness, and hardening, which ultimately reduces its physical properties [3,4]. Research has shown that UV light from sunlight is the primary contributor to the aging of polymer materials [3,5]. Therefore, incorporating UV-shielding materials is crucial to improve the stability of PVC film composite materials.

In recent years, there has been a growing interest in the development and utilization of inorganic-organic hybrid nanomaterials, such as hydrotalcite, due to their unique properties that arise from the interaction between the inorganic matrix and the organic material. Layered double hydroxides (LDHs), also known as hydrotalcite-like compounds or anionic clays, are a diverse group of layered materials that can be either natural or synthetic. LDHs are characterized by the general formula: $[M^{2+}{}_{1-x}M^{3+}{}_{x}(OH)_2]^{x+}(A^{n-})_{x/n}\cdot mH_2O$, where M^{2+} and M^{3+} represent divalent and trivalent metal cations, respectively, and A^{n-} represents exchangeable anions with negative charge n, which can be organic or inorganic in nature [6].

By regulating its plate and interlayer structure, hydrotalcite can be customized for various functional purposes. It has important applications in various fields including catalysis, environmental protection [7–10], electrochemistry [11,12], and polymer additives [13,14].

The use of organic UV absorbers and inorganic nanoparticles to delay the photoaging degradation of polymeric materials, such as PVC, is common. However, these materials have certain limitations, such as high lipophilicity [15] and a narrow range of UV wavelengths shielding ability [16]. Moreover, the UV absorption ability of anionic surfactant intercalated hydrotalcite, containing a conjugated structure and dodecyl chain length as UV-shielding materials, is insufficiently studied. Existing studies mainly focus on the adsorption capacity of such anionic surfactants after intercalation of hydrotalcite. They can adsorb heavy metals or organic pollutants through hydrogen bonding and electrostatic interactions after anion insertion into hydrotalcite, and physical adsorption by increased layer spacing [9,17–21].

Although numerous studies have examined the effect of single UV wavelength range on the photoaging of PVC films [22,23], there is a lack of comparison between the results obtained and different wavelength ranges. Additionally, when adding intercalated hydrotalcite powders to polymers, it is important to consider their compatibility and interlayer crystalline water content, a factor that has not been thoroughly explored in previous studies.

To address these gaps in knowledge, this study investigates the mechanism of MADS-LDH fillers to inhibit the photoaging degradation of PVC film composite materials under different UV wavelength range irradiation. This is based on the existing oxidation reaction mechanism of photo-oxidation reaction generating carbonyl during PVC aging [24]. As a result, the objectives of this study are to: (1) study and compare the UV absorption range and absorption intensity between CO_3-LDH and MADS-LDH; (2) analyze and discuss the desorption behavior of adsorbed water and crystalline water during the thermal decomposition of MADS-LDH; (3) analyze the morphology and pore size of MADS-LDH fillers, and their effect on structure and compatibility of PVC; (4) investigate the influence of different wavelength ranges of UV light on the MADS-LDH/PVC film composite materials' irradiation aging behavior, and the mechanism by which MADS-LDH inhibits the photoaging degradation of MADS-LDH/PVC film composite materials.

2. Materials and Methods

2.1. Materials

$ZnCl_2$ (98.0%), $AlCl_3\cdot 6H_2O$ (97.0%), and NaOH (96.0%) were supplied by Shanghai Macklin Biochemical Co., Ltd. (Shanghai, China), Sodium dodecyl diphenyl ether disulfonate (45% aqueous solution) was purchased from Shanghai Aladdin Biochemical Technology Co., Ltd. (Shanghai, China), and ethanol (99.7%) was supplied by Anhui Zesheng Technology Co., Ltd. (Anqing, China). Further, 1, 2-dichloroethane (99.5%) was purchased from Meryer (Shanghai) Biochemical Technology Co., Ltd. (Shanghai, China). PVC (99.0%) and was supplied by Zhangmutou Ruixiang Polymer Materials Business Department, Dongguan City (Dongguan, China). Deionized water was decarbonized by boiling for 5 min before use.

2.2. Preparation of MADS Anion-Intercalated LDH

MADS-LDH was prepared by co-precipitation method as follows: a mixture of 0.625 M $ZnCl_2$ and 0.313 M $AlCl_3\cdot 6H_2O$ (Zn/Al molar ratio = 2:1) was slowly added to 100 mL of MADS solution (0.125 M) with stirring in N_2 atmosphere. The pH of the reaction solution was adjusted to 9.0 by dropwise addition of 1 M NaOH alkaline solution. The resulting slurry was crystallized at 70 °C for 24 h. The white precipitates were filtered and separated. They were rinsed with hot decarbonated deionized water and anhydrous ethanol until reaching a state of neutrality. Finally, the sample was subjected to a 24 h drying process at 100 °C, followed by grinding and sieving. Instead of MADS, Na_2CO_3 was utilized as the precursor. Meanwhile, the preparation of CO_3-LDH was exactly following the same method.

2.3. Preparation of LDHs/PVC Composite Materials

Prepared MADS-LDH of 1.0 g dry weight was added to a round bottom flask containing 100 mL of 1,2-dichloroethane for dispersion and stirring for 30 min to ensure uniform dispersion. Then, 10.0 g of PVC was added to the same round-bottom flask and stirred at 25 °C for 2 h. The resulting dispersion system of MADS-LDH and PVC was defoamed and poured onto a flat and smooth glass plate. The dispersion system solution was immediately scraped with a wet applicator and placed in a fume hood for static drying. The resulting film was then cut into dimensions of length × width × height= 30 mm × 20 mm × 0.1 mm was and used for subsequent testing. The reference samples of pure PVC and 10.0 wt% CO_3-LDH/PVC film materials were also prepared using the same method described above.

2.4. UV Photoaging Test

The UV photoaging tests were performed in a custom-made UV-aging chamber that was equipped with individually controllable UVA (315~400 nm, 8 W), UVB (280~315 nm, 8 W), and UVC (200~280 nm, 8 W) lamps. The chamber was designed to have a cooling system to eliminate the influence of thermal aging during the photoaging test of PVC films. The PVC films were subjected to photoaging treatment three times with a total UV exposure time of 144 h.

2.5. Characterization

The X-ray diffraction (XRD) patterns were analyzed using a Shimadzu XRD-6000 diffractometer (Shimadzu Corporation, Japan) with monochromatic Cu Kα radiation (λ = 0.15406 nm, 40 kV, 40 mA) from 2 to 70° at a scan speed of $10° \cdot min^{-1}$. Fourier transform infrared (FT-IR) spectra were recorded on a Bruker Vector 22 infrared spectrophotometer (Bruker Corporation, Billerica, MA, USA) with a scan range of 4000~400 cm^{-1} and a resolution of 1 cm^{-1}. The surface and internal morphology of the sample were examined using scanning electron microscope (TESCAN MIRA LMS SEM, TESCAN Orsay Holding, a.s., Brno, Czech Republic) and transmission Electron Microscope (Hitachi HT7700 TEM, Hitachi High Technologies Corporation, Tokyo, Japan). Thermogravimetric analysis (TGA) was carried out on a PCT-IA instrument (Thermal Analysis Instruments, Inc., New Castle, DE, USA) in the range of 25~800 °C with a heating rate of 10 $°C \cdot min^{-1}$ under N_2 atmosphere. BET analysis was performed on a Micromeritics ASAP 2460 instrument (Micromeritics Instrument Corporation, Norcross, GA, USA). The ultraviolet–visible (UV-vis) absorption spectra were obtained using a Shimadzu UV-2501 PC instrument (Shimadzu Corporation, Kyoto, Japan).

3. Results and Discussion

3.1. XRD Analysis

The XRD patterns of powder CO_3-LDH and MADS-LDH are shown in Figure 1. These two samples have characteristic diffraction peaks of LDHs materials. From Table 1, the planar spacing of LDHs (003), (006), and (009) showed good multiplicative relationships, with (d_{003}) = 2 (d_{006}) = 3 (d_{009}) [25], confirming the successful preparation of LDHs materials. The XRD spectrum LDH of MADS-LDH (Figure 1) shows only one set of typical (00l) (l = 3, 6, 9) Bragg reflections, indicating the successful intercalation of MADS into the interlayer of Zn_2Al-LDH with only a single crystalline phase. The basal spacings (d_{003}) of CO_3-LDH and MADS-LDH are approximately 0.75 and 2.87 nm, respectively. Subtracting the LDH plate layer thickness of 0.48 nm [26], the interlayer spacings of CO_3-LDH and MADS-LDH are approximately 0.27 and 2.39 nm, respectively. Additionally, we propose a possible orientation for MADS intercalated in Zn_2Al-LDH interlayer (Figure 2).

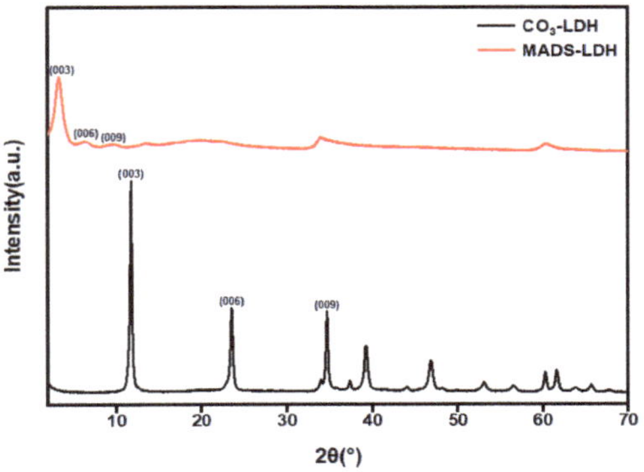

Figure 1. XRD patterns of CO_3-LDH and MADS-LDH.

Table 1. Crystal spacing for CO_3-LDH and MADS-LDH.

Parameter (nm)	CO_3-LDH	MADS-LDH
d_{003}	0.75	2.87
d_{006}	0.38	1.41
d_{009}	0.26	0.93
d_{110}	0.15	0.15

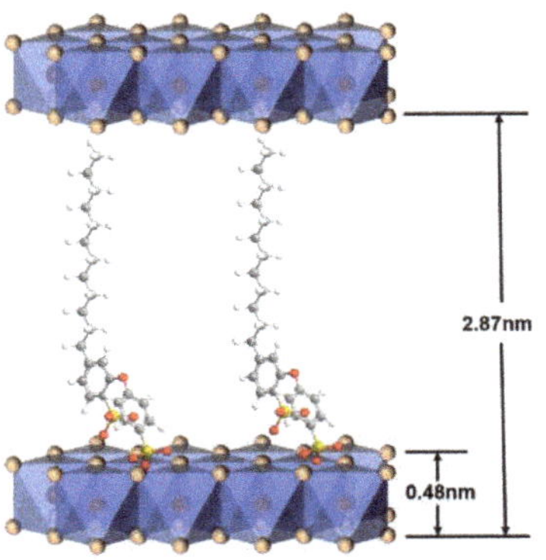

Figure 2. The probable orientation of MADS intercalated in Zn_2Al-LDH interlayer.

It has been reported by Yanjun Lin et al. [27] that an increase in LDHs layer spacing can improve the stability of PVC films. This is because a larger interlayer spacing facilitates the entry of Cl^- into the LDHs interlayer channels, thereby inhibiting the autocatalytic dechlorination decomposition of PVC by Cl^-. Therefore, MADS-LDH, with a layer spacing of 2.39 nm, is expected to improve the UV aging resistance of PVC.

3.2. FT-IR Analysis

The resulting FT-IR spectra of CO_3-LDH, MADS-LDH, and MADS are shown in Figure 3. The broad absorption band near 3430 cm^{-1} in the spectra of CO_3-LDH and MADS-LDH can be attributed to the O-H stretching vibration of the LDH plate layer and interlayer water molecules [18,20]. The intense band at very low wavenumber (428 cm^{-1})

was only observed for CO_3-LDH and MADS-LDH, which is attributed to the Zn-O-bound for LDH structures [28,29]. The absence of CO_3^{2-} groups at 1360 cm^{-1} indicates that there were no CO_3^{2-} anions in the interlayer of MADS-LDH [30]. The C-H of alkyl chains formed strong peaks at 2958, 2928, and 2858 cm^{-1}, while the C-C of diphenoxy produced strong peaks at 1590 and 1482 cm^{-1}, and the -O- of diphenoxy resulted in a peak at 1245 cm^{-1} [31,32].

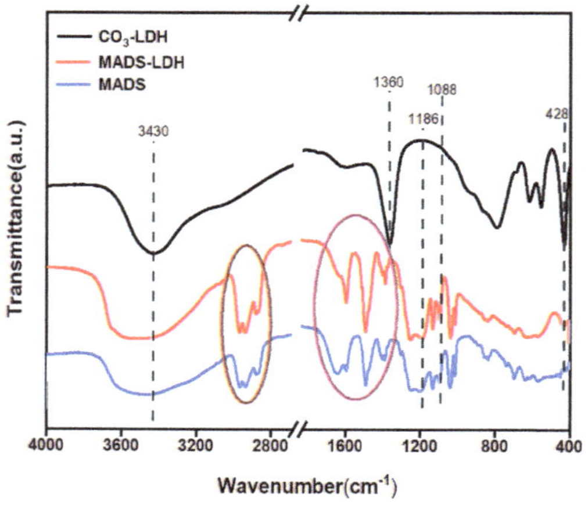

Figure 3. FT-IR spectra of CO_3-LDH,MADS-LDH and MADS.

In the FT-IR spectra of MADS-LDH, most of the characteristic absorption bands of MADS were observed, indicating that MADS was successfully intercalated into the interlayer of Zn_2Al-LDH. However, some subtle differences were observed between MADS-LDH and MADS. The asymmetric and symmetric vibrational absorption peaks of the -SO_3^- groups of MADS were shifted from 1190 and 1092 cm^{-1} to 1186 and 1088 cm^{-1}, respectively, due to the host-guest interaction between the Zn_2Al-LDH layer and interlayer MADS anion [31]. These shifts suggest that the environment experienced by MADS anion in the interlayer gallery of LDHs is distinct from that of UV absorber. The successful preparation of a new organic-inorganic UV shielding material, MADS-LDH, is confirmed by XRD and FT-IR analyses. Similarly, the XRD and FT-IR spectra of CO_3-LDH (Figures 1 and 3) verified the successful intercalation of CO_3^{2-} into Zn_2Al-LDH.

3.3. SEM and TEM Analysis

Figure 4 illustrates the SEM images of CO_3-LDH and MADS-LDH. The obtained CO_3-LDH exhibited flaky morphological features, while the SEM image of MADS-LDH showed aggregation of irregular flaky particles. This aggregation is attributed to the intercalation of MADS anions, which leads to the aggregation of Zn_2Al-LDH crystals [33]. To further confirm the successful intercalation of MADS anion into Zn_2Al-LDH, transmission electron microscopy (TEM) was used, and the obtained TEM image of MADS-LDH was displayed in Figure 5. Clusters with irregularly shaped MADS-LDH nanosheets stacked in the TEM images were observed. The surfactant intercalated samples exhibit a rather different morphology. The agglomeration of MADS-LDH resulted in a decrease in its crystallinity, which is consistent with the XRD results.

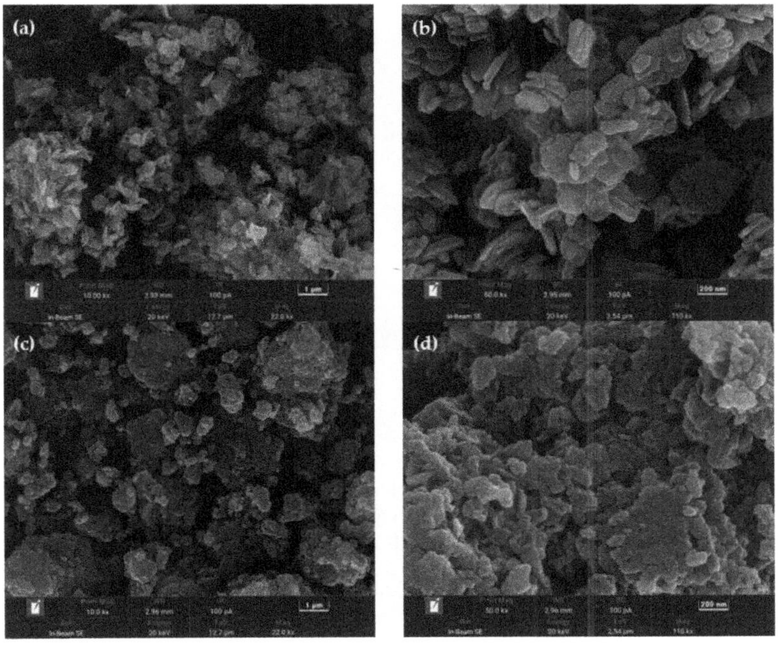

Figure 4. SEM images of (**a**,**b**) CO_3-LDH and (**c**,**d**) MADS-LDH.

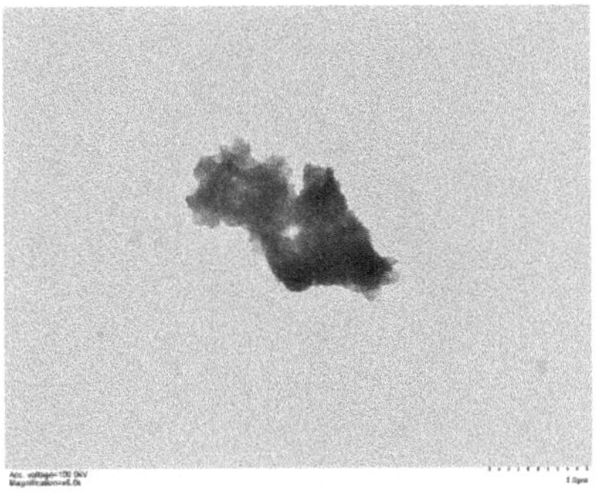

Figure 5. TEM image of MADS-LDH.

3.4. BET Analysis

Figure 6 shows the N_2 adsorption–desorption isotherms of CO_3-LDH and MADS-LDH. The nitrogen adsorption method revealed that both CO_3-LDH and MADS-LDH are mesoporous materials with type IV isotherms and H3 hysteresis loops [34]. The BET surface areas were calculated to be 30.430 m^2/g for CO_3-LDH and 23.927 m^2/g for MADS-LDH [35]. The decrease in the specific surface area of MADS-LDH after MADS anion intercalation is consistent with the increase in the unit mass of MADS-LDH due to agglomeration [36], which further confirms the successful intercalation of MADS anion into Zn_2Al-LDH.

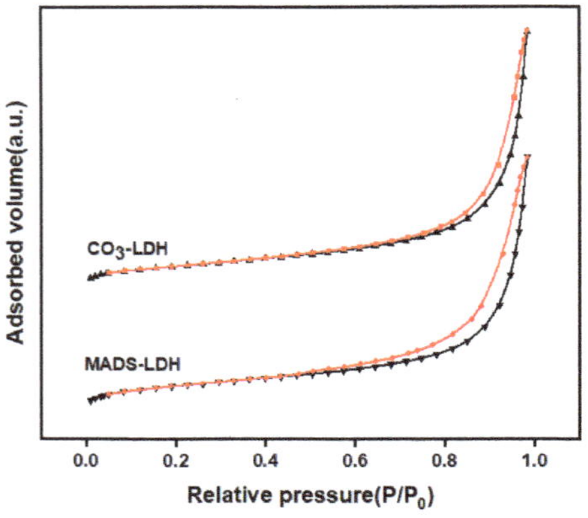

Figure 6. N_2 adsorption–desorption isotherms of CO_3-LDH and MADS-LDH.

Figure 7 presents the pore diameter distribution of CO_3-LDH and MADS-LDH. The results show that MADS-LDH has a narrower pore diameter distribution with a maximum effective pore diameter of 4.6 nm compared to CO_3-LDH. These findings are consistent with the adsorption isotherm results, which indicate that both CO_3-LDH and MADS-LDH have a mesoporous structure with pore diameters ranging from 2 nm to 50 nm. These findings are crucial in the development of new UV-shielding materials based on MADS-LDH, as they provide insights into the specific surface area and porosity of the materials.

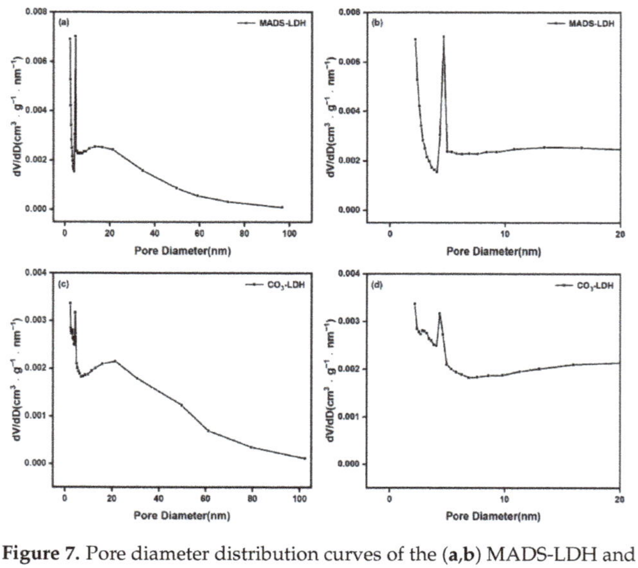

Figure 7. Pore diameter distribution curves of the (**a**,**b**) MADS-LDH and (**c**,**d**) CO_3-LDH.

Table 2 show Constitutive properties of CO_3-LDH and MADS-LDH. As shown in Table 2, the specific surface area of CO_3-LDH (S_{BET} = 30.430 m^2/g) is much larger than that of MADS-LDH (S_{BET} = 23.927 m^2/g). Meanwhile, the pore volume V_p and the pore diameter D_p of both were almost equal. This indicates that the MADS anion intercalation into the interlayer of Zn_2Al-LDH has the greatest effect on its specific surface area.

Table 2. Constitutive properties of CO_3-LDH and MADS-LDH.

Sample	S_{BET} (m^2/g)	V_p (cm^3/g)	D_p (nm)
CO_3-LDH	30.430	0.119	16.610
MADS-LDH	23.927	0.117	16.187

3.5. TGA Analysis

Thermogravimetric analysis (TGA) was utilized to assess the thermal stability of the prepared LDHs. Figure 8 displays the TG curves of CO_3-LDH, MADS-LDH, and MADS. The mass loss stage for MADS occurred within the range of 25~765 °C, with 10% mass loss temperature ($T_{0.1}$) and 50% mass loss temperature ($T_{0.5}$) at 58 and 156 °C, respectively. In comparison, the weight loss rate of LDHs was substantially lower than that of MADS within the thermal decomposition temperature of 25~800 °C, indicating a better thermal stability.

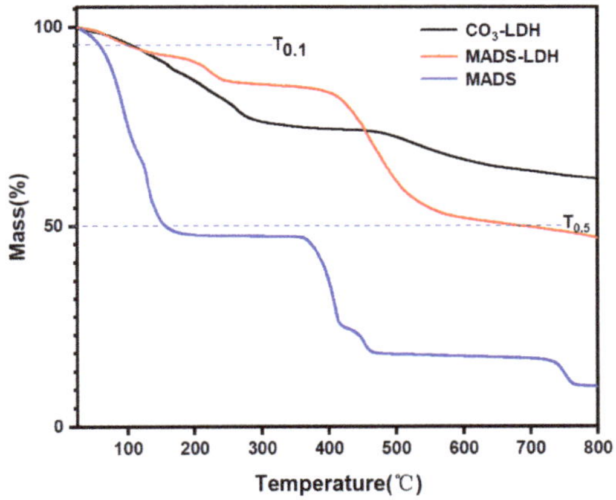

Figure 8. TG curves of CO_3-LDH, MADS-LDH, and MADS.

Specifically, the weight loss rate of MADS-LDH between 120~452 °C was lower than that of CO_3-LDH. At this stage, the weight loss for LDHs is primarily caused by the desorption of water of crystallization between its layers [18]. The hydrophobic environment created by the alkyl chain of the MADS anion within the MADS-LDH interlayer made it difficult for water molecules to enter and form crystalline water [36]. Furthermore, the weight loss rate of MADS-LDH between 452~800 °C was higher than that of CO_3-LDH due to the thermal decomposition of organic carbon constituted of the MADS anion present in the MADS-LDH interlayer [37]. The detailed $T_{0.1}$ and $T_{0.5}$ for MADS and LDHs are listed in Table 3, where ΔT represents the difference between MADS and LDHs at the temperature of $T_{0.1}$ or $T_{0.5}$. Notably, the $T_{0.1}$ and $T_{0.5}$ values of MADS-LDH were at least 41 °C higher than those of MADS, indicating the superior thermal stability of MADS-LDH over MADS alone.

Table 3. Summary of TG analysis results (unit: °C).

Sample	$T_{0.1}$	$\Delta T_{0.1}$	$T_{0.5}$	$\Delta T_{0.5}$
MADS	58	–	156	–
CO_3-LDH	104	46	>800	>644
MADS-LDH	99	41	685	529

3.6. UV Absorption of CO_3-LDH, MADS-LDH and MADS

The UV-vis absorption spectra of CO_3-LDH, MADS-LDH, and MADS are presented in Figure 9. CO_3-LDH displayed moderate UV absorption in the wavelength range of 200~228 nm and 300~400 nm, with low UV absorption in the range of 228~300 nm. In comparison with CO_3-LDH, MADS exhibited stronger absorption in the range of 228~300 nm. After intercalation, MADS-LDH showed a wider absorption band in the range of 200~400 nm and stronger UV absorption in the range of 228~320 nm. The absorbance of MADS-LDH reached a maximum of 0.835, which was about 28% higher than that of MADS. These results could be attributed to supramolecular interactions, such as hydrogen bonding, electrostatic and gravitational forces, and van der Waals forces, between the cations on the LDHs plate layer and the interlayer MADS anion [38]. The π-π conjugation interactions between the aromatic groups of the MADS-LDH interlayer MADS anion resulted in a lower excitation energy, leading to a red-shifted and broadened absorption band [39,40]. Therefore, the Zn_2Al-LDH not only inherits the strong UV absorbing capabilities of MADS but also broadens the absorption band through π-π conjugate interactions between the interlayer guest anions, resulting in a new material with superior UV shielding performance.

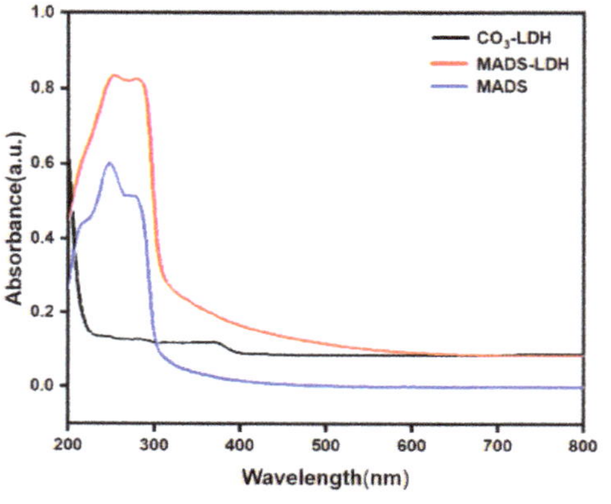

Figure 9. UV absorbance curves of CO_3-LDH, MADS-LDH and MADS.

3.7. Effect of LDHs on the Structure of PVC Film Composite Materials

The structural and dispersion effects of CO_3-LDH and MADS-LDH on PVC film composite materials were illustrated using X-ray diffraction (XRD) patterns. The XRD patterns of PVC, CO_3-LDH/PVC, and MADS-LDH/PVC film composite materials are presented in Figure 10. For the MADS-LDH/PVC film composite materials, the absence of characteristic diffraction peaks of MADS-LDH in the range of 2~10° on the test scale, as well as the increased half-peak width of the XRD diffraction peaks, indicated that the PVC macromolecular chains were intercalated within the pore channels of MADS-LDH [41]. This is related to the formation of a hydrophobic environment within the interlayer space of hydrotalcite by the alkyl chains on the interlayer MADS anion after intercalation [41,42], and the excellent compatibility of PVC with the alkyl chains on the MADS anion. During film preparation, hydrophobic solvated PVC molecules, which are soluble in 1,2-dichloroethane, preferred to enter the MADS-LDH interlayer [43], leading to the disappearance of the (003) characteristic diffraction peak of MADS-LDH. Conversely, the interlayer of CO_3-LDH constitutes a hydrophilic environment that is not conducive to the entry of hydrophobic PVC molecules into the interlayer. Consequently, all the characteristic X-ray diffraction peaks of CO_3-LDH were present in the XRD pattern of CO_3-LDH/PVC.

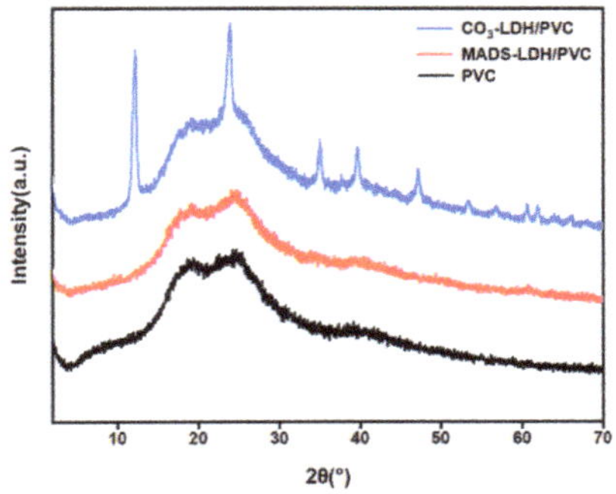

Figure 10. XRD patterns of PVC, CO_3-LDH/PVC, and MADS-LDH/PVC film composite materials.

3.8. Film Color Change of LDHs/PVC Film Composite Materials during Photoaging

Table 4 presents a comparison of the photostability of PVC film, CO_3-LDH/PVC, and MADS-LDH/PVC film composite materials. During UVB band irradiation aging, no significant color changes were observed in any of the films.

Table 4. Film color change process of PVC, CO_3-LDH/PVC, and MADS-LDH/PVC films during UVB, UVC, and UV ultraviolet light aging, respectively.

		UVB			UVC			UV		
Time / Sample	0 h	48 h	96 h	144 h	48 h	96 h	144 h	48 h	96 h	144 h
PVC										
MADS-LDH/PVC										
CO_3-LDH/PVC										

However, under UVC band irradiation aging, PVC and CO_3-LDH/PVC film composite materials experienced significant color changes, turning orange-red after 96 h of aging and finally brown after 144 h. In contrast, the MADS-LDH/PVC film composite materials had a longer aging time before showing significant color changes and did not undergo an abrupt jump in color as observed in the other films.

During the aging process of UV band irradiation, the PVC film changed color rapidly. It turned brown after 48 h of aging, turns dark brown after aging to 96 h, and finally photoages to black after 144 h. Similarly, CO_3-LDH/PVC films underwent a similar change process during the aging process, but with relatively light aging. The color change during aging of CO_3-LDH/PVC film was more evident than that of MADS-LDH/PVC film. The color change process of MADS-LDH/PVC films when aged under UV band irradiation for 144 h was similar to the change process when aged under UVC band irradiation conditions. These findings suggest that the MADS-LDH/PVC films exhibit better photostability under UVC and UV band conditions.

The improved photostability of the MADS-LDH/PVC film composite materials can be further verified by the increase in carbonyl index values and relative degradation rates, as shown in Figure 11 and Table 5. Specifically, PVC film composite materials containing MADS-LDH exhibited good photostability.

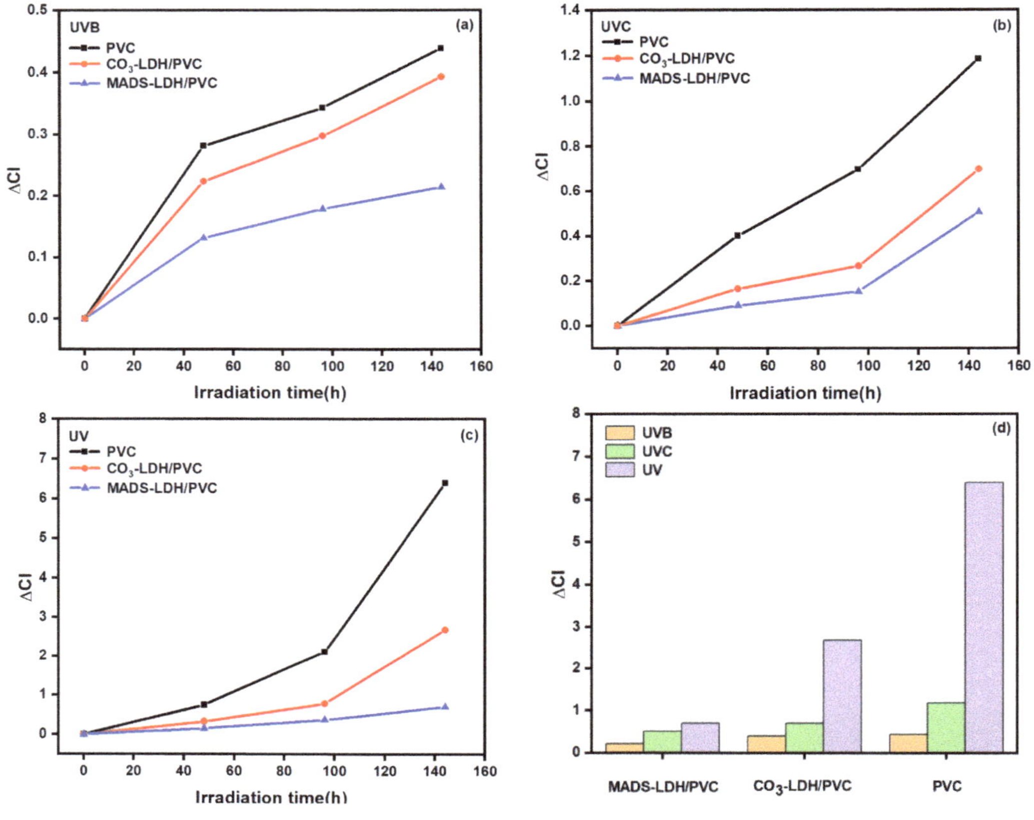

Figure 11. (**a**–**c**) The increase in carbonyl index (ΔCl) with aging time for PVC, CO_3-LDH/PVC, and MADS-LDH/PVC films during UVB (**a**), UVC (**b**), and UV (**c**) aging, respectively; (**d**) ΔCl of UVB, UVC and UV on PVC, CO_3-LDH/PVC, and MADS-LDH/PVC films after aging for 144 h, respectively.

Table 5. Carbonyl index (ΔCl) and relative degradation rate (RDR) of pristine PVC, CO_3-LDH/PVC, and MADS-LDH/PVC films after irradiation aging for 144 h.

	UVB		UVC		UV	
Sample	**ΔCl**	**RDR**	**ΔCl**	**RDR**	**ΔCl**	**RDR**
PVC	0.439	100%	1.185	100%	6.401	100%
CO_3-LDH/PVC	0.393	89.522%	0.697	58.819%	2.687	41.978%
MADS-LDH/PVC	0.215	48.975%	0.507	42.785%	0.705	11.014%

Moreover, by comparing the color changes during the aging process of the films, it could be inferred that the degree of influence of UVB, UVC, and UV band on the aging of MADS-LDH/PVC film composite materials was in the order of UV > UVC > UVB, which was consistent with the results obtained in Figures 11d and 12d below.

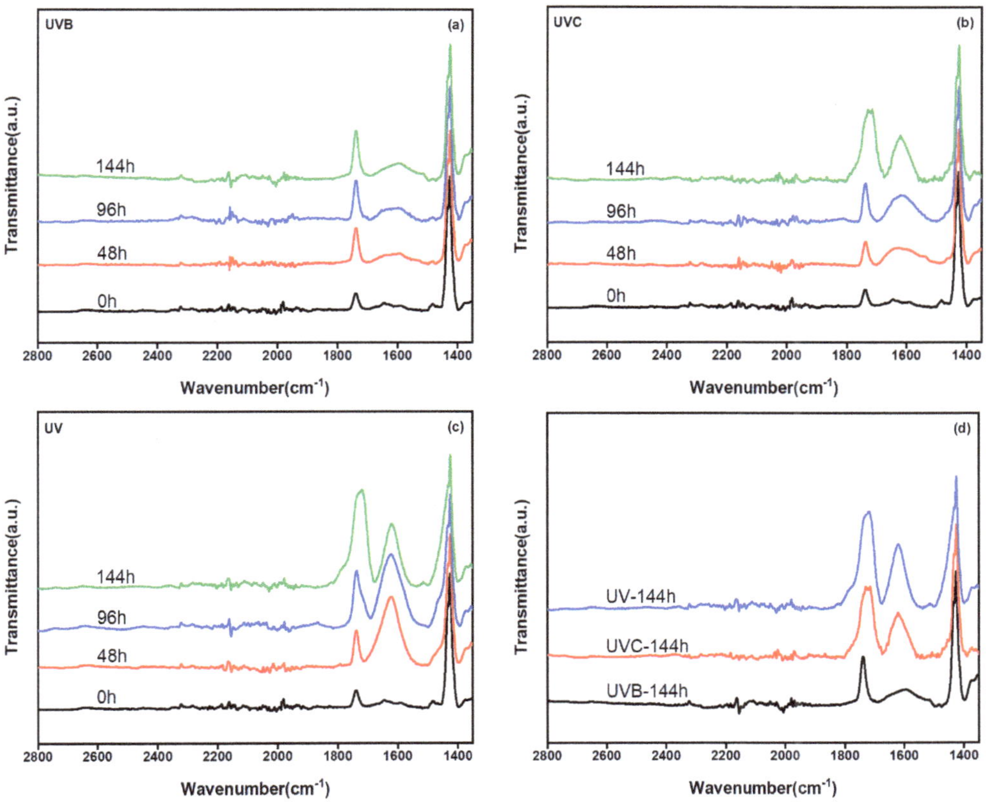

Figure 12. (**a**–**c**) FT-IR patterns of MADS-LDH/PVC films at different aging times under UVB, UVC, and UV photoaging conditions, respectively; (**d**) FT-IR patterns of UVB, UVC, and UV on MADS-LDH/PVC films after 144 h aging, respectively.

3.9. Light Stability of LDHs/PVC Film Composite Materials

To evaluate the UV-light aging resistance of MADS-LDH/PVC film composite materials, the surface chemical properties of pristine PVC, CO_3-LDH/PVC, and MADS-LDH/PVC film composite materials were analyzed and compared during UV photoaging. Previous studies have shown that carbonyl compounds are the primary molecules produced on the surface of PVC films during photo-oxidative aging [44]. PVC undergoes de-hydrogen chloride and photo-oxidative aging under UV-light, which results in the production of C=O compounds [45].

UV accelerated photoaging tests were conducted to assess the photostability of the pristine PVC and LDHs/PVC film composite materials. The degree of aging during photoaging was determined by analyzing the variation of ΔCI values with irradiation time [42,46]. When exposed to UV light radiation, PVC undergoes photoaging through oxidation reactions, which leads to the production of numerous carbonyl groups [44]. The ΔCI value represents the increase in the number of carbonyl groups induced by UV light [28]. Figure 11a–c display the curves of irradiation time vs. carbonyl index increase for pristine PVC, CO_3-LDH/PVC, and MADS-LDH/PVC film composite materials under UVB, UVC, and UV band ultraviolet light conditions, respectively. The increase in carbonyl index (ΔCI) during aging was calculated using the peak area method, with the absorption peak area A_{1430} at 1430 cm^{-1} as the internal standard. This method provided a more intuitive way of evaluating the degree of photoaging. The calculation equations were as follows:

$$\Delta CI = \frac{(A_{1730,t} - A_{1730,0})}{A_{1430}} \tag{1}$$

The absorption peak areas at 1730 cm^{-1} at the aging times t and 0 were denoted as $A_{1730,t}$ and $A_{1730,0}$, respectively, to calculate the carbonyl index increase. Figure 11a–c shows the ΔCI values of both the pristine PVC and LDHs/PVC film composite materials increased with increasing photoaging time, indicating the occurrence of photooxidation reactions with different oxidation rates. The ΔCI value of the LDHs/PVC film composite materials was smaller during aging than that of the pristine PVC, suggesting that the LDH samples could reduce the formation of peroxides during PVC aging and improve its UV stability. Moreover, the ΔCI curves of the MADS-LDH films had the smallest values during the aging process, indicating a slower oxidation rate and better UV shielding properties compared to CO_3-LDH.

The relative degradation rates (RDR) were calculated using the following equations, and the results are presented in Table 5. After 144 h of UVB, UVC, and UV band irradiation aging, the MADS-LDH/PVC films showed RDR values of 48.975%, 42.785%, and 11.014%, respectively, compared to the original PVC. Furthermore, analysis of Table 5 and Figure 11d indicated that the magnitude of the aging effect on all three films follows the order of UV > UVC > UVB.

$$\text{Relative degradation rate} = \left(\Delta CI_{LDH/PVC}/\Delta CI_{PVC}\right) \times 100\% \tag{2}$$

where ΔCI_{PVC} and $\Delta CI_{LDH/PVC}$ are the increase values of carbonyl index of the original PVC and LDHs/PVC film composite materials after irradiation aging process for 144 h, respectively.

Figure 12a–c present the FT-IR spectra of MADS-LDH/PVC film composite materials subjected to different photoaging times under UVB, UVC, and UV wavelength aging conditions. It is known that photoaging of pristine PVC under UVB irradiation produced a considerable number of carbonyl groups through oxidation reactions. The C-H anti-stretching vibrational absorption peak at 1430 cm^{-1} in the R-CH_2-R′ structure of the PVC molecular chain was accompanied by the peaks in the range of 1620 cm^{-1} and 1730 cm^{-1}, which were associated with the C=C stretching vibrational absorption peaks in the R-C=C-R′ polyene structure and C=O in the R-C=O-R′ carbonyl structure [47]. Both C=C and C=O stretching vibration absorption peaks in the sample were produced by the UV photoaging process. Therefore, the carbonyl group's absorption intensity at 1730 cm^{-1} was selected to represent the degree of photoaging. As shown in Figure 12a–c, the intensity of the carbonyl absorption peak at 1730 cm^{-1} increased to different degrees with increasing photoaging time, indicating different levels of photoaging degradation of the MADS-LDH/PVC film composite materials. Comparing with Figure 12d, it can be seen that the effect of UV, UVC, and UVB band irradiation conditions on the aging of MADS-LDH/PVC film composite materials followed the order of UV > UVC > UVB, which was consistent with the results obtained in Figure 11d above.

4. Conclusions

This study focused on the preparation and characterization of MADS-LDH and its potential to improve the photoaging resistance of PVC films. The mechanism of MADS-LDH inhibition of photoaging degradation of MADS-LDH/PVC film composite materials was also investigated. The results indicate that MADS-LDH effectively inhibits the production rate of carbonyl groups during photoaging, thereby improving the photoaging resistance of MADS-LDH/PVC film composite materials. Based on the experimental results obtained in this study, the following conclusions can be drawn:

(1) The successful preparation of MADS-LDH was confirmed through XRD and FT-IR analysis, which presented a d_{003} peak position at 3.08°, a proportional relationship between planar spacing in (d_{003}) = 2 (d_{006}) = 3 (d_{009}), and a single set of Bragg reflections (00*l*) (*l* = 3, 6, 9). Most of the characteristic absorption bands of MADS were detected in the FT-IR spectra of MADS-LDH.
(2) The agglomeration of MADS-LDH nanosheet particles through stacking was observed using SEM and TEM. BET and TGA analysis indicated that the pore size of MADS-LDH is mesoporous with a homogeneous pore size distribution, and that the MADS anion intercalation created a hydrophobic environment within the layer, reducing the formation of interlayer crystalline water in MADS-LDH.
(3) MADS-LDH/PVC film composite materials exhibited the least color change and the lowest degree of aging under both UVC and UV band irradiation compared to pristine PVC and CO_3-LDH/PVC films. The color of PVC films changed significantly under both UVC and UV band irradiation, with similar changes in CO_3-LDH/PVC films, but the degree of aging was relatively low.
(4) The addition of MADS-LDH inhibited the generation of carbonyl groups in the MADS-LDH/PVC film composite materials during photoaging, resulting in lower values of both carbonyl index (ΔCl) and relative degradation rate (RDR) compared to pristine PVC and CO_3-LDH/PVC films. The degree of influence of UVB, UVC, and UV bands on the photoaging of PVC film and LDHs/PVC film composite materials was found to be UV > UVC > UVB.

Author Contributions: Conceptualization, E.Z. and Y.L.; Methodology, E.Z. and X.C.; Validation, J.H. and X.L.; Data curation, H.Y. and Y.Z.; Writing—original draft preparation, E.Z.; writing—review and editing, E.Z.; Visualization, E.Z. All authors have read and agreed to the published version of the manuscript.

Funding: This paper describes research activities mainly requested and supported by the Natural Science Foundation of China under grant numbers 22278086, supported by the Science and Technology Program of Guangzhou under grant numbers 202102021043, supported by the Demonstration Base for Joint Training of Postgraduates between Guangdong University of Technology and GCH Technology Co., Ltd., supported by the GCH Technology Co., Ltd., under grant numbers 607210383.

Institutional Review Board Statement: Not applicable.

Informed Consent Statement: Not applicable.

Data Availability Statement: Not applicable.

Conflicts of Interest: The authors declare no conflict of interest.

References

1. Kann, Y.; Billingham, N.C. Chemiluminescence is shedding light on degradation and stabilisation of plasticised poly (vinyl chloride). *Polym. Degrad. Stab.* **2004**, *85*, 957–966. [CrossRef]
2. Belhaneche-Bensemra, N.; Ouazene, N. Study of the influence of atmospheric pollutants on the natural ageing of rigid polyvinyl chloride. *Macromol. Symp.* **2002**, *180*, 181–190. [CrossRef]
3. Decker, C. Photostabilization of poly (vinyl chloride) by protective coatings. *J. Vinyl Addit. Technol.* **2001**, *7*, 235–243. [CrossRef]
4. Belhaneche-Bensemra, N. Influence of atmospheric pollutants on the natural and artificial aging of rigid poly (vinyl chloride). *J. Vinyl Addit. Technol.* **2002**, *8*, 45–54. [CrossRef]
5. Benavides, R.; Castillo, B.M.; Castañeda, A.O.; López, G.M.; Arias, G. Different thermo-oxidative degradation routes in poly (vinyl chloride). *Polym. Degrad. Stab.* **2001**, *73*, 417–423. [CrossRef]
6. Mohapi, M.; Sefadi, J.S.; Mochane, M.J.; Magagula, S.I.; Lebelo, K. Effect of LDHs and Other Clays on Polymer Composite in Adsorptive Removal of Contaminants: A Review. *Crystals* **2020**, *10*, 957. [CrossRef]
7. Nazir, M.A.; Khan, N.A.; Cheng, C.; Shah, S.S.A.; Najam, T.; Arshad, M.; Sharif, A.; Akhtar, S.; Rehman, A.u. Surface induced growth of ZIF-67 at Co-layered double hydroxide: Removal of methylene blue and methyl orange from water. *Appl. Clay Sci.* **2020**, *190*, 105564. [CrossRef]
8. Nazir, M.A.; Najam, T.; Jabeen, S.; Wattoo, M.A.; Bashir, M.S.; Shah, S.S.A.; ur Rehman, A. Facile synthesis of Tri-metallic layered double hydroxides (NiZnAl-LDHs): Adsorption of Rhodamine-B and methyl orange from water. *Inorg. Chem. Commun.* **2022**, *145*, 110008. [CrossRef]

9. Haleem, A.; Shafiq, A.; Chen, S.Q.; Nazar, M. A Comprehensive Review on Adsorption, Photocatalytic and Chemical Degradation of Dyes and Nitro-Compounds over Different Kinds of Porous and Composite Materials. *Molecules* **2023**, *28*, 1081. [CrossRef]
10. Jamshaid, M.; Nazir, M.A.; Najam, T.; Shah, S.S.A.; Khan, H.M.; Rehman, A.u. Facile synthesis of Yb^{3+}-Zn^{2+} substituted M type hexaferrites: Structural, electric and photocatalytic properties under visible light for methylene blue removal. *Chem. Phys. Lett.* **2022**, *805*, 139939. [CrossRef]
11. Balbin Tamayo, A.I.; Esteva Guas, A.M.; Pupim Ferreira, A.A.; Aucélio, R.Q.; Huertas Flores, J.O. Influence of molar fraction of Mg/Al, on the electrochemical behavior of hycrotalcite-epoxy-graphite composites. *Mater. Chem. Phys.* **2020**, *253*, 123392. [CrossRef]
12. Scavetta, E.; Berrettoni, M.; Giorgetti, M.; Tonelli, D. Electrochemical characterisation of Ni/Al X hydrotalcites and their electrocatalytic behaviour. *Electrochim. Acta* **2002**, *47*, 2451–2461. [CrossRef]
13. Shen, L.; Chen, Z.; Kou, J. High-quality modification of general polypropylene by the synergistic effect of zinc adipate, hydrotalcite, and polypropylene (SP179). *J. Appl. Polym. Sci.* **2023**, *140*, e53704. [CrossRef]
14. Yan, J.; Yang, Z. Intercalated hydrotalcite-like materials and their application as thermal stabilizers in poly (vinyl chloride). *J. Appl. Polym. Sci.* **2017**, *134*, 44896. [CrossRef]
15. Huang, Y.; Law, J.C.-F.; Lam, T.-K.; Leung, K.S.-Y. Risks of organic UV filters: A review of environmental and human health concern studies. *Sci. Total Environ.* **2021**, *755*, 142486. [CrossRef] [PubMed]
16. Gao, H.; Yao, A.; Shi, Y.; Noor, N.; Zeb, A.; Li, M.; Li, H. Preparation and properties of hierarchical Al–Mg layered double hydroxides as UV resistant hydrotalcite. *Mater. Chem. Phys.* **2020**, *256*, 123630. [CrossRef]
17. Zeng, R.; Tang, W.; Zhou, Q.; Liu, X.; Liu, Y.; Wang, S.; Chen, Z.; Yi, N.; Wang, Z.; Chen, J. Efficient adsorption of Pb (II) by sodium dodecyl benzene sulfonate intercalated calcium aluminum hydrotalcites: Kinetic, isotherm, and mechanisms. *Environ. Sci. Pollut. Res. Int.* **2022**, *29*, 46161–46173. [CrossRef]
18. Bouraada, M.; Lafjah, M.; Ouali, M.; Demenorval, L. Basic dye removal from aqueous solutions by dodecylsulfate- and dodecyl benzene sulfonate-intercalated hydrotalcite. *J. Hazard. Mater.* **2008**, *153*, 911–918. [CrossRef]
19. Li, Y.; Bi, H.Y.; Shen, S.L. Removal of bisphenol A from aqueous solution by a dodecylsulfate ion-intercalated hydrotalcite-like compound. *Environ. Technol.* **2012**, *33*, 1367–1373. [CrossRef]
20. Bouraada, M.; Ouali, M.S.; de Ménorval, L.C. Dodecylsulfate and dodecybenzenesulfonate intercalated hydrotalcites as adsorbent materials for the removal of BBR acid dye from aqueous solutions. *J. Saudi Chem. Soc.* **2016**, *20*, 397–404. [CrossRef]
21. Milagres, J.L.; Bellato, C.R.; Ferreira, S.O.; de Moura Guimaraes, L. Preparation and evaluation of hydrocalumite-iron oxide magnetic intercalated with dodecyl sulfate for removal of agrichemicals. *J. Environ. Manag.* **2020**, *255*, 109845. [CrossRef] [PubMed]
22. Zhang, X.; Pi, H.; Guo, S. The mechanism for inorganic fillers accelerating and inhibiting the UV irradiation aging behaviors of rigid poly (vinyl chloride). *J. Appl. Polym. Sci.* **2011**, *122*, 2869–2875. [CrossRef]
23. Ma, X.; Gao, H.; Lu, Y.; Liu, X.; Dang, L.; Xu, S. Application of β-diketone boron complex as an ultraviolet absorber in polyvinyl chloride film. *Mater. Res. Express* **2020**, *7*, 076403. [CrossRef]
24. Yousufzai, A.; Zafar, M.J.; Hasan, S.U. Radical degradation of polyvinyl chloride. *Eur. Polym. J.* **1972**, *8*, 1231–1236. [CrossRef]
25. Yu, J.J.; Jiang, Z.; Zhu, L.; Hao, Z.P.; Xu, Z.P. Adsorption/Desorption Studies of NOx on Well-Mixed Oxides Derived from Co−Mg/Al Hydrotalcite-like Compounds. *J. Phys. Chem. B* **2006**, *110*, 4291–4300. [CrossRef]
26. Li, L.; Ma, R.; Ebina, Y.; Fukuda, K.; Takada, K.; Sasaki, T. Layer-by-Layer Assembly and Spontaneous Flocculation of Oppositely Charged Oxide and Hydroxide Nanosheets into Inorganic Sandwich Layered Materials. *J. Am. Chem. Soc.* **2007**, *129*, 8000–8007. [CrossRef]
27. Lin, Y.; Wang, J.; Evans, D.G.; Li, D. Layered and intercalated hydrotalcite-like materials as thermal stabilizers in PVC resin. *J. Phys. Chem. Solids* **2006**, *67*, 998–1001. [CrossRef]
28. Zhang, Q.; Song, K.; Zhao, J.; Kong, X.; Sun, Y.; Liu, X.; Zhang, Y.; Zeng, Q.; Zhang, H. Hexanedioic acid mediated surface–ligand-exchange process for transferring $NaYF_4$: Yb/Er (or Yb/Tm) up-converting nanoparticles from hydrophobic to hydrophilic. *J. Colloid Interface Sci.* **2009**, *336*, 171–175. [CrossRef]
29. Sampath, S.K.; Cordaro, J.F. Optical properties of zinc aluminate, zinc gallate, and zinc aluminogallate spinels. *J. Am. Ceram. Soc.* **1998**, *81*, 649–654. [CrossRef]
30. Mannepalli, L.K.; Dupati, V.; Vallabha, S.J.; Sunkara, V.M. Synthesis of substituted guanidines using Zn–Al hydrotalcite catalyst. *J. Chem. Sci.* **2013**, *125*, 1339–1345. [CrossRef]
31. Bai, L.; Liu, X.; Jiao, T.; Wang, Y.; Huo, Y.; Niu, J. Surface and Interfacial Properties of Mono and Didodecyl Diphenyl Ether Disulfonates. *Tenside Surfactants Deterg.* **2018**, *55*, 302–311. [CrossRef]
32. Liu, X.; Niu, J.; Wang, X. Synthesis, Surface and Interfacial Properties of Dodecyl Diphenyl Oxide Disulfonate with Different Counterions. *J. Surfactants Deterg.* **2015**, *18*, 675–679. [CrossRef]
33. Tao, Q.; Yuan, J.; Frost, R.L.; He, H.; Yuan, P.; Zhu, J. Effect of surfactant concentration on the stacking modes of organo-silylated layered double hydroxides. *Appl. Clay Sci.* **2009**, *45*, 262–269. [CrossRef]
34. Wang, X.; Wu, P.; Huang, Z.; Zhu, N.; Wu, J.; Li, P.; Dang, Z. Solar photocatalytic degradation of methylene blue by mixed metal oxide catalysts derived from ZnAlTi layered double hydroxides. *Appl. Clay Sci.* **2014**, *95*, 95–103. [CrossRef]
35. Li, Z.-X.; Zeng, H.-Y.; Gohi, B.F.C.A.; Ding, P.-X. Preparation of CeO_2-decorated organic-pillared hydrotalcites for the UV resistance of polymer. *Appl. Surf. Sci.* **2020**, *507*, 145110. [CrossRef]

36. Sharma, D.; Sakthivel, A.; Michelraj, S.; Muthurasu, A.; Ganesh, V. Surfactant Intercalated Mono-metallic Cobalt Hydrotalcite: Preparation, Characterization, and its Bi-functional Electrocatalytic Application. *ChemistrySelect* **2020**, *5*, 9615–9622. [CrossRef]
37. Mahjoubi, F.Z.; Khalidi, A.; Elhalil, A.; Barka, N. Characteristics and mechanisms of methyl orange sorption onto Zn/Al layered double hydroxide intercalated by dodecyl sulfate anion. *Sci. Afr.* **2019**, *6*, e00216. [CrossRef]
38. Wang, C.-X.; Pu, M.; Zhang, P.-H.; Gao, Y.; Yang, Z.-Y.; Lei, M. Structure Simulation and Host–Guest Interaction of Histidine-Intercalated Hydrotalcite–Montmorillonite Complex. *Minerals* **2018**, *8*, 198. [CrossRef]
39. Ma, R.; Tang, P.; Feng, Y.; Li, D. UV absorber co-intercalated layered double hydroxides as efficient hybrid UV-shielding materials for polypropylene. *Dalton Trans.* **2019**, *48*, 2750–2759. [CrossRef]
40. Feng, Y.; Li, D.; Wang, Y.; Evans, D.G.; Duan, X. Synthesis and characterization of a UV absorbent-intercalated Zn–Al layered double hydroxide. *Polym. Degrad. Stab.* **2006**, *91*, 789–794. [CrossRef]
41. Kohno, Y.; Asai, S.; Shibata, M.; Fukuhara, C.; Maeda, Y.; Tomita, Y.; Kobayashi, K. Improved photostability of hydrophobic natural dye incorporated in organo-modified hydrotalcite. *J. Phys. Chem. Solids* **2014**, *75*, 945–950. [CrossRef]
42. Miotto Menino, N.; da Silveira Salla, J.; do Nascimento, M.S.; Dallago, R.M.; Peralta, R.A.; Moreira, R.F. High-performance hydrophobic magnetic hydrotalcite for selective treatment of oily wastewater. *Environ. Technol.* **2021**, *44*, 1426–1437. [CrossRef] [PubMed]
43. Zhang, Y.; Yuan, Z.-P.; Qin, Y.; Dai, J.; Zhang, T. Comparative Studies on Hydrophilic and Hydrophobic Segments Grafted Poly (vinyl chloride). *Chin. J. Polym. Sci.* **2018**, *36*, 604–611. [CrossRef]
44. Owen, E.D. Photodegradation of Polyvinyl chloride. In *Ultraviolet Light Induced Reactions in Polymers*; American Chemical Society: Washington, DC, USA, 1976; Volume 25, pp. 208–219.
45. Owen, J. *Degradation and Stabilisation of PVC*; Springer Science & Business Media: Cham, Switzerland, 2012.
46. Zhang, X.; Pi, H.; Guo, S.; Fu, J. Influence of ultraviolet absorbers on the ultraviolet-irradiation behaviour of rigid poly vinyl chloride. *Plast. Rubber Compos.* **2016**, *45*, 352–361. [CrossRef]
47. Liu, H.; Dong, L.; Xie, H.; Wan, L.; Liu, Z.; Xiong, C. Ultraviolet light aging properties of PVC/$CaCO_3$ composites. *J. Appl. Polym. Sci.* **2013**, *127*, 2749–2756. [CrossRef]

coatings

MDPI

Article

Rheological Properties of Composite Inorganic Micropowder Asphalt Mastic

Tengteng Guo [1], Haijun Chen [1], Deqing Tang [1], Shengquan Ding [1], Chaohui Wang [2], Decai Wang [1], Yuanzhao Chen [1,*] and Zhenxia Li [1]

[1] School of Civil Engineering and Communication, North China University of Water Resources and Electric Power, Zhengzhou 450045, China; guotth@ncwu.edu.cn (T.G.); chenhaijun@ncwu.edu.cn (H.C.); tdq18738952643@163.com (D.T.); dsqcad@163.com (S.D.); wangdecai@ncwuedu.cn (D.W.); zhenxiali2009@ncwu.edu.cn (Z.L.)

[2] School of Highway, Chang'an University, Xi'an 710064, China; wchh0205@chd.edu.cn

* Correspondence: cyz740513@ncwu.edu.cn

Abstract: Graphene Tourmaline Composite Micropowder (hereinafter referred to as GTCM) modified asphalt was prepared by the ball milling method. The effects of different temperatures and different frequencies on the high-temperature performance of composite-modified asphalt were evaluated by dynamic shear rheological test, and the viscoelastic properties of composite-modified asphalt under different stresses and different temperatures were analyzed. The low-temperature rheological properties of GTCM-modified asphalt were analyzed by bending beam rheological test, and its mechanism was analyzed by Fourier transform infrared spectroscopy (FTIR) test. The results show that the temperature sensitivity and anti-aging resistance of GTCM-modified asphalt are significantly higher than that of tourmaline-modified asphalt. The improvement effect gradually increases with the increase in graphene powder content, and its addition does not change the viscoelastic properties of asphalt. The complex shear modulus and phase angle of GTCM-modified asphalt at appropriate temperatures are more conducive to tourmaline-modified asphalt and matrix asphalt, which can improve the rutting resistance of asphalt. In the same type, with the increase in composite modified micropowder content, the rutting resistance of modified asphalt is better. The improvement of rutting resistance of GTCM-0.5, GTCM-1.0 and GTCM-1.5-modified asphalt can reach 12.95%, 10.12% and 24.25%, respectively; the improvement range is more complicated due to temperature and frequency changes. The GTCM-modified asphalt has good low-temperature crack resistance. The creep stiffness modulus of GTCM-modified asphalt decreases with the increase in load time under different types and dosages, and its stiffness modulus is smaller than that of tourmaline-modified asphalt and mineral powder asphalt mastic. The creep rate increases with the extension of load time, which is greater than that of tourmaline-modified asphalt and mineral powder asphalt mastic. When the load was 60 s, the creep stiffness modulus of GTCM-0.5, GTCM-1.0 and GTCM-1.5-modified asphalt decreased by 5.75%, 6.97% and 13.73%, respectively, and the creep rate increased by 1.37%, 2.52% and 4.35%, respectively. After adding GTCM or tourmaline to the matrix asphalt, no new functional groups were produced due to the chemical reaction with the asphalt.

Keywords: pavement material; tourmaline; graphene; asphalt mastic; rheological properties

Citation: Guo, T.; Chen, H.; Tang, D.; Ding, S.; Wang, C.; Wang, D.; Chen, Y.; Li, Z. Rheological Properties of Composite Inorganic Micropowder Asphalt Mastic. *Coatings* **2023**, *13*, 1068. https://doi.org/10.3390/coatings13061068

Academic Editor: Valeria Vignali

Received: 21 May 2023
Revised: 6 June 2023
Accepted: 6 June 2023
Published: 8 June 2023

1. Introduction

The overload problem of asphalt pavement is escalating in severity due to the nation's fast growth in car ownership. It is difficult for matrix asphalt to solve the problem of the long fatigue life of the pavement. However, the performance of traditional modified asphalt has gradually failed to meet the current requirements of asphalt pavement, resulting in frequent diseases of asphalt pavement and the actual service life cannot reach the design life, resulting in a serious waste of resources and economic losses [1–3]. At present, SBS modified asphalt and crumb rubber modified asphalt are the most widely used types of modified asphalt. Due to their inherent component incompatibility, easy oxidation degradation and density difference, there are still some inevitable problems such as low road performance, poor durability, and insufficient stability [4,5], which seriously restrict the service life of pavement laid with such materials. Therefore, it is urgent to develop high-performance and long-life asphalt pavement materials.

Due to its unique material properties, the modification of asphalt by inorganic compounds has emerged as a research hotspot in the field of asphalt modification in light of the issues with conventionally treated asphalt. wide sources, and simple production equipment. To a certain extent, inorganic ingredients can enhance the overall performance of asphalt mastic and asphalt mixture, hence enhancing the quality of asphalt pavement [6]. Compared with organic-modified asphalt, inorganic powder-modified asphalt has the characteristics of a simple production process, low price, and excellent performance. Therefore, many scholars at home and abroad have carried out a series of studies on inorganic powder-modified asphalt. The commonly used inorganic micropowder materials are carbon black, fiber, diatomite, cement, hydrated lime, silica fume, layered silicate, nano-calcium carbonate, etc. Inorganic micropowder-modified asphalt is by adding inorganic micropowder materials into asphalt, through physical adsorption and chemical reaction, etc., to improve the road performance of asphalt and asphalt mixture, improve the performance of asphalt pavement, and improve the performance of asphalt mastic to a certain extent.

Graphene has been widely used in composite materials, nano-electronic devices, catalyst carriers, sensors, and energy storage due to its excellent thermal conductivity, high strength and large specific surface area [7–10]. The existing results show that the high-temperature rutting resistance and high-temperature stability of asphalt can be improved by means of the high thermal conductivity, high specific surface area and excellent mechanical properties of graphene [11–15]. Amirkhanian S et al. [16,17] studied the interaction between graphene oxide and asphalt and its road performance. The results show that graphene oxide does not chemically react with asphalt, but only a simple physical blending, but the adhesion between them is good. Graphene oxide-modified asphalt will reduce the ductility of asphalt and increase the viscosity. It significantly enhances the rutting resistance of modified asphalt within the range of 30 to 80 °C. M. Marasteanu et al. [18,19] studied the performance changes of graphene-modified asphalt before and after aging. Graphene modifiers can lessen asphalt's aging rate, postpone the production of carbonyl and sulfoxide groups during aging, and increase the resilience of asphalt to fatigue cracking. Christiansen S. [20] proposed a method for graphene to enhance the strength of foamed asphalt, indicating that graphene oxide can improve the mechanical properties of foamed asphalt. Wang Chaohui et al. [21–24] systematically studied the effects of tourmaline on the high-temperature performance, low-temperature crack resistance, water temperature qualitative and fatigue resistance of asphalt and its mixture, as well as the functions of thermal companion emission reduction, flame retardant smoke suppression and air purification. The results show that tourmaline has a good improvement effect on asphalt pavement performance and environmental efficacy. Haibo Ding et al. [25] prepared polymer-modified asphalt containing different proportions of tourmaline modifier and found that a small amount of modifier could improve the mechanical properties of asphalt binder, and the modification effect of tourmaline ion powder was the best. Kim S. J. et al. [26] prepared a TiO_2 photoanode with different content of tourmaline by using the spontaneous polarization characteristics of tourmaline and compared it with the original TiO_2 photoanode by electrochemical impedance spectroscopy, current density, and voltage. Compared with the original TiO_2 photoanode, the electron lifetime and power conversion efficiency of the TiO_2 photoanode with 3% tourmaline were increased by about 42% and 20%, respectively. Kang S. J. et al. [27] added tourmaline particles into nylon fibers doped with TiO_2 particles and found that tourmaline can play a synergistic role with TiO_2 to improve the ability of photocatalytic degradation of organic pollutants. Moreno-Navarro et al. [28] tested the rheological and thermal properties of binders prepared with different amounts of graphene sheets. It was found that the addition of graphene to asphalt can produce a greater elastic response. The impact of tourmaline content on the photocatalytic activity of TiO2-graphene was investigated by Baeissa et al. [29]. Tourmaline's spontaneous polarization effect can enhance photocatalytic activity, and the composite material with the highest performance has a G-to-T ratio of 1% to 2.5%.

In summary, although graphene and tourmaline are two environmentally friendly materials that belong to different substances, they can exert excellent adsorption capacity and play a major role in environmental protection. When used alone as modifiers, they can greatly enhance the asphalt's rheological characteristics and fatigue resistance. However, there are few studies on the combination of the two in the road field. Therefore, this paper prepared GTCM asphalt mastic by graphene/tourmaline, and systematically studied the rheological properties of inorganic micropowder asphalt mastic, to produce composite asphalt-modified materials with cheap cost, environmental protection, excellent quality, and long life.

2. Materials and Tests

2.1. Materials

2.1.1. GTCM

The composite inorganic powder refers to a powder mixture made by ball milling of graphene and tourmaline. Among them, graphene (Figure 1) is provided by Suzhou Carbonfeng Graphene Technology Co., Ltd. (Suzhou, China). The main physical properties are shown in Table 1.

(a) (b)

Figure 1. Graphene. (**a**) photo. (**b**) SEM picture.

Table 1. Technical specifications of graphene.

Performance Parameter	Appearance	Purity (wt%)	Thickness (nm)	Specific Surface Area (m^2/g)	Layer Diameter (µm)
measured value	black powder	95	1.35	89	47

The appearance of the tourmaline used is a black powder (Figure 2), 2000 mesh, and the main components are shown in Table 2.

(a)

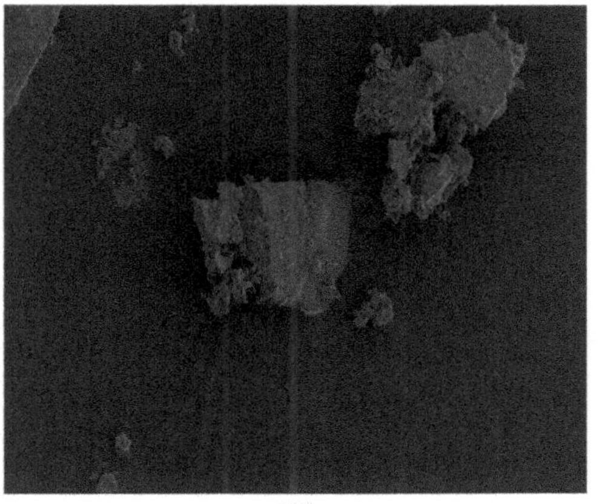

(b)

Figure 2. Tourmaline. (**a**) photo. (**b**) SEM picture.

Table 2. The main components and content of tourmaline.

Essential Component	SiO_2	Al_2O_3	B_2O_3	MgO	FeO	Fe_2O_3	CaO	TiO_2	Na_2O	H_2O
content (%)	33.54	31.98	11.21	0.49	2.88	16.32	0.01	0.28	0.72	2.51

GTCM (Figure 3) is based on tourmaline as the main material. By adding a certain mass fraction of graphene to the tourmaline powder material, the two are combined with a certain process to form a composite material with a certain process. It is expected that the two will play a synergistic role and play a synergistic role in enhancing the performance of asphalt and environmental efficacy. In view of the high price of graphene with superior quality sold in the market, the amount of graphene should be appropriately controlled when preparing the composite material. The ball milling method is used to obtain [30], in which the mass ratio of graphene to tourmaline is 0.5%, 1.0% and 1.5%, respectively. They are denoted as GTCM-0.5, GTCM-1.0 and GTCM-1.5, respectively.

(a)

(b)

Figure 3. GTCM. (**a**) photo. (**b**) SEM picture.

2.1.2. Asphalt

AH-70 # asphalt is used as asphalt, and Table 1 displays its technical indications.

2.2. Preparation of GTCM Asphalt Mastic

The performance of modified asphalt is directly impacted by the preparation procedure used to create the inorganic micropowder modified asphalt mastic. To ensure the dispersion of inorganic powder, modified asphalt was prepared using the following method:

(1) Weigh the dried asphalt after quantitative dehydration. Based on the quality of asphalt and the content of GTCM, titanate coupling agent TC-131 and quantitative dry GTCM were weighed and well combined. When the asphalt is heated to good fluidity, the uniformly mixed powder material is poured into it, and one side is added for manual stirring for 5 min.
(2) The high-speed shearing instrument was started. First, after shearing and dispersing at 1000 rpm for ten minutes, the speed was increased to 3000 rpm for thirty minutes. The rotor was taken out of the mixing barrel by adjusting the speed, and the temperature of asphalt in the modification process was controlled at 150 °C.
(3) To get rid of the air bubbles in the GTCM-modified asphalt, it was manually stirred with a stirring rod for 10 min before being placed in storage.

2.3. Basic Performance Test Method

According to the relevant provisions of "Test Specifications for Asphalt and Asphalt Mixtures for Road Construction" (JTG E20-2011), the penetration index (PI) and equivalent softening point (T_{800}) of asphalt were also evaluated. The penetration, softening point, and ductility of asphalt were examined. Asphalt was aged using the Rotating Film Oven Heating Test (RTFOT) (Ai Yao Scientific Instrument Co., Ltd., Shanghai, China) according to ASTM D2872. Aged asphalt was tested for penetration and softening point, and the residual penetration ratio and softness point increase were computed.

2.4. Test Method for Rheological Properties

2.4.1. Test Method for Dynamic Shear Rheological Properties

The H-PTD200 dynamic shear rheometer was produced by Anton Paar (Graz, Austria), as shown in Figure 4. The temperature control range of the instrument is −40~+200 °C, supporting the speed control range of 10^{-6} μrad/s~314 rad/s, equipped with a variety of parallel plates and fixtures, and supporting the test of asphalt samples under various modes such as temperature scanning and frequency scanning.

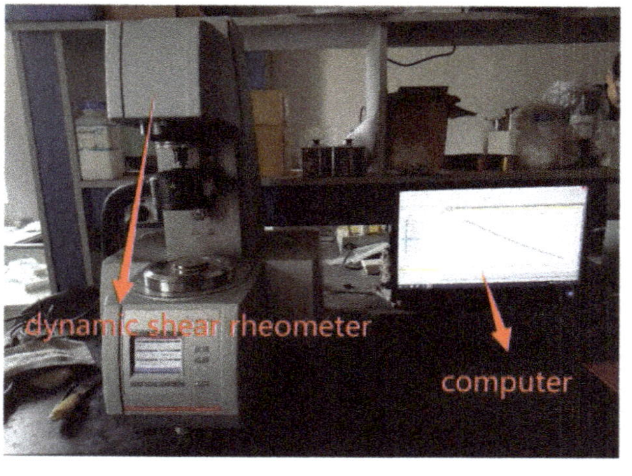

Figure 4. Dynamic shear rheometer.

In the experiment, the specimen was removed from the mold and placed between two parallel plates with either an 8 or 25 mm diameter. The distance between the plates was adjusted to meet the experimental requirements. A scraper was used to remove extra asphalt outside the plates. The upper plate spun around the central axis at a predetermined angular rate while the lower square circular plate remained fixed during the test. A sensor recorded the test results.

Dynamic shear rheological tests were conducted on GTCM-modified asphalt at varying temperatures and frequencies, with different types and dosages. The resulting data allowed for a comprehensive and objective evaluation of the dynamic shear rheological properties of the modified asphalt. Table 3 shows the specific test plan which includes a control strain of 1% and a frequency of 10 rad/s, with a temperature range of 30~80 °C. Additionally, a frequency scanning control strain of 10% will be used with a temperature of 60 °C and a frequency range of 0.1~100 rad/s.

Table 3. DSR test scheme of GTCM asphalt mastic.

Order Number	Types	Dosage (%)
1	asphalt	-
2	T asphalt mastic	20
3	GTCM-0.5 asphalt mastic	20
4	GTCM-1.0 asphalt mastic	10; 20; 30
5	GTCM-1.5 asphalt mastic	20

2.4.2. Low-Temperature Rheological Property Test Method

The TE-BBR-F type low temperature bending beam rheometer produced by CANNON company (Peschiera Borromeo (MI), Italy) was used. As shown in Figure 5, according to the ASTM D6648-01 test method, the test temperature of −12 °C and the stress continuous action of 100 g weight for 240 s were selected to study the low-temperature creep properties of composite inorganic powder-modified asphalt under different types and dosages. At the same time, two test temperatures of −18 and −24 °C and 100 g weight were selected to apply stress for 60 s. To thoroughly assess the low-temperature performance of GTCM-modified asphalt, the bending creep test of several types of GTCM-modified asphalt was conducted. The law of change with temperature. During the test, tourmaline-modified asphalt, mineral powder asphalt mastic and matrix asphalt were used as control tests. The specific test scheme is shown in the dynamic shear rheological performance test method.

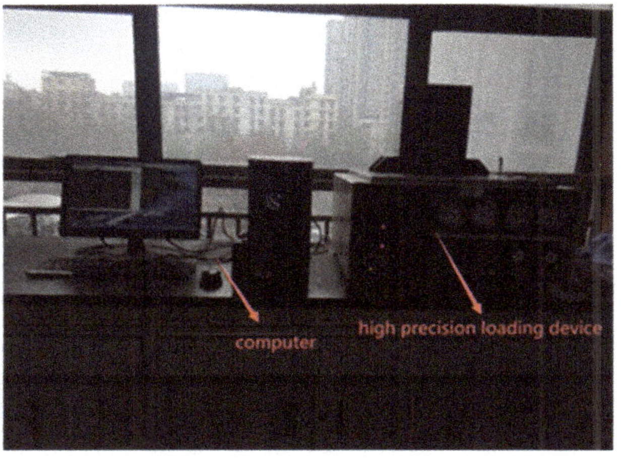

Figure 5. Asphalt low temperature bending beam rheometer.

2.5. FTIR Spectral Analysis Method

The infrared spectra of samples were measured by TENSOR27 FTIR Fourier transform infrared spectrometer produced by Thermo Nicolet Corporation from Madison, WI, USA, as shown in Figure 6. The instrument can measure the spectral range of 8300~30 cm^{-1}. SNR better than 40,000: 1 (peak–peak); resolution less than or equal to 4 cm^{-1}; the test site is the Key Laboratory of Applied Surface and Colloidal Chemistry, College of Chemical Engineering, Shaanxi Normal University, Xi'an, China.

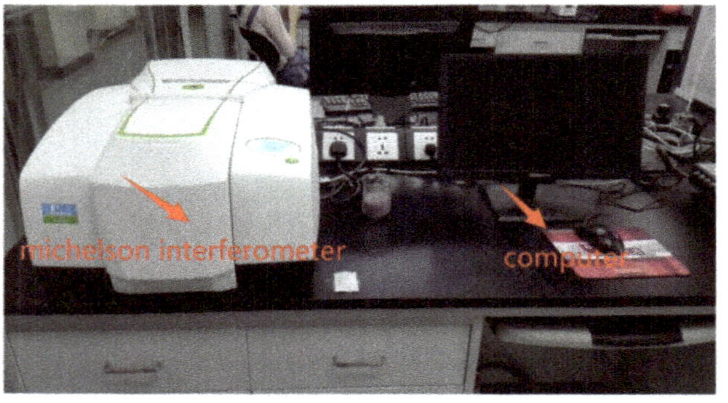

Figure 6. Fourier transform infrared spectrometer.

3. Result and Discussion

3.1. Basic Performance

To clarify the content of different types of GTCM in asphalt and its influence on asphalt performance, inorganic micropowder modified asphalt with different contents has been prepared, and a thorough investigation of the temperature sensitivity, high temperature, and anti-aging characteristics of modified asphalt. Table 4 presents the outcomes.

Table 4. Main technical indexes of matrix asphalt.

Test Project		Unit	Test Results
Penetration (25 °C)		0.1 mm	64
Penetration index PI		-	−1.37
Ductility (10 °C)		cm	38.3
Softening point (R&B)		°C	47.9
60 °C Kinetic viscosity		Pa·s	226
After TFOT	Quality change	%	0.12
	Penetration ratio	%	74.1
	Ductility (10 °C)	cm	12.4

According to Table 5.

(1) The temperature sensitivity of GTCM-modified asphalt is superior to tourmaline micropowder-modified asphalt under the same circumstances, and the improvement impact is more pronounced the more graphene micropowder is present. Compared to tourmaline micropowder modified asphalt, the temperature sensitivity of composite micropowder modified asphalt is increased by up to 87.2%.

(2) The high-temperature stability of asphalt changed with GTCM is superior to asphalt modified with tourmaline micropowder under the same circumstances. The improvement effect is more significant with higher graphene micropowder content. Compared to tourmaline micropowder modified asphalt, the GTCM-modified asphalt showed a 6.35% improvement in 25 °C penetration, 7.62% improvement in softening point, and 7.17% improvement in equivalent softening point.

(3) Under the same circumstances, the anti-aging performance of asphalt modified with composite micropowder is superior to that of asphalt modified with tourmaline micropowder. Additionally, the higher the content of graphene micropowder, the more significant the improvement effect. In comparison to tourmaline micropowder modified asphalt, the GTCM-modified asphalt showed an improvement of 8.60% in 25 °C penetration residual ratio and 57.58% in softening point increment.

Table 5. Basic performance test results of inorganic powder modified asphalt.

Asphalt Type	Dosage (%)	Needle Penetration (100 g, 5 s, 0.1 mm)					Residual Needle Penetration Ratio (%)	Softening Point (°C)			
		Before Aging				After Aging		Before Aging	Equivalent Softening Point T_{800}	After Aging	Increment
		15 °C	25 °C	30 °C	PI	25 °C					
Matrix asphalt	-	25	75	138	−1.317	43	57.2	41.5	45.5	49.5	8.0
T asphalt	10	23	65	116	−1.031	40	61.8	49.0	48.0	52.0	3.0
	20	22	61	107	−0.909	39	63.7	51.0	49.	52.5	1.5
	30	20	58	99	−0.895	38	64.3	52.5	50.0	53.5	1.0
GTCM-0.5	10	24	64	115	−0.826	41	63.7	49.5	49.0	52.0	2.5
	20	22	61	103	−0.669	40	65.3	52.0	50.0	53.0	1.0
	30	21	57	95	−0.536	38	67.6	53.0	51.5	54.5	1.5
GTCM-1.0	10	24	63	109	−0.581	42	66.2	50.5	50.0	52.5	2.0
	20	23	60	98	−0.369	40	66.9	52.5	51.5	53.5	1.0
	30	22	55	91	−0.244	40	71.3	55.5	53.0	56.5	1.0
GTCM-1.5	10	24	62	107	−0.446	42	67.6	51.5	50.5	53.0	1.5
	20	24	59	100	−0.228	41	69.0	53.0	52.0	54.0	1.0
	30	23	55	93	−0.115	40	72.9	56.5	53.5	57.5	1.0

The high-temperature stability of asphalt changed with GTCM is superior to asphalt modified with tourmaline micropowder under the same circumstances, that is, GTCM-modified asphalt had much better temperature sensitivity, high-temperature resistance, and anti-aging capabilities than tourmaline-modified asphalt, and as the amount of graphene micropowder rose, the improving impact steadily grew, which has been proved in the existing research [21,30,31].

3.2. Dynamic Shear Rheological Property Analysis

3.2.1. Temperature Scanning

The temperature scanning test of GTCM-modified asphalt was carried out. The outcomes are displayed in Figures 7–10, and the rheological properties of GTCM-modified asphalt in the continuous temperature range are analyzed.

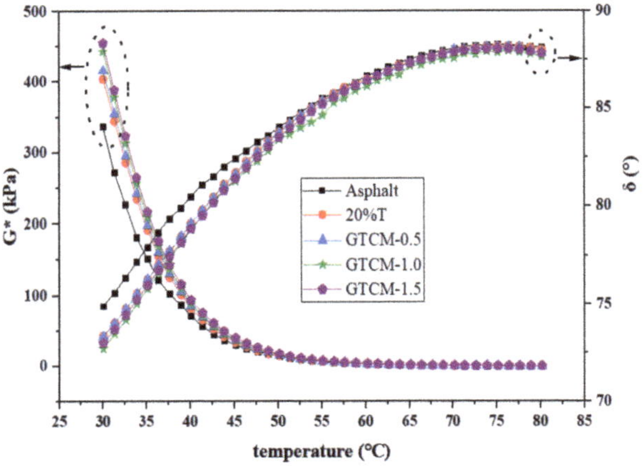

Figure 7. Curves of complex shear modulus and phase angle of different types of GTCM-modified asphalt with temperature.

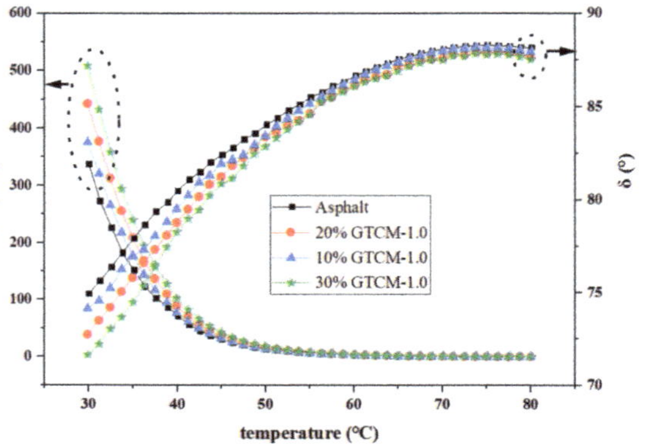

Figure 8. Curves of complex shear modulus and phase angle of GTCM with different contents changing with temperature.

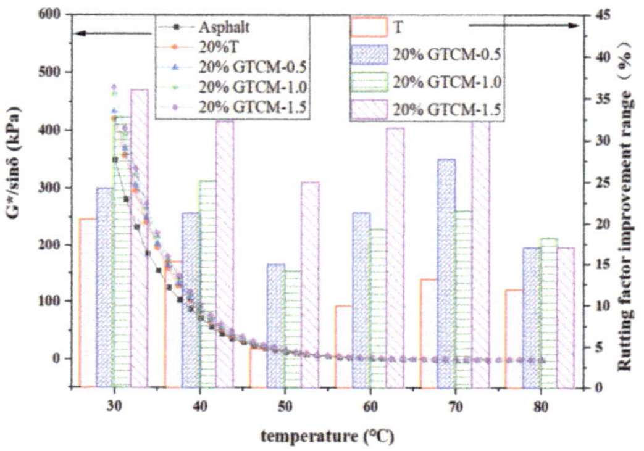

Figure 9. Rutting factor of different types of GTCM-modified asphalt and its improvement trend with temperature.

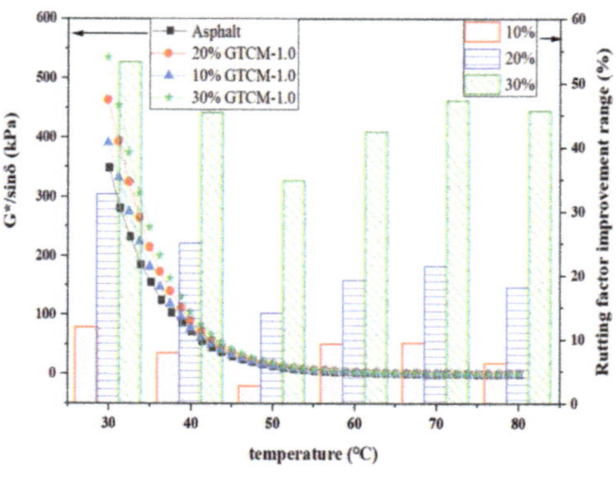

Figure 10. Rutting factor of different types of GTCM-modified asphalt and its improvement trend with temperature.

It is shown in Figure 7.

(1) Similar to matrix asphalt and tourmaline micropowder modified asphalt, the complicated shear modulus change trend of various types of GTCM-modified asphalt is the temperature under the same dose. Temperature increases cause an exponential

decline in the complex modulus, indicating that the GTCM is used for modified asphalt. Asphalt's viscoelastic characteristics remained the same, but the complex modulus of different types of GTCM-modified asphalt at the same temperature was higher than that of tourmaline micropowder modified asphalt and matrix asphalt.

(2) Similar to matrix asphalt and tourmaline-modified asphalt, the phase angle of GTCM-modified asphalt has the same content. Additionally, the phase angle increases with temperature. During dynamic mechanical testing, the phase angle δ of stress and strain hysteresis indicates the ratio of viscosity and elasticity of viscoelastic materials. The phase angle of completely elastic material is 0°, and the phase angle of completely viscous material is 90°, while the phase angle of viscoelastic material is in the range of 0–90°. The phase angle of various asphalt kinds is within 88° as the temperature rises from 30 °C to 80 °C. It also shows that the addition of GTCM to asphalt does not change the viscoelastic properties of asphalt.

It is shown in Figure 8.

(1) The complex shear modulus and phase angle fluctuation trends of modified asphalt varying GTCM concentrations and temperature are essentially identical to those of matrix asphalt. The temperature rises cause an exponential decline in the complex modulus and an increase in the phase angle. It shows that the viscoelastic properties of asphalt are not changed after the GTCM material is used for modified asphalt in the range of 30%.
(2) The complex modulus between matrix asphalt and GTCM-modified asphalt is noticeably different in the region of 30 to 50 °C, and the higher the content, the higher the complex shear modulus. When the temperature exceeds 50 °C, the content has less of an impact on the complex modulus of GTCM-modified asphalt, and the complex modulus between different contents is very small, but the overall trend is still 30%. The complex modulus of GTCM-modified asphalt is the largest.
(3) In the range of 30~55 °C, the phase angle between different content of GTCM-modified asphalt and matrix asphalt is obvious, and the larger the content, the smaller the phase angle. When the temperature exceeds 55 °C, the content has less of an impact on the phase angle of GTCM-modified asphalt. The phase angle between different content composite modified asphalt is small, but the overall trend is still 30%. The phase angle of GTCM-modified asphalt is the smallest.

It is shown in Figure 9.

(1) Under the condition of the same content, the rutting factor of different kinds of GTCM is similar to that of tourmaline-modified asphalt and matrix asphalt, which falls down dramatically as the temperature rises, but the rutting coefficient of GTCM to asphalt is larger than that of tourmaline and matrix asphalt, indicating that the rutting factor of GTCM to asphalt is better than that of tourmaline composite powder.
(2) According to the results, between 30 and 50 °C, the rutting factors of matrix asphalt and modified asphalt are very different, while the rutting factors of different types of GTCM-modified asphalt are small. In the temperature range of more than 50 °C, the rut factor of various types of asphalt has little difference and is at a low level.
(3) When the temperature rises, the improvement of the rutting factor decreases first, then increases, and finally increases again. At different temperatures, the improvement of the rutting coefficient of different types of GTCM-modified asphalt is different. Below 50 °C, with the addition of GTCM, the improvement of the rutting factor is gradually increasing. In the temperature range of 50~70 °C, with the addition of graphene powder, the improvement of the rutting factor shows a trend of decreasing first and then increasing. At the temperature of 50~70 °C, with the addition of graphene powder, the improvement degree of the rutting factor shows a trend of increasing first and then decreasing, indicating that GTCM-1.5-modified asphalt has the best anti-rutting ability when the temperature does not exceed 70 °C, while GTCM-1.0 has the best anti-rutting ability at the temperature of 70~80 °C.

It is shown in Figure 10.

(1) The change trend of the rut factor of GTCM-modified asphalt with temperature is basically the same as that of matrix asphalt, and the rut factor decreases exponentially with the increase in temperature. The rutting factor of modified asphalt increases with increasing GTCM content at the same temperature, suggesting that adding more GTCM can help asphalt become more resistant to rutting.
(2) The improvement range of the rutting factor of modified asphalt with different content of GTCM is basically the same as that of temperature, which decreases first, then increases and then increases with the increase in temperature. At the same temperature, the improvement range of the rutting factor of modified asphalt with GTCM increases with the increase in content.
(3) At 60 °C, the rut factor of matrix asphalt is 2.97 kPa, and the rut factor of GTCM-modified asphalt is 3.25 kPa at 10% dosage, which is 9.43% higher than that of matrix asphalt. In comparison to matrix asphalt, the rutting factor of GTCM-modified asphalt with a 20% component is 19.53% greater at 3.55 kPa. The 30% GTCM-modified asphalt has a rutting factor of 4.23 kPa, which is 42.42% more than matrix asphalt.

3.2.2. Frequency Scanning Results and Analysis

To clarify the viscoelastic properties of GTCM-modified asphalt under different load frequencies, it is necessary to study the rheological properties of GTCM-modified asphalt under different load frequencies. In Figures 11–13, the precise test results are displayed.

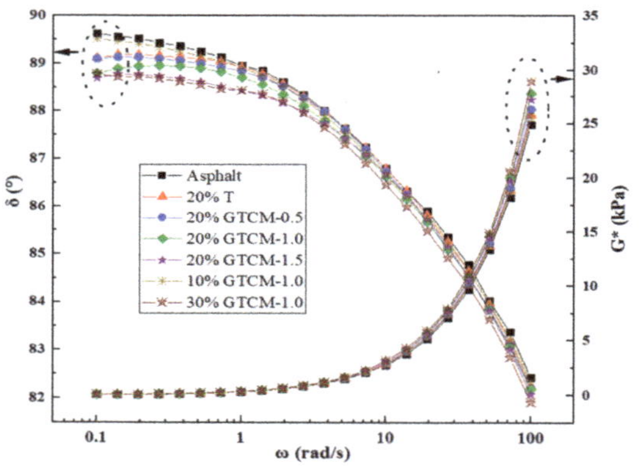

Figure 11. Curves of complex modulus and phase angle of GTCM-modified asphalt with frequency under different types and contents.

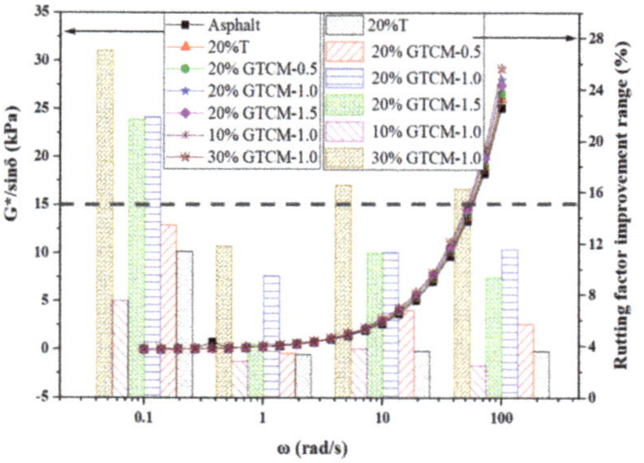

Figure 12. Rutting factor of GTCM-modified asphalt and its improvement trend with frequency under different types and contents.

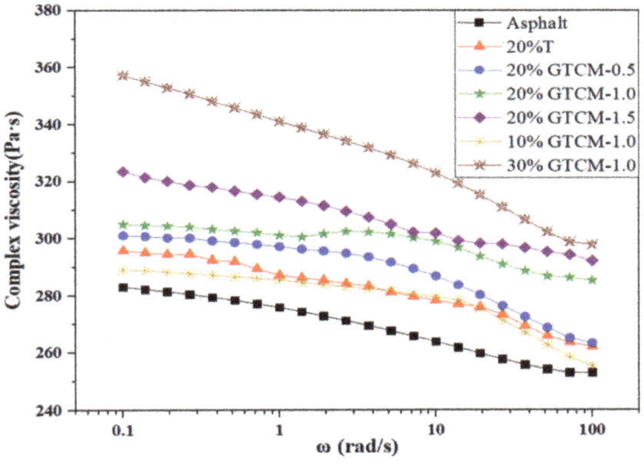

Figure 13. The curve of complex viscosity of GTCM-modified asphalt with frequency under different types and contents.

It is shown in Figure 11.

(1) The complex shear modulus and phase angle fluctuation trends of various types of GTCM-modified asphalt with frequency are essentially identical to those of matrix asphalt and tourmaline micropowder-modified asphalt. The complex modulus increases exponentially with the increase of frequency, and the phase angle decreases gradually with the increase of frequency, indicating that the viscoelastic properties of asphalt are not changed after the GTCM is used for modified asphalt. However, the phase angle of various types of GTCM-modified asphalt is smaller than that of tourmaline micropowder-modified asphalt and matrix asphalt, and the complex modulus of these modified asphalt types is larger at the same frequency.
(2) The complex shear modulus and phase angle of GTCM-modified asphalt has the same trend with frequency under the same type and different content. At the same frequency, the complex shear modulus increases with increasing GTCM amount while the phase angle decreases.

It is shown in Figure 12.

(1) Under the same dosage, the rutting factor of different types of GTCM-modified asphalt increases exponentially with the increase of frequency, which is basically consistent with the trend of the rutting factor of tourmaline micropowder-modified asphalt and matrix asphalt with frequency. However, under the same conditions, the finding that matrix asphalt and tourmaline micropowder-modified asphalt have lower rutting factors than GTCM-modified asphalt suggests that GTCM has the best impact on enhancing the rutting resistance of asphalt.
(2) The rut factor of GTCM-modified asphalt with the same type and different content is basically the same as the frequency change trend, it gets bigger when the frequency gets bigger. At the same frequency, the larger the content of GTCM, the greater the rut factor, indicating that increasing the content helps to improve the rut resistance of GTCM-modified asphalt.
(3) When the frequency is 10 rad/s, the rut factor of tourmaline micropowder modified asphalt is 2.78 kPa. The rut factors of GTCM-0.5, GTCM-1.0 and GTCM-1.5-modified asphalt are 2.87, 3.00 and 2.99 kPa, respectively, which are 3.24%, 7.91% and 7.55% higher than that of tourmaline micropowder modified asphalt.
(4) When the frequency is 10 rad/s, the rut factor of the matrix asphalt is 2.69 kPa, and the rut factors of the graphene tourmaline composite powder-modified asphalt at 10%, 20%, and 30% are 2.79, 3.00, and 3.14 kPa, respectively. Compared with the matrix asphalt, it increased by 3.72%, 14.20%, and 16.73%, respectively.

It is shown in Figure 13.

(1) Under various kinds and contents, the complex viscosity of GTCM-modified asphalt falls as frequency increases. When compared to matrix asphalt, GTCM-modified asphalt has a greater complex viscosity, and is higher than that of tourmaline micropowder-modified asphalt at the same content, indicating that the shear performance of GTCM-modified asphalt is better than that of tourmaline micropowder-modified asphalt matrix asphalt.
(2) At the same frequency, by comparing the complex viscosity of different types of GTCM-modified asphalt, It can be shown that the complex viscosity of composite-modified asphalt increases with increasing graphene micropowder amount, that is, the shear resistance of GTCM-1.5-modified asphalt is the best.
(3) When the frequency is 10 rad/s, the complex viscosity of tourmaline micropowder modified asphalt is 278.28 Pa·s. The complex viscosity of GTCM-0.5, GTCM-1.0 and GTCM-1.5-modified asphalt is 286.81, 298.92 and 301.85 Pa·s, respectively, which is 3.07%, 7.42% and 8.46% higher than that of tourmaline micropowder modified asphalt.
(4) When the frequency is 10 rad/s, the complex viscosity of the matrix asphalt is 263.81 Pa·s. The complex viscosity of the GTCM-modified asphalt at 10%, 20% and 30% is 279.42, 298.92 and 322.9 Pa·s, respectively, which is 5.92%, 13.31% and 22.40% higher than that of the matrix asphalt.

Existing studies have shown that GO can improve the rutting resistance of 90 A and SBS MA in the temperature range of 30–80 °C [16]. The existing research shows that tourmaline can significantly improve the high-temperature rheological properties of asphalt. The dynamic shear rheological properties of tourmaline-modified asphalt steadily improve with increasing tourmaline amount. When the tourmaline content exceeds 14%, The effect of tourmaline content on the enhancement of asphalt's high-temperature performance is insignificant [32]. Through comprehensive comparative analysis of the rheological properties of GTCM-modified asphalt at different temperatures and frequencies, it can be concluded that the anti-rutting performance of GTCM-modified asphalt is better than that of tourmaline micropowder-modified asphalt. Compared with the same content of tourmaline micropowder-modified asphalt, The anti-rutting performance of GTCM-0.5, GTCM-1.0 and GTCM-1.5-modified asphalt can be increased by 12.95%, 10.12% and 24.25%, respectively.

3.3. Analysis of Rheological Properties at Low Temperature

The low-temperature creep tests of GTCM-modified asphalt with different types and contents were carried out at −12, −18 and −24 °C by low-temperature bending beam rheometer. The test outcomes are displayed in Figures 14–16.

(1) The stiffness modulus (−12 °C) of GTCM-modified asphalt under different types and contents.

It is shown in Figure 14.

Under various kinds and doses, the creep stiffness modulus of GTCM-modified asphalt falls as loading time increases, as with matrix asphalt, tourmaline micropowder-modified asphalt, and mineral powder asphalt mastic, the change pattern is essentially the same. However, during the 240 s loading period, the stiffness modulus of graphene micropowder reinforced tourmaline-modified asphalt is smaller than that of tourmaline micropowder-modified asphalt and mineral powder asphalt mastic, indicating that the low-temperature performance of GTCM-modified asphalt is better than that of tourmaline micropowder modified asphalt.

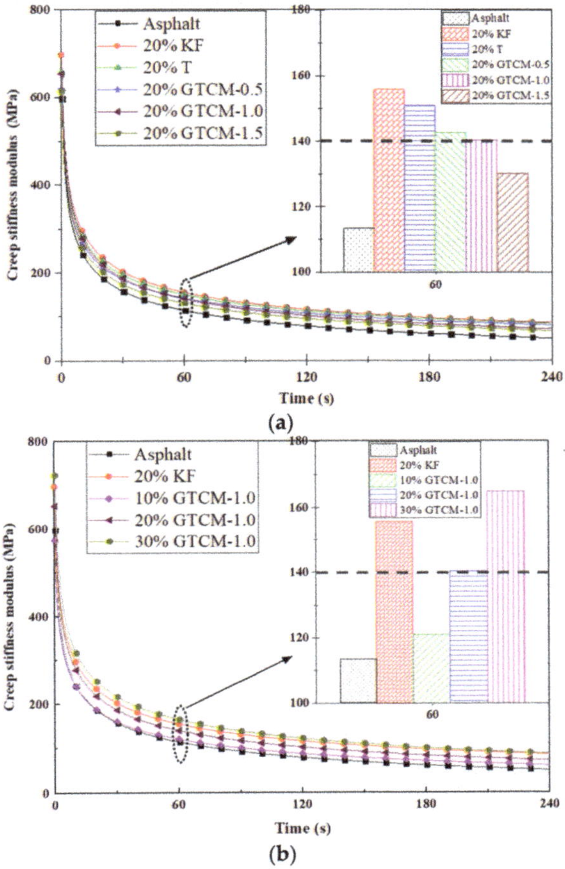

Figure 14. Stiffness modulus of GTCM-modified asphalt with different types and contents. (**a**) Different kinds. (**b**) Different dosage.

The creep stiffness modulus of GTCM-modified asphalt with the same type and different content also decreases with the extension of load action time. The different content only changes the creep stiffness modulus of asphalt and does not affect the rheological properties of asphalt under low-temperature load. The larger the content of GTCM, the larger the creep stiffness modulus of modified asphalt, indicating that too much content will adversely affect the low-temperature performance of asphalt.

When the loading time is 60 s, the creep stiffness modulus of tourmaline powder modified asphalt is 150.95 MPa, and the creep stiffness modulus of GTCM-0.5-modified asphalt is 142.74 MPa, which is 5.75% higher than that of tourmaline powder modified asphalt. The creep stiffness modulus of GTCM-1.0-modified asphalt is 140.43 MPa, which is 6.97% higher than that of tourmaline powder-modified asphalt. The creep stiffness modulus of GTCM-1.5-modified asphalt is 130.22 MPa, which is 13.73% higher than that of tourmaline powder modified asphalt, that is when the loading time is 60 s, the low-temperature performance of GTCM-1.5-modified asphalt is the best.

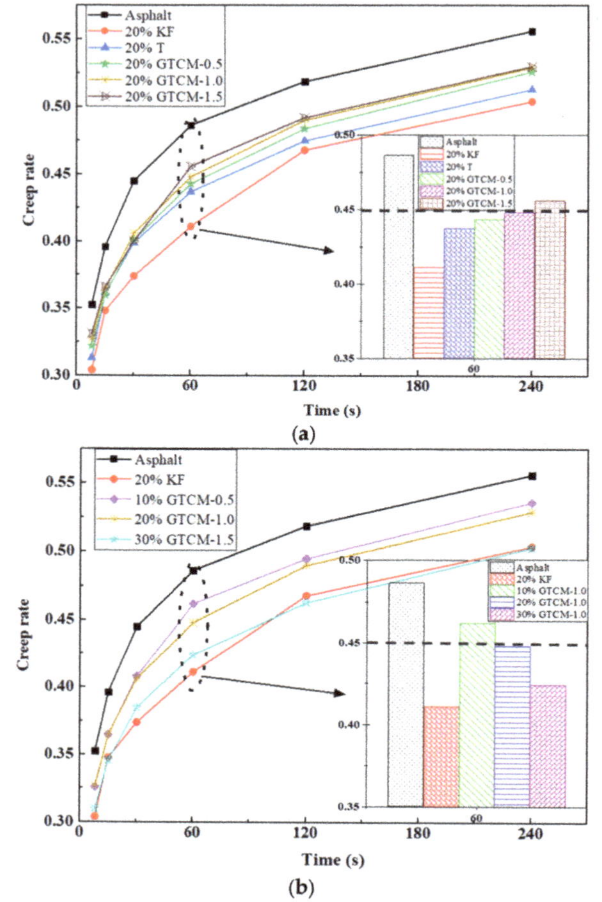

Figure 15. Creep rate of GTCM-modified asphalt under different kinds and contents. (**a**) Different kinds. (**b**) Different dosage.

(2) Creep rate (−12 °C) of GTCM-modified asphalt under different types and contents It is shown in Figure 15.

Under different types and contents, the creep rate of GTCM-modified asphalt gradually increases with the extension of load time, and its change trend is basically the same as that of matrix asphalt, tourmaline micropowder-modified asphalt and mineral powder asphalt mastic. However, in the 240 s load period, the creep rate of GTCM-modified asphalt is greater than that of tourmaline micropowder-modified asphalt and mineral powder asphalt mastic, indicating that the low-temperature performance of GTCM-modified asphalt is better than that of tourmaline micropowder modified asphalt.

The creep rate of GTCM-modified asphalt with the same type and different content also increases with the extension of load action time. The different content only changes the creep rate of asphalt and does not affect the rheological properties of asphalt under low temperature load. The larger the content of GTCM, the smaller the creep rate of modified asphalt, indicating that too much content will have an adverse effect on the low-temperature performance of modified asphalt.

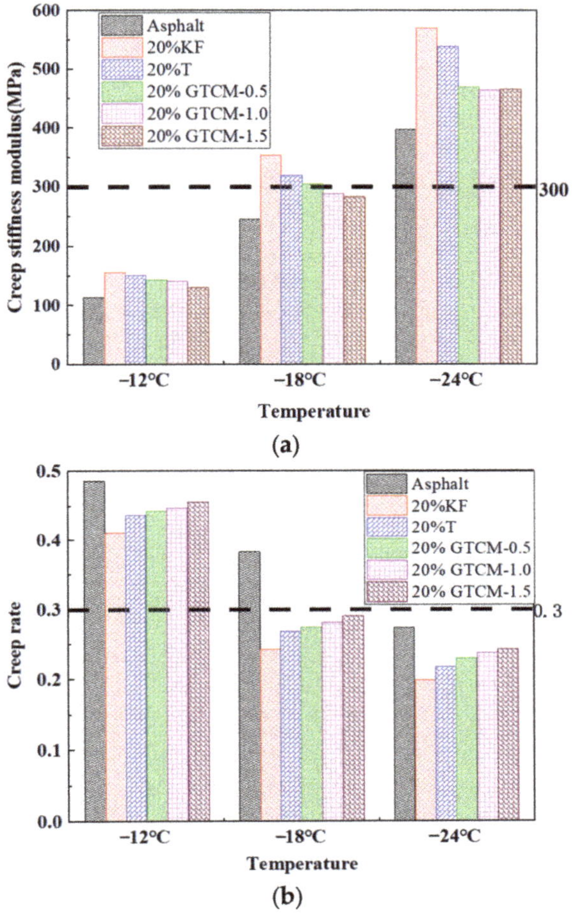

Figure 16. Creep stiffness modulus and creep rate of GTCM-modified asphalt at different temperatures, (**a**) Creep stiffness modulus and (**b**) Creep rate.

When the loading time is 60 s, the creep rate of tourmaline powder-modified asphalt is 0.437, and the creep rate of GTCM-0.5-modified asphalt is 0.443, which is 1.37% higher than that of tourmaline powder-modified asphalt. The creep rate of GTCM-1.0-modified asphalt is 0.448 MPa, which is 2.52% higher than that of tourmaline powder-modified asphalt. The creep rate of GTCM-1.5-modified asphalt is 0.456, which is 4.35% higher than that of tourmaline powder-modified asphalt. When the loading time is 60 s, the low-temperature performance of GTCM-1.5-modified asphalt is the best.

(3) The stiffness modulus and creep rate of GTCM-modified asphalt at different temperatures (60 s)

It can be clearly seen from Figure 16.

With a drop in temperature, the creep stiffness modulus of GTCM-modified asphalt steadily rises. At the same temperature, although the creep stiffness modulus of GTCM-modified asphalt is higher than that of matrix asphalt, it is lower than the creep stiffness modulus of tourmaline micropowder-modified asphalt and mineral powder asphalt mastic under the same dosage.

With a drop in temperature, the creep rate of GTCM-modified asphalt steadily reduces. The creep rate of GTCM-modified asphalt is higher than that of tourmaline micropowder modified asphalt and mineral powder asphalt mastic at the same dose, yet it is less than that of matrix asphalt at the same temperature.

When the ambient temperature is −12 and −18 °C, the creep stiffness modulus of the GTCM-modified asphalt is lower than 300 MPa, and the creep rate is only greater than 0.3 at −12 °C, indicating that the GTCM-modified asphalt has good low-temperature crack resistance at −12 °C.

In summary, the low-temperature crack resistance of GTCM-modified asphalt is better than that of tourmaline micropowder-modified asphalt, and the greater the content of graphene micropowder, the more obvious the improvement effect on low-temperature performance. Relevant research shows that tourmaline powder-modified asphalt has lower flexibility and weaker stress relaxation ability than matrix asphalt. Tourmaline reduces the low-temperature crack resistance of asphalt to a certain extent [33]. Adding graphene oxide can effectively absorb the load. Thus, the flexibility and deformation ability of asphalt are enhanced, and the low-temperature cracking resistance of asphalt binder is improved [21]. The reason may be that the flexibility of graphene is better than that of tourmaline. Because graphene has a far bigger specific surface area than tourmaline, it is distributed throughout asphalt at a considerably higher density than tourmaline, which improves the flexibility of asphalt in low-temperature environments. At the same time, asphalt molecules may partially permeate the graphene lattice because of its stable two-dimensional lamellar structure, and some stress is absorbed by graphene during loading, which weakens the direct effect of stress on asphalt, thereby improving the low-temperature crack resistance of asphalt.

3.4. Analysis of Fourier Infrared Spectrum

The infrared spectra of matrix asphalt, tourmaline-modified asphalt, GTCM-0.5-modified asphalt, GTCM-1.0-modified asphalt and GTCM-1.5-modified asphalt were tested, respectively. The test results are shown in Figure 17.

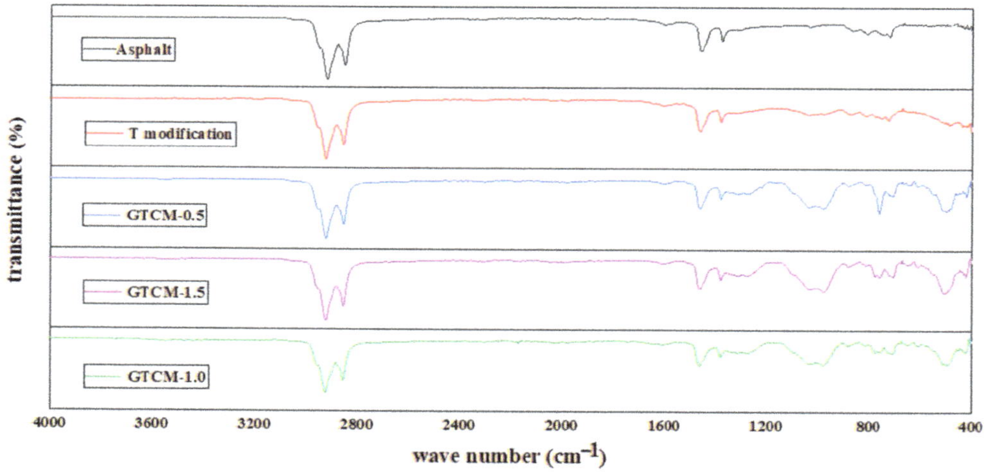

Figure 17. FTIR spectrum of GTCM-modified asphalt.

From the FTIR spectra of different types of asphalt in Figure 17, it can be seen that there are some differences in the absorption peaks of different types of asphalt in the infrared spectrum. In the FTIR spectrum of matrix asphalt, there are strong absorption peaks at 2920.05 and 2850.96 cm^{-1}, which are antisymmetric and symmetric stretching vibrations of CH_2 in long-chain alkanes and cycloalkanes. The absorption peaks of C = C and C–C skeleton vibration in the benzene ring of aromatic compounds were 1601.98, 1456.26 and 1376.34 cm^{-1}. The absorption peak of S = O stretching vibration of dibenzyl sulfoxide at 1031.10 cm^{-1}. The four smaller characteristic peaks at 864~721 cm^{-1} are the C–H out-of-plane bending vibration on the benzene ring of aromatic compounds.

The difference between the FTIR spectra of the tourmaline-modified asphalt and the GTCM-modified asphalt and the matrix asphalt mainly appears in the low-frequency region, which is reflected in the absorption peaks with low sharp strength in the range of 720~400 cm^{-1}. The tourmaline-modified asphalt has obvious absorption peaks at 483.64, and 433.04 cm^{-1}. GTCM-0.5-modified asphalt showed obvious absorption peaks at 704.48, 634.90, 493.69 and 417.97 cm^{-1}. The absorption peaks of the low-frequency region in the FTIR spectrum of GTCM-1.0-modified asphalt mainly appeared at 641.50, 504.64, 491.36 and 419.06 cm^{-1}. The different absorption peaks of GTCM-1.5-modified asphalt FTIR spectrum are mainly distributed at 704.50, 634.98, 606.84, 505.78 and 418.85 cm^{-1}. Compared with the tourmaline FTIR spectrum, the primary source of the low-frequency region's small and sharp absorption peaks is the lattice vibration of some oxides such as ZnO, Fe_2O_3, Fe_3O_4, Al_2O_3, MgO, TiO_2, and SiO_2. This part of the absorption peak in the modified asphalt FTIR spectrum mainly comes from tourmaline. After adding tourmaline or GTCM material to asphalt, asphalt does not undergo a chemical reaction that would create new functional groups, but only increases the infrared activity of asphalt material in a lower frequency range.

4. Conclusions

In this paper, GTCM-modified asphalt is taken as the object. Firstly, the influence of different temperatures and different frequencies on the high-temperature performance of GTCM-modified asphalt is evaluated by dynamic shear rheological test, and the viscoelastic properties of GTCM-modified asphalt under different stress and different temperatures are analyzed. Secondly, the low-temperature rheological properties of GTCM-modified asphalt were analyzed by bending beam rheological test. Finally, the mechanism was analyzed by FTIR. The main conclusions are as follows:

(1) Compared to tourmaline-modified asphalt, GTCM-modified asphalt has much better temperature sensitivity, high temperature resistance, and anti-aging capabilities. The improved impact will also rise when the graphene powder content is increased, and the temperature sensitivity of GTCM-modified asphalt can be increased by up to 87.2%. The improvement of penetration, softening point, and equivalent softening point at 25 °C can be up to 6.35%, 7.62% and 7.17%, respectively. The improvement of penetration residual ratio and softening point increment at 25 °C can be up to 8.60% and 57.58%, respectively.

(2) The viscoelasticity of asphalt was unaffected by the addition of GTCM. The anti-rutting performance of GTCM-modified asphalt is better than that of tourmaline-modified asphalt. At the same temperature, it increases with the increase in composite powder content. At different temperatures, the modification effect of the composite powder is still optimal. Compared with the same amount of tourmaline-modified asphalt, the anti-rutting performance of GTCM-0.5, GTCM-1.0 and GTCM-1.5-modified asphalt can be improved by 12.95%, 10.12% and 24.25%, respectively.

(3) Under various load frequencies, GTCM-modified asphalt with various contents performs better than tourmaline-modified asphalt in terms of anti-rutting and shear resistance, and its performance improves as the amount of graphene powder and GTCM rises. When the load was 60 s, the creep stiffness modulus of GTCM-0.5, GTCM-1.0 and GTCM-1.5-modified asphalt decreased by 5.75%, 6.97% and 13.73%, respectively, and the creep rate increased by 1.37%, 2.52% and 4.35%, respectively.

(4) GTCM has stronger low-temperature crack resistance than tourmaline, and the improvement in low-temperature performance is increasingly noticeable as graphene powder gets larger. However, the high content of GTCM will have a negative impact on its low-temperature performance. When the ambient temperature is 12 °C, GTCM-modified asphalt has good low-temperature crack resistance.

(5) FTIR analysis showed that after adding GTCM material or tourmaline to the matrix asphalt, no new functional groups were produced due to the chemical reaction with the asphalt.

Author Contributions: T.G.: Conceptualization, Project administration, Supervision, Writing-Review and Editing. H.C.: Conceptualization, Formal analysis, Methodology, Visualization. D.T.: Supervision, Project administration, Data curation, Formal analysis. S.D.: Conceptualization, Writing-Original Draft, Supervision, Investigation. C.W.: Conceptualization, Supervision, Writing-Original Draft. D.W.: Conceptualization, Project administration, Supervision, Investigation. Y.C.: Funding acquisition, Investigation. Z.L.: Project administration, Resources. All authors have read and agreed to the published version of the manuscript.

Funding: This research was sponsored by the Henan Province 2021 Science and Technology Key Project 'Performance and Synergistic Mechanism of Graphene/Tourmaline Composite Modified Asphalt' (Grant No. 212102310089), North China University of Water Resources and Hydropower Research Project (Grant No. 201811035) and Fundamental Research Funds for the Central Universities (Grant No. 300102219314). That sponsorship and interest are gratefully acknowledged. Open Fund Project of Henan Province Engineering Technology Research Center for Environment-friendly and High-performance Pavement Materials (Henan Science foundation (2020) No.18).

Institutional Review Board Statement: Not applicable.

Informed Consent Statement: Not applicable.

Data Availability Statement: Not applicable.

Conflicts of Interest: The authors declare no conflict of interest.

References

1. Gedik, A. A review on the evaluation of the potential utilization of construction and demolition waste in hot mix asphalt pavements. *Resour. Conserv. Recycl.* **2020**, *161*, 104956. [CrossRef]
2. Xiao, F.; Su, N.; Yao, S.; Amirkhanian, S.; Wang, J. Performance grades, environmental and economic investigations of reclaimed asphalt pavement materials. *J. Clean. Prod.* **2019**, *211*, 1299–1312. [CrossRef]
3. Zhang, R.; Tang, N.; Zhu, H.; Zeng, J.; Bi, Y.; Xi, Y. Environmental and economic comparison of semi-rigid and flexible base asphalt pavement during construction period. *J. Clean. Prod.* **2022**, *340*, 130791. [CrossRef]
4. Khare, P.; Machesky, J.; Soto, R.; He, M.; Presto, A.A.; Gentner, D.R. Asphalt-related emissions are a major missing nontraditional source of secondary organic aerosol precursors. *Sci. Adv.* **2020**, *6*, eabb9785. [CrossRef] [PubMed]
5. Sun, G.; Zhu, X.; Zhang, Q.; Yan, C.; Ning, W.; Wang, T. Oxidation and polymer degradation characteristics of high viscosity modified asphalts under various aging environments. *Sci. Total Environ.* **2022**, *813*, 152601. [CrossRef]
6. Wang, J. Preparation and modification mechanism of inorganic micropowder modified asphalt. Master's Thesis, Wuhan University of Technology, Wuhan, China, 2009.
7. Liang, Y.; Tang, X.; Zhu, Q.; Han, J.; Wang, C. A review: Application of tourmaline in environmental fields. *Chemosphere* **2021**, *281*, 130780. [CrossRef]
8. Zeng, Q.; Liu, Y.; Liu, Q.; Liu, P.; He, Y.; Zeng, Y. Preparation and modification mechanism analysis of graphene oxide modified asphalts. *Constr. Build. Mater.* **2020**, *238*, 117706. [CrossRef]
9. Jastrzębski, K.; Kula, P. Emerging Technology for a Green, Sustainable Energy-Promising Materials for Hydrogen Storage, from Nanotubes to Graphene—A Review. *Materials* **2021**, *14*, 1–24. [CrossRef]
10. Zheng, F.; Dong, F.; Lv, Z.; Li, H.; Zhou, L.; Chen, Y.; Luo, X. A novel g-C3N4/tourmaline composites equipped with plasmonic MoO3−x to boost photocatalytic activity. *Colloid Interface Sci. Commun.* **2021**, *43*, 100434. [CrossRef]
11. Han, M.; Muhammad, Y.; Wei, Y.; Zhu, Z.; Huang, J.; Li, J. A review on the development and application of graphene based materials for the fabrication of modified asphalt and cement. *Constr. Build. Mater.* **2021**, *285*, 122885. [CrossRef]
12. Yu, R.; Wang, Q.; Wang, W.; Xiao, Y.; Wang, Z.; Zhou, X.; Zhang, X.; Zhu, X.; Fang, C. Polyurethane/graphene oxide nanocomposite and its modified asphalt binder: Preparation, properties and molecular dynamics simulation. *Mater. Des.* **2021**, *209*, 109994. [CrossRef]
13. Zhu, J.; Zhang, K.; Liu, K.; Shi, X. Performance of hot and warm mix asphalt mixtures enhanced by nano-sized graphene oxide. *Constr. Build. Mater.* **2019**, *217*, 273–282. [CrossRef]
14. Chen, Q.; Wang, C.; Wen, P.; Sun, X.; Guo, T. Performance evaluation of tourmaline modified asphalt mixture based on grey target decision method. *Constr. Build. Mater.* **2019**, *205*, 137–147. [CrossRef]
15. Liu, J.; Hao, P.; Jiang, W.; Sun, B. Rheological properties of SBS modified asphalt incorporated polyvinylpyrrolidone stabilized graphene nanoplatelets. *Constr. Build. Mater.* **2021**, *298*, 123850. [CrossRef]

16. Amirkhanian, S.; Li, Y.; Wu, S. Investigation of the graphene oxide and asphalt interaction and its effect on asphalt pavement performance. *Constr. Build. Mater.* **2018**, *165*, 572–584.
17. Wu, S.; Zhao, Z.; Li, Y.; Pang, L.; Amirkhanian, S.; Riara, M. Evaluation of aging resistance of graphene oxide modified asphalt. *Appl. Sci.* **2017**, *7*, 702. [CrossRef]
18. Le, J.; Marasteanu, M.; Turos, M. Experimental investigation of properties of graphene nanoplatelet-modified asphalt binders and mixtures. *TRB 95th Annu. Meet. Compend. Pap.* **2016**, *16*, 3088.
19. Marasteanu, M.O.; Le, J.L.; Turos, M.I. Graphene Nano-Platelet (Gnp) Reinforced Asphalt Binders and Mixtures. In Proceedings of the 6th Eurasphalt & Eurobitume Congress, Prague, Czech Republic, 1–3 June 2016. [CrossRef]
20. Christiansen, S.; Restrepo, D.; Stoltz, R.; Bullington, J. Graphite oxide entrainment in cement and asphalt composite. U.S. Patent EP3194337, 2017.
21. Wang, C.; Wang, M.; Chen, Q.; Zhang, L. Basic performance and asphalt smoke absorption effect of environment-friendly asphalt to improve pavement construction environment. *J. Clean. Prod.* **2022**, *333*, 130142. [CrossRef]
22. Wang, C.; Zhang, D.; Chen, H.; Zhang, H.; Xiao, X.; Liu, Z. Preparation and Properties of Silicon-Modified Epoxy Grouting Material for Repairing Microcracks. *J. Mater. Civ. Eng.* **2022**, *34*, 04021479. [CrossRef]
23. Chen, Q.; Wang, C.; Qiao, Z.; Guo, T. Graphene/tourmaline composites as a filler of hot mix asphalt mixture: Preparation and properties. *Constr. Build. Mater.* **2020**, *239*, 117859. [CrossRef]
24. Chen, Q.; Wang, C.H.; Li, Y.W.; Feng, L.; Huang, S. Performance Development of Polyurethane Elastomer Composites in Different Construction and Curing Environments. *Constr. Build. Mater.* **2023**, *365*, 130047. [CrossRef]
25. Ding, H.; Rahman, A.; Li, Q.; Qiu, Y. Advanced mechanical characterization of asphalt mastics containing tourmaline modifier. *Constr. Build. Mater.* **2017**, *150*, 520–528. [CrossRef]
26. Kim, S.J.; Ko, H.S.; Jeong, G.H.; Yun, J.J.; Park, K.H.; Han, E.M. The effect of tourmaline additives in TiO_2 photoanode for high-efficiency dye sensitized solar cells. *Mol. Cryst. Liq. Cryst.* **2016**, *629*, 200–205. [CrossRef]
27. Kang, S.J.; Tijing, L.D.; Hwang, B.S.; Jiang, Z.; Kim, H.Y.; Kim, C.S. Fabrication and photocatalytic activity of electrospun nylon-6 nanofibers containing tourmaline and titanium dioxide nanoparticles. *Ceram. Int.* **2013**, *39*, 7143–7148. [CrossRef]
28. Moreno-Navarro, F.; Sol-Sánchez, M.; Gámiz, F.; Rubio-Gámez, M.C. Mechanical and thermal properties of graphene modified asphalt binders. *Constr. Build. Mater.* **2018**, *180*, 265–274. [CrossRef]
29. Baeissa, E.S. Green synthesis of methanol by photocatalytic reduction of CO2, under visible light using a graphene and tourmaline co-doped titania nanocomposites. *Ceram. Int.* **2014**, *40*, 12431–12438. [CrossRef]
30. Wang, C.; Chen, Q.; Guo, T.; Li, Q. Environmental effects and enhancement mechanism of graphene/tourmaline composites. *J. Clean. Prod.* **2020**, *262*, 121313. [CrossRef]
31. Guo, T.; Fu, H.; Wang, C.; Chen, H.; Chen, Q.; Wang, Q.; Chen, Y.; Li, Z.; Chen, A. Road Performance and Emission Reduction Effect of Graphene/Tourmaline-Composite-Modified Asphalt. *Sustainability* **2021**, *13*, 8932. [CrossRef]
32. Wang, C.; Li, Y.; Li, R. Preparation and properties of low-carbon multi-functional tourmaline modified asphalt. *China Highw. J.* **2013**, *26*, 34–41. [CrossRef]
33. Huang, X. Research on Technical Performance and Modification Mechanism of Tourmaline Powder Modified Asphalt. Master's Thesis, Changsha University of Science and Technology, Changsha, China, 2018. [CrossRef]

coatings

MDPI

Article

A Comparative Study for Creep and Recovery Behavior Characterization of Modified Bitumens Using the MSCR Test

Guodong Zeng [1,2], Jianjing Zhang [3], Hongming Huang [1,4], Xin Xiao [3] and Chuanqi Yan [5,6,*]

1 Foshan Road and Bridge Supervision Station Co., Ltd., Foshan 528041, China
2 Foshan Transportation Science and Technology Co., Ltd., Foshan 528041, China
3 School of Transportation, Civil Engineering and Architecture, Foshan University, Foshan 528225, China
4 School of Traffic and Transportation Engineering, Changsha University of Science & Technology, Changsha 410004, China
5 Highway Engineering Key Laboratory of Sichuan Province, Chengdu 610031, China
6 School of Civil Engineering, Southwest Jiaotong University, Chengdu 610031, China
* Correspondence: ycq@swjtu.edu.cn

Abstract: In recent decades, the application of modified bitumens has experienced tremendous growth. However, due to the varying modification mechanism of different modifiers, the creep and recovery properties of modified bitumen have not been comprehensively understood. This study aims to evaluate the creep and recovery properties of several representative modified bitumens using the multi-stress creep recovery (MSCR) test. The MSCR test can highlight the unique delayed elasticity of modified bitumen and it uses a high stress level, which is more comparable to the field. In particular, this test also aims to identify the effects of different aging conditions. To do so, a total of 15 bitumens, including 7 elastomeric-modified bitumens, 5 non-elastomeric-modified bitumens, and 3 plain bitumens, were prepared and examined. Furthermore, 10 different aging conditions were considered. The results suggest that the generation mechanism of elasticity varies for different modified bitumens. There are two types of elasticities, which are energy elasticity and entropy elasticity, and their differences need more attention in the road bitumen material community. Aging changes the percentages of contributions from energy elasticity and entropy elasticity to the bitumen's overall recovery performance. The increase in "bad" energy elasticity may compensate for part of the "good" entropy elasticity loss, but overall, the bitumen's recovery rate is decreasing and the ratio of energy elasticity is increasing, which might hinder the bitumen's road performance.

Keywords: bitumen; PMB; elasticity; creep and recovery

Citation: Zeng, G.; Zhang, J.; Huang, H.; Xiao, X.; Yan, C. A Comparative Study for Creep and Recovery Behavior Characterization of Modified Bitumens Using the MSCR Test. *Coatings* **2023**, *13*, 1445. https://doi.org/10.3390/coatings13081445

Academic Editor: Aomar Hadjadj

Received: 22 July 2023
Revised: 11 August 2023
Accepted: 12 August 2023
Published: 16 August 2023

1. Introduction

As a result of the increased demand placed on highways from higher traffic volumes and heavier loads, bitumen with better rheological properties is gaining more attention [1]. Bitumens modified with polymers and other additives show improved rheological properties such as higher modulus and higher elasticity, which can significantly enhance the pavement deformation resistance at high temperatures. In recent decades, the application of modified bitumens has experienced tremendous growth. There are numerous modifiers available, and the Styrene-Butadiene-Styrene (SBS) polymer is among one of the most used ones [2]. SBS polymer can form a three-dimensional rubbery network within the bitumen, which significantly benefits pavement performance [3].

Besides using SBS, a more economical approach to modifying the bitumen could be possible via the use of small amounts of modifiers. In this sense, polyphosphoric acid (PPA) might be a good alternative [4]. Some studies have suggested that small amounts of PPA show a comparable effect as other polymers [5]. Yang et al. [6] evaluated the possibility of a composite modifying PPA with SBS and bio-oil and achieved good performances. Besides SBS and PPA, ethyl vinyl acetate (EVA) [7,8] and natural rock bitumen [9] are also commonly used modifiers to improve the bitumen performance. Wentao et al. [10] reviewed the recent progress of EVA bitumen and claimed the utilization of EVA sees notable growth due to its rigid three-dimensional network structure and special functional groups.

The application of recycled and environmentally friendly materials is also trending in the road bitumen material community. Two representative examples are crumb rubber particles and bio-oil. Abundant researchers have reported that crumb rubber can improve the bitumen's rheological properties by increasing the bitumen viscosity, modulus, and elasticity [11,12]. Bio-oil, as a renewable material, can enhance the bitumen's fatigue and low-temperature properties [13]. Some also claimed that bio-oil facilitates the swelling of the polymer and thus helps with the compatibility between the polymer and base bitumen [14].

Bitumen is a typical viscoelastic material, and some of them (especially modified ones) exhibit notable elasticities. Many studies have indicated there are obvious positive correlations between the elasticity of bitumen and its mechanical properties [15,16], as good elasticities improve the bitumen's overall properties such as better rutting resistance and better fatigue resistance. To better evaluate the modified bitumen regarding its elasticity and creep/recovery behaviors, D'angelo [17] proposed a test named multi-stress creep recovery (MSCR). The MSCR test uses the well-established creep and recovery test concept from polymer testing to evaluate the bitumen's elasticity and potential for permanent deformation [18]. The MSCR test is operated in rotational mode using 1 s creep load followed by 9 s recovery, and then the compliance and recovery rate are recorded for analysis. MSCR has been quickly accepted as one of the most popular experiments for modified bitumens. Yanlei's [19] study used MSCR as a useful method to optimize the preparation procedures of various modified bitumens.

The MSCR test has two main advantages. First, it utilizes a relatively high stress level (3.2 kPa or even higher), which is comparable with the field. Second, the MSCR test leaves much time for the material to recover. Therefore, the effects of time-dependent entropy elasticity or delayed elasticity from elastomeric modifiers are revealed. These two aspects allow the MSCR to accurately quantify the performance of both plain bitumen and modified bitumen. Moreover, since the strain and stress data during the MSCR can be recorded with a high resolution, researchers can fit the creep and recovery behavior using physical modeling or numerical simulation to achieve in-depth and quantitative analysis [20,21].

Besides the influence of modification, aging is another essential factor that changes the bitumen's creep and recovery properties. It has been widely accepted that aging can negatively influence the performance of asphalt mixtures. In recent research carried out by Haichuan et al. [22], dynamic modulus and uniaxial fatigue tests were performed to investigate the stiffness and cracking properties of specimens after different aging durations, and the results suggest that aging can notably impair the mixtures' fatigue resistance. However, currently, there is no clear conclusion on how aging will change the bitumen's elasticity. Some claimed that thermal aging could stiffen the bitumen and push it to a more elastic end [23]. Li et al. [24] indicated that asphaltenes had a skeleton function so the modified bitumen with a high content of asphaltene presented a strong elastic response after long-term aging. However, on the other hand, numerous studies have reported that polymer-modified bitumen could be vulnerable to thermal aging and polymer oxidative degradation will weaken the modified bitumen's elasticity [25,26]. To overcome the negative consequences of aging, Lijun et al. [27] even explored the possibility of lignin, quercetin, and gallic acid as antioxidants in asphalt.

This study aims to evaluate the creep and recovery properties of several representative modified bitumens using the MSCR test. It can highlight the unique delayed elasticity of modified bitumen and it uses a high stress level, which is more comparable to the field. In particular, this test aims to investigate the effects of aging and different testing temperatures. To achieve this aim, a total of 15 bitumens, including 7 elastomeric-modified bitumens, 5 non-elastomeric-modified bitumens, and 3 plain bitumens, were prepared and examined.

Furthermore, 10 different aging conditions were considered. After the MSCR test, the recovery rate and non-recoverable compliance were recorded for analysis.

2. Experimental Design

2.1. Materials

2.1.1. Raw Materials

In this study, a total of 15 bitumens, including 7 elastomeric-modified bitumens, 5 non-elastomeric-modified bitumens, and 3 plain bitumens, were investigated and compared. Modified bitumens were prepared with either ES base bitumen, SK base bitumen, or ES base bitumen coupled with 10% bio-oil. The basic information on raw materials for bitumen preparation is summarized in Table 1. The raw materials were selected based on common selections in the industry.

Table 1. Properties of raw materials for bitumen preparation.

Material	Basic Properties
Plain bitumen	Two Perfomance-Grading (PG) 58–22 plain bitumen are utilized in this study. One is from the United States of America, under the brand name of ESSO, and the other is from South Korea under the brand name of SK
EVA	24% vinyl acetate content, with a melting index of 19
SBS	Linear SBS polymer with a S/B ratio of 30/70
Crumb rubber	30 mesh recycled truck radial tire crumbs, natural/synthesis rubber content: 54%
PPA	85% P_2O_5 content
Bio oil	Treated waste cooking oil with a viscosity of 37.2 m^2/s (100 °C)
Natural rock bitumen	Density: 1.066 g/cm^3, bitumen content: 98.4%, ash content: 0.72%, 1.18 mm passing: 100%

2.1.2. Bitumen Modification and Preparation

In this study, 13 modified bitumens were prepared and examined. They were all laboratory prepared by mixing the plain bitumen with selected modifiers. The specific mixing processes varied based on the type of modifier used. For EVA, PPA, and rock bitumen-modified bitumens, plain bitumen was first heated to 160 °C to achieve fluidity and then the desired modifiers were added to the hot bitumen. Subsequently, the blend was sheared for 20 min (4000 rpm) and then stirred for 60 min (800 rpm). For the SBS-modified bitumen (SBSMB) and crumb rubber-modified bitumen, mixing was carried out at 180 °C, and the shearing stage was extended to 120 min to achieve better swelling and homogeneity. The stability of modified bitumens used in this study has been verified using the cigar-tube separation test.

2.1.3. Laboratory Aging

According to AASHTO T350, a standard MSCR test requires RTFO aging before testing to simulate the short-term aging during asphalt mixture mixing and paving. To discuss the influence of aging, this study used 10 different laboratory aging processes including RTFO aging. The applied aging processes are summarized in Table 2.

2.1.4. Summary of Tested Bitumen Samples

A summary of the experimental matrix is presented in Table 3. The content of the modifiers was selected based on the common ranges of their applications in the industry.

Table 2. Summary of aging processes used in this study.

Aging Level	Standard	Explanation
VG	None	Virgin, unaged
R163, R193	AASHTO T240	Rolling Thin Film Oven Test (RTFOT), conducted at either 163 °C or 193 °C
M163, M193	Bahia's work [28]	Modified Rolling Thin Film Oven Test (MRTFOT), where a steel rod is put into the glass bottle to create extra shearing forces to spread the bitumen film. It was conducted at either 163 °C or 193 °C
PAV, 2PAV, 4PAV	AASHTO R28	Standard Pressure Aging Vessel (PAV), double PAV, four times PAV
S135, S163, S193	AASHTO PP2	Short-term Oven Aging (STOA), conducted at either 135 °C, 163 °C or 193 °C, then aged bitumens were recovered from the aged mixture following AASHTO T164

Table 3. Summary of tested samples.

Type	No	Bitumen ID	Modifier	Aging Level
Plain bitumen	1	ES	None	VG, R163, R193, 2PAV, 4PAV
	2	SK	None	VG, R163, 4PAV
	3	ES: 10%Bio	10% Bio oil	VG
Non-elastomeric modified bitumen	4	ES: 12%Rock	12% Rock bitumen	VG, R163
	6	ES: 0.8%PPA	0.8% PPA	VG, R163
	7	ES: 2%PPA	2% PPA	VG, R163
	8	ES: 2%EVA	2% EVA	VG
	9	ES: 4%EVA	4% EVA	VG
Elastomeric modified bitumen	10	ES: 4.5%SBS	4.5%SBS	VG, R163, R193, PAV, 2PAV
	11	ES: 7.5%SBS	7.5%SBS	VG, R163, M163, M193, R193, 2PAV, 4PAV, S135, S163, S193
	12	ES: 4.5%SBS + 10%Bio	4.5%SBS + 10%Bio	VG
	13	ES: 7.5%SBS + 10%Bio	7.5%SBS + 10%Bio	VG
	14	ES: 6%SBS	6%SBS	VG
	15	SK: 7.5%SBS	7.5%SBS	VG, R163, R193, 2PAV
	16	ES: 10% Crumb rubber	10% Crumb rubber	VG
	17	ES: 20% Crumb rubber	20% Crumb rubber	VG

In this study, samples at different aging levels are coded as "Bitumen ID [Aging level]". For instance, "SK: 7.5%SBS" and "SK: 7.5%SBS [2PAV]" refer to unaged SK-based 7.5% SBS-modified bitumen and double PAV-aged SK-based 7.5% SBS-modified bitumen, respectively ("VG", which represents the unaged status, is omitted).

2.2. MSCR Testing Procedure

The TA DHR-3 type DSR was used to perform the MSCR test. The 25 mm parallel plate with a 1 mm gap was selected. The bitumen was first heated to 165 °C to remove any thermal history, and then the liquid bitumen sample was dripped into a silicon mold to prepare the sample for MSCR testing.

According to AASHTO T350, the MSCR test was operated in rotational mode using a 1 s creep load followed by a 9 s recovery for each cycle. Twenty creep and recovery cycles were run at 0.1 kPa creep stress followed by ten cycles at 3.2 kPa creep stress. Figure 1 presents a typical creep and recovery cycle. For each cycle, two parameters, the percent recovery (R) and non-recoverable creep compliance (J_{nr}), are calculated following Equations (1) and (2):

$$R = \frac{\varepsilon_p - \varepsilon_u}{\varepsilon_p} \times 100\% \quad (1)$$

$$J_{nr} = \frac{\varepsilon_u}{\sigma} \quad (2)$$

where ε_p represents the peak strain, ε_u represents the unrecovered strain, and σ is the stress level (either 0.1 kPa or 3.2 kPa).

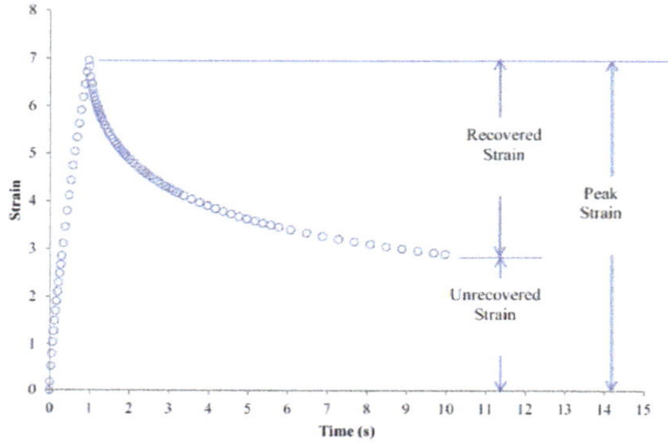

Figure 1. A typical creep and recovery cycle in the MSCR test.

Then, the average percent recovery at 0.1 kPa and 3.2 kPa is calculated and expressed as R0.1 and R3.2, respectively. The average non-recoverable creep compliance at 0.1 kPa and 3.2 kPa is expressed as J_{nr}0.1 and J_{nr}3.2. This study only investigates the measurements of 3.2 kPa (R3.2 and J_{nr}3.2) because a higher stress level is considered to be more comparable with the field. Some European agencies are even running MSCR at 10 kPa. The test was conducted at 4 different temperatures (64 °C, 70 °C, 76 °C, and 82 °C), hence the influence of temperature could be discussed.

2.3. MSCR Curve

The AASHTO T350 specification also includes a technique to further evaluate the bitumen creep and recovery behavior. This technique is referred to as the MSCR curve or polymer curve [29]. An MSCR curve is established by plotting R3.2 against J_{nr}3.2 (shown in Figure 2). A standard curve depicted as Equation (3) is used to examine whether the bitumen has enough elastic response for elastomeric-modified bitumen.

$$y = 29.37x^{-0.2633} \quad (3)$$

where y is R3.2 and x is J_{nr}3.2.

According to D'angelo's work [17], this standard curve is an empirical decision based on numerous modified bitumen samples. This curve will assure that the polymer-modified bitumens have well-developed polymer networks. The MSCR curve is not commonly employed in the literature, but authors find it could be a useful tool to analyze the bitumen creep and recovery properties. In this study, the MSCR curve is applied for more in-depth analysis.

A flowchart is shown in Figure 3 to help understand the methodology of this study.

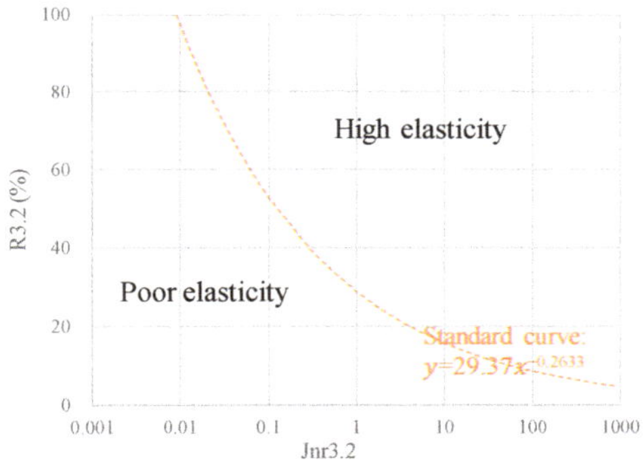

Figure 2. Diagram of the MSCR curve and standard curve to determine the minimum R3.2 for measured $J_{nr}3.2$.

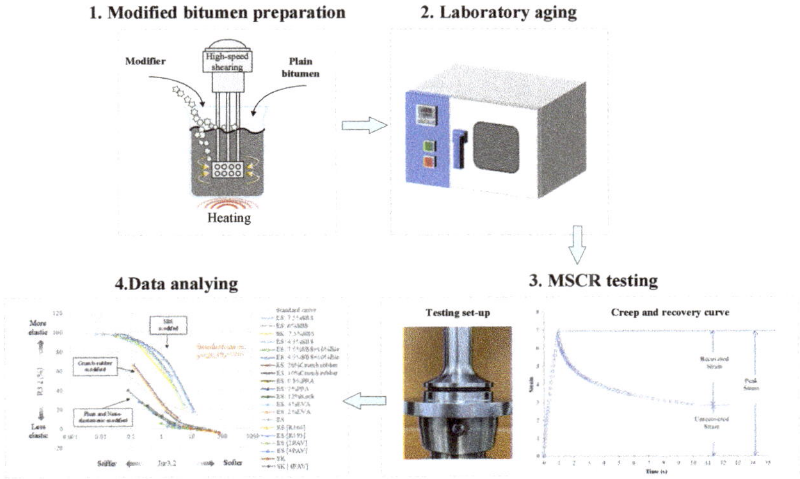

Figure 3. Methodology of this study.

3. Results and Discussion

3.1. Comparison between Different Modifiers

In this section, all seven elastomeric-modified bitumens, five non-elastomeric-modified bitumens, and three plain bitumens are compared in terms of their R3.2 and $J_{nr}3.2$. Only unaged virgin bitumens are discussed in this section. The influence of aging will be discussed later. The testing results of R3.2 and $J_{nr}3.2$ are summarized in Figures 4 and 5, respectively.

Figure 4 shows that elastomeric-modified bitumens have the highest recovery rate, followed by non-elastomeric-modified bitumens and plain bitumens. Specific to each modifier, SBS shows the highest recovery rate, followed by crumb rubber and rock bitumen. EVA and PPA do not show any evident improvement in the recovery rate. Bio-oil notably diminishes bitumen's recovery rate as bio-oil-modified SBSMBs are less elastic than conventional SBSMB with the same SBS content.

The summary of $J_{nr}3.2$ is shown in Figure 5. It shows that elastomeric-modified bitumens have the best permanent deformation resistance (lowest $J_{nr}3.2$), followed by the non-elastomeric-modified bitumens and plain bitumens. Specific to each modifier, SBS shows the smallest $J_{nr}3.2$, followed by crumb rubber and rock bitumen. EVA and PPA modification slightly decrease the $J_{nr}3.2$ for plain bitumen. Bio-oil notably diminishes bitumen's deformation resistance as bio-modified bitumens show much more increased

$J_{nr}3.2$ than conventional plain bitumen-based bitumens. These observations are consistent with previous studies that show that SBS polymer-modified asphalt has good rutting resistance [30] and bio-oil may hinder the bitumen rutting resistance [31].

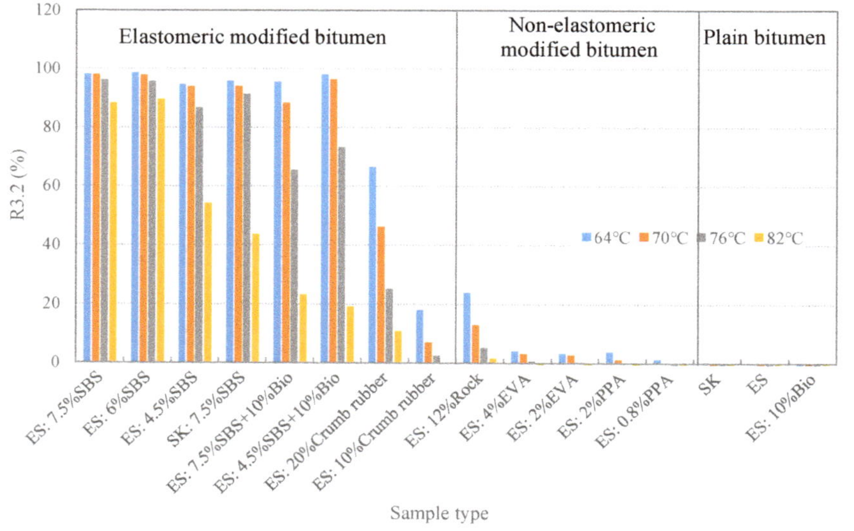

Figure 4. R3.2 from MSCR testing at different temperatures (64 °C, 70 °C, 76 °C, and 82 °C).

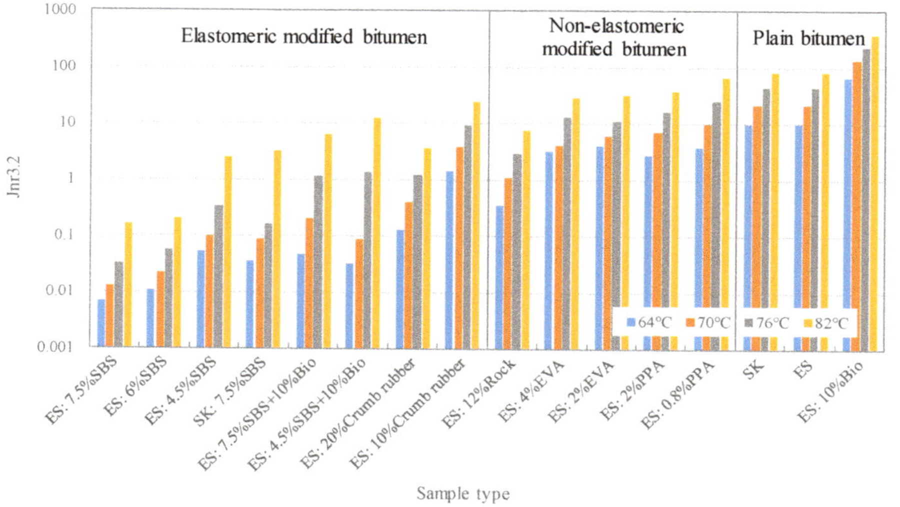

Figure 5. $J_{nr}3.2$ from MSCR testing at different temperatures (64 °C, 70 °C, 76 °C, and 82 °C).

It is worth mentioning that SK: 7.5%SBS only shows comparable $J_{nr}3.2$ and R3.2 with ES: 4.5%SBS. This observation highlights the significant influence of the base bitumen source on the rheological properties of SBS-modified bitumen. The SBS polymer swells better in an aromatic fraction. Since ES bitumen contains a higher proportion of the aromatic fraction, it is often considered an ideal base bitumen for modification.

3.2. MSCR Testing Results for Plain Bitumens

A separate discussion on R3.2 or $J_{nr}3.2$ is not sufficient to fully explain the creep and recovery behavior of bitumens. In this section, the MSCR curve is applied for more in-depth analysis. Moreover, laboratory-aged bitumens are examined.

An MSCR curve for plain bitumen is established in Figure 6. Three plain bitumens, which are ES, SK, and ES: 10%Bio, and their aged replicates are evaluated. For each bitumen at each aging level, results from four different temperatures (64 °C, 70 °C, 76 °C, and 82 °C) are plotted together as a series, highlighting the temperature-induced trends. Noted for the purpose of conciseness, the testing temperature is not labeled in the graph. To tell which data point corresponds to which temperature, the data points of 64 °C are always at the left-top end of the series line because 64 °C yields the highest stiffness and highest recovery rate, followed by data points of 70 °C, 76 °C, and 82 °C.

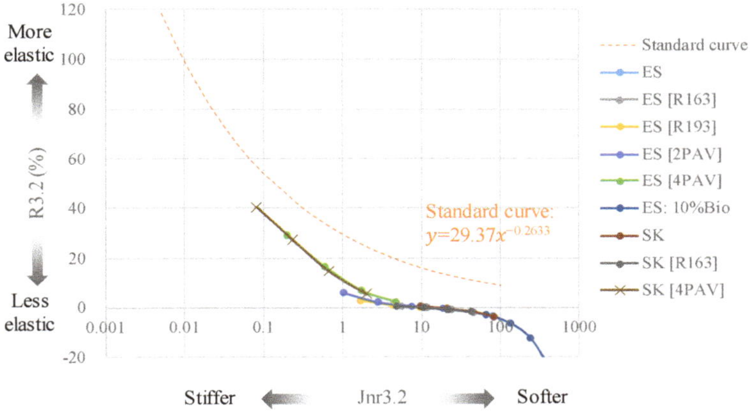

Figure 6. MSCR curve for plain bitumens of different aging conditions.

From Figure 6, it is very interesting to see that all points converge onto a smooth "master curve", irrespective of the bitumen type or aging level. This master curve is lower than the standard curve, indicating these plain bitumens are not elastic enough to be categorized as elastomer-modified asphalt. Aging and reduced testing temperatures will bring the recovery rates to a higher level but will not make the curve cross the standard curve.

This smooth master curve suggests that there is a strong correlation between the plain bitumen's recovery rate and its stiffness. Due to its relatively low molecular weight and absence of a long-chain polymer network, plain bitumen relies predominantly on molecular internal energy changes for its elasticity [32]. This type of elasticity, referred to as energy elasticity, exhibits a significant dependence on bitumen stiffness. In other words, variables such as the testing temperature and aging level affect the bitumen recovery rate by changing its stiffness.

3.3. MSCR Testing Results for Non-Elastomeric-Modified Bitumens

Three non-elastomeric modifiers, which are rock bitumen, PPA, and EVA, are evaluated in this section. The MSCR curve for non-elastomeric-modified bitumens is drawn in Figure 7 following the same pattern as Figure 6. Results show that these non-elastomeric-modified bitumens behave very much like plain bitumen. Similarly, a smooth and continuous master curve is observed, and it almost overlaps with the one generated from plain bitumens. Their master curves both start from (0.1, 40) and cross the $J_{nr}3.2$ axis around 10. This similarity can be comprehended as a non-elastomeric modifier also acquiring its elasticity mainly from internal energy changes. According to Cheng's work [33], adding rock bitumen will increase the asphaltene/heavy molecule ratio and result in an increase in stiffness. The mechanism is quite similar to aging because aging will also bring more heavy molecules into the system. With the increase in heavy molecular and chemical fractions, the intermolecular interaction becomes stronger, and thus the bitumen becomes stiffer and shows an increased elastic response.

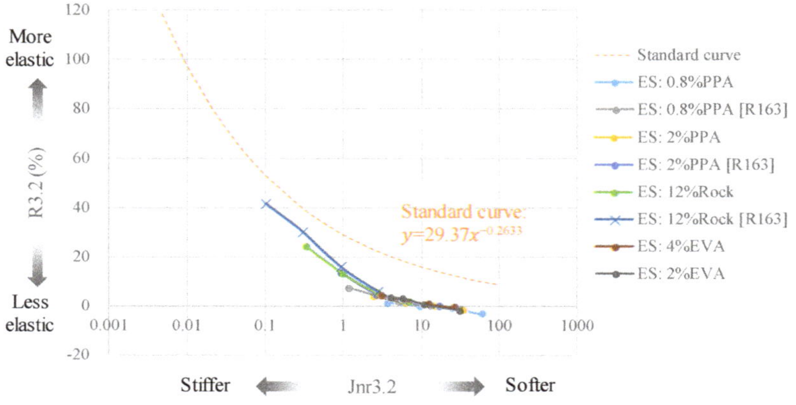

Figure 7. MSCR curve for non-elastomeric-modified bitumens.

3.4. MSCR Testing Results for Elastomeric-Modified Bitumens

Two commonly used elastomeric modifiers, namely SBS and crumb rubber, are tested in this study. Note that only unaged modified bitumens are discussed in this section because it is found that in terms of aging, elastomeric-modified bitumens' responses are much more complicated and thus will be discussed in a separate section.

Following the same fashion, a master curve is established for elastomeric-modified bitumens. The results are shown in Figure 8. For comparison, plain bitumens and non-elastomeric-modified bitumens are also included in Figure 8.

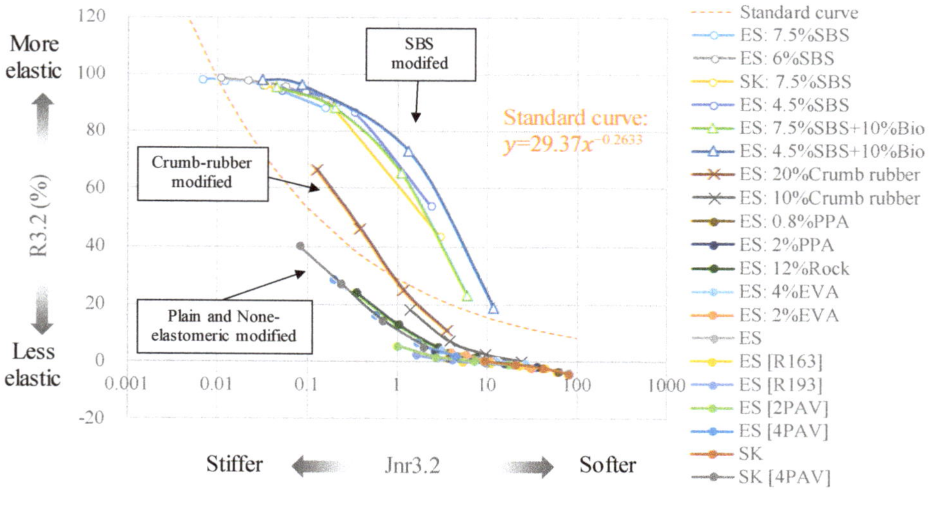

Figure 8. MSCR curve for all kinds of modified bitumens.

According to Figure 8, SBSMB shows the most improved elastic response. Crumb rubber-modified bitumen also shows higher elasticity than plain bitumen, but the improvement is not comparable with SBSMB. Moreover, 20% crumb rubber-modified bitumen meets the standard curve at 64 °C and 70 °C. However, further increases in temperature will make it unqualified. In addition, 10% crumb rubber-modified bitumen fails the standard curve at all testing temperatures. The relatively lower recovery rate of crumb rubber-modified bitumen is possibly explained by the poor compatibility between crumb rubber and plain bitumen [34].

It can be seen that SBSMB shows more elasticity (R3.2) than plain/non-elastomeric bitumen at any specific stiffness (J_{nr}3.2). This highlights the different origins of elasticity between these bitumens. For plain bitumen and non-elastomeric-modified bitumen, the elasticity is mainly attributed to internal energy change, whereas SBSMB acquires its elasticity from both the internal energy change (energy elasticity) and the rubbery polymer network (entropy elasticity).

According to thermodynamics, the elastic behavior of materials is governed by Helmholtz's free energy [35], which is expressed as Equation (4) and has two parts: The energy term and the entropy term.

$$F = U - TS \tag{4}$$

where F is free energy, U is (internal) energy, T is temperature, and S is entropy.

When deformed, molecules "want" to go back to the state with minimal free energy, that is, to minimize their internal energy and maximize their entropy. Therefore, depending on whether the energy or the entropy term dominates the free energy, the elasticity can be classified as energy-elasticity and entropy-elasticity. While low-molecular-weight materials such as plain bitumen usually show energy elasticity, high-molecular-weight materials such as elastomer and elastomer-modified bitumen can exhibit a considerable portion of entropy elasticity.

Due to the fundamental difference in the generating mechanism, entropy elasticity has a much higher yield strain and is less temperature-sensitive than energy elasticity. In fact, entropy elasticity is positively proportional to the temperature increase because a higher ambient temperature facilitates molecular thermal motions [36]. Both energy elasticity and entropy elasticity lead to improved recovery rates but energy elasticity does not always translate to better field performance. Stiff bitumen that shows considerable energy elasticity could also be vulnerable to fatigue and cracking damage.

In terms of applications in bitumen concrete pavement, entropy elasticity might be considered a "better" kind of elasticity. It improves the bitumen's overall properties such as better rutting resistance (higher recovery rate), better fatigue resistance (higher yield strain), and better low-temperature cracking resistance (less temperature-sensitive). Many studies have indicated there are obvious correlations between the elasticity of bitumen and its mechanical properties [15,16]. However, no wide-accepted performance standard has been established based on bitumen's elasticity properties. This might be because insufficient attention has been paid to differentiating the origin of elasticity (energy or entropy). This research field requires more in-depth studies.

3.5. Temperature Sensitivity of Recovery Rate

The difference between energy elasticity and entropy elasticity in terms of temperature sensitivity is examined in this section. The influence of temperature on the bitumen recovery rate (R3.2) is shown in Figure 9. To highlight the difference, only six representative bitumens are listed in this section, and the testing temperature range is broadened to 40 °C~82 °C with a 6 °C gap.

Based on the available data, stiff samples such as non-elastomeric-modified bitumen (ES:12%Rock) and seriously aged bitumen (SK [4PAV]) show the most obvious temperature sensitivity, highlighting the strong temperature sensitivity of energy elasticity. This sensitivity is explained by the fact that stiff material has higher intermolecular forces, which translate to higher internal force and consequently higher temperature sensitivity. SBSMB on the other hand does not show much temperature sensitivity because it does not rely on stiffness to gain its elasticity. Instead, entropy elasticity is the main source, and it is much less temperature-sensitive. SBSMB does not show a noticeable decrease in the recovery rate until 76 °C. Bio-oil-modified SBSMB (marked with squares) show a higher temperature sensitivity than conventional ones, and this is attributed to the viscous nature of bio-oil. This is favorable with related studies showing that bio-oil can hinder the elasticity of bitumens [37].

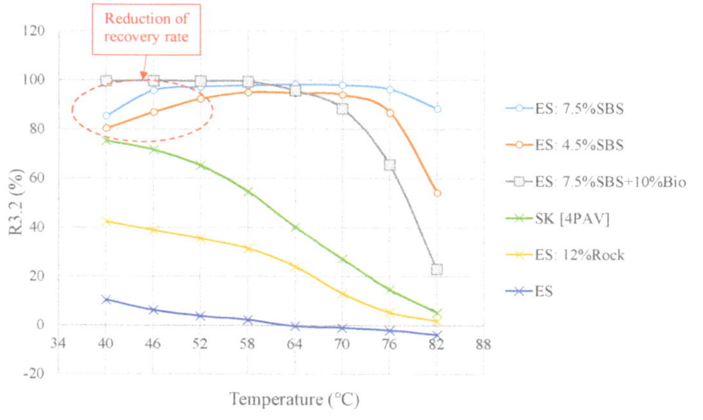

Figure 9. Influence of temperature on bitumen recovery rate.

From Figure 9, it is also interesting to notice that SBSMB exhibits an R3.2 reduction at lower temperatures (marked in red circle). This might be explained by two reasons. First, entropy elasticity is fundamentally attributed to the thermal motion of polymer chains, and thus a lower ambient temperature will weaken the molecular thermal motion, as well as the entropy elasticity. Second, bitumen stiffens at lower temperatures, and it hinders the relaxation of polymer chains. Either way, this is a noteworthy phenomenon because the MSCR test is always conducted at relatively high temperatures (usually PG temperatures) while the pavement spends most of its service life at intermediate temperatures. Recovery rate measurements at intermediate temperatures may require more attention.

3.6. *Effects of Aging on SBS-Modified Bitumens*

The influence of aging on SBSMB is discussed in this section. Depending on the difference in base bitumen source and SBS content, three representative SBSMB are investigated in this section, which are ES: 7.5%SBS, SK: 7.5%SBS, and ES: 4.5%SBS. The results for aged ES: 7.5%SBS bitumens are presented in Figure 10.

Based on Figure 10, the aging behavior of SBSMB notably differs from plain bitumen or non-elastomeric-modified bitumens in two aspects. First, after aging, SBSMB's recovery rates tend to decrease rather than increase. Second, the data points of aged SBSMB no longer form a smooth master curve. This phenomenon suggests that the aging process changed the percentages of contributions from energy elasticity and entropy elasticity to the SBSMB's overall creep and recovery performance. During aging, SBSMB experienced oxidative stiffening coupled with polymer degradation [38], which led to an increase in energy elasticity and a decrease in entropy elasticity. The increase in "bad" energy elasticity may compensate for part of the "good" entropy elasticity loss, but overall, the bitumen's recovery rate is decreasing and the ratio of energy elasticity is climbing. This variation is not preferable from the perspective of field performance, especially for anti-cracking and anti-fatigue performance.

The results for aged SK: 7.5%SBS bitumens are presented in Figure 11. A similar outcome is seen for SK: 7.5%SBS as a smooth master curve is no longer observed and the recovery rate tends to decrease after aging.

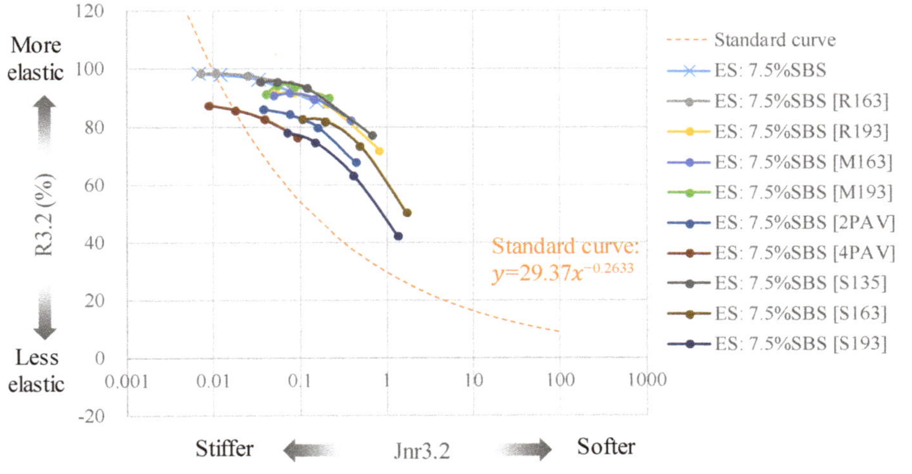

Figure 10. MSCR curve for ES: 7.5%SBS bitumen of different aging conditions.

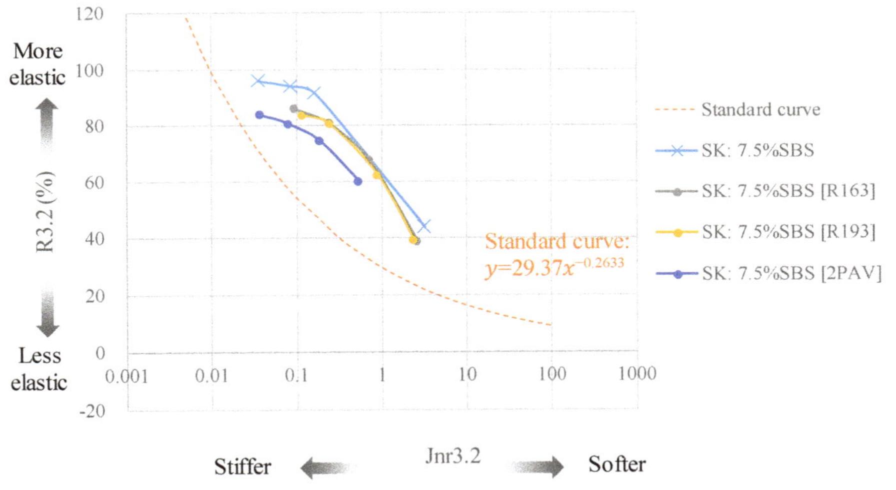

Figure 11. MSCR curve for aged SK: 7.5%SBS bitumens at 3.2 kPa.

The results for aged ES: 4.5%SBS bitumens are presented in Figure 12. Similarly, no smooth master curve is seen. Due to the relatively lower SBS dosage and the consequently lower degree of entropy elasticity, it can be seen that part of the data points for the PAV and 2PAV aged ES: 4.5%SBS samples have failed the standard curve. The PAV and 2PAV aged bitumens behave very much like seriously aged plain bitumens. This is a coupled result of oxidative stiffing and polymer degradation. The SBS polymer has a readily oxidizable nature, thus SBSMA is vulnerable to aging. Due care should be taken to prevent excessive aging of SBSMA in case it becomes stiff and brittle like aged plain bitumen.

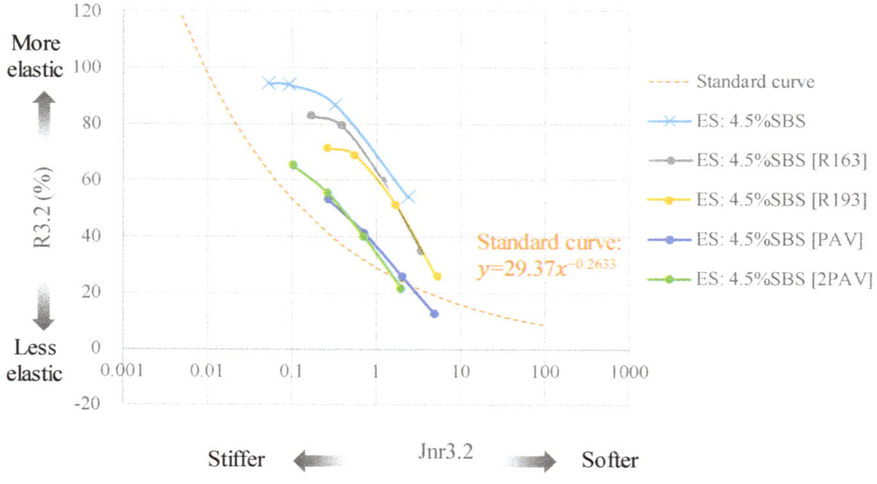

Figure 12. MSCR curve for aged ES: 4.5%SBS bitumens at 3.2 kPa.

4. Conclusions

This study aims to evaluate the creep and recovery properties of 15 kinds of representative modified bitumens used in the field based on the MSCR test. The MSCR test can highlight the unique delayed elasticity of modified bitumen and it uses a high stress level, which is more comparable to the field. The bitumens' temperature sensitivity was examined, and their aging behavior was evaluated based on the consideration of 10 different laboratory aging conditions. The detailed conclusions were as follows:

- For plain bitumen and non-elastomeric-modified bitumen, the elasticity is mainly attributed to the internal energy change (energy elasticity), whereas elastomeric-modified bitumens acquire their elasticity from both energy elasticity and entropy elasticity.
- The difference between energy elasticity and entropy elasticity needs more attention in the road bitumen material community. Compared with energy elasticity, entropy elasticity is less temperature-sensitive and has a higher yield strain. It is considered a "better" kind of elasticity from the field performance perspective.
- High strain creep and recovery tests such as MSCR could be a promising technique to analyze bitumen's elasticity properties. An interesting "master curve" is observed for different bitumens and can be utilized to discriminate energy elasticity from entropy elasticity.
- The aging process changes the percentages of contributions from energy elasticity and entropy elasticity to the bitumen's overall recovery performance. The increase in "bad" energy elasticity may compensate for part of the "good" entropy elasticity loss, but overall, the bitumen's recovery rate is decreasing and the ratio of undesirable energy elasticity is climbing, which might hinder the bitumen's performance.

Author Contributions: Methodology, G.Z. and J.Z.; Formal analysis, H.H.; Investigation, G.Z., J.Z., H.H. and X.X.; Resources, H.H.; Data curation, X.X.; Writing—original draft, C.Y.; Writing—review & editing, C.Y.; Visualization, G.Z.; Supervision, H.H.; Project administration, C.Y.; Funding acquisition, C.Y. All authors have read and agreed to the published version of the manuscript.

Funding: This research was funded by the National key R&D program of China (2022YFB2602603), Hong Kong scholars program (XJ2022040), National Natural Science Foundation of China (52008353), and Sichuan Youth Science and Technology Innovation Research Team (2021JDTD0023, 2022JDTD0015).

Institutional Review Board Statement: Not applicable.

Informed Consent Statement: Not applicable.

Data Availability Statement: Not applicable.

Conflicts of Interest: The authors declare no conflict of interest.

References

1. Lv, Q.; Huang, W.; Sadek, H.; Xiao, F.; Yan, C. Investigation of the rutting performance of various modified asphalt mixtures using the Hamburg Wheel-Tracking Device test and Multiple Stress Creep Recovery test. *Constr. Build. Mater.* **2019**, *206*, 62–70. [CrossRef]
2. Shan, L.; He, H.; Wagner, N.J.; Li, Z. Nonlinear rheological behavior of bitumen under LAOS stress. *J. Rheol.* **2018**, *62*, 975–989. [CrossRef]
3. Yan, C.; Huang, W.; Lin, P.; Zhang, Y.; Lv, Q. Chemical and rheological evaluation of aging properties of high content SBS polymer modified asphalt. *Fuel* **2019**, *252*, 417–426. [CrossRef]
4. Behnood, A.; Olek, J. Rheological properties of asphalt binders modified with styrene-butadiene-styrene (SBS), ground tire rubber (GTR), or polyphosphoric acid (PPA). *Constr. Build. Mater.* **2017**, *151*, 464–478. [CrossRef]
5. Nuñez, J.Y.M.; Domingos, M.D.I.; Faxina, A.L. Susceptibility of low-density polyethylene and polyphosphoric acid-modified asphalt binders to rutting and fatigue cracking. *Constr. Build. Mater.* **2014**, *73*, 509–514. [CrossRef]
6. Yang, X.; Liu, G.; Rong, H.; Meng, Y.; Peng, C.; Pan, M.; Ning, Z.; Wang, G. Investigation on mechanism and rheological properties of Bio-asphalt/PPA/SBS modified asphalt. *Constr. Build. Mater.* **2022**, *347*, 128599. [CrossRef]
7. Hussein, I.A.; Iqbal, M.H.; Al-Abdul-Wahhab, H.I. Influence of M w of LDPE and vinyl acetate content of EVA on the rheology of polymer modified asphalt. *Rheol. Acta* **2005**, *45*, 92–104. [CrossRef]
8. Sengoz, B.; Isikyakar, G. Evaluation of the properties and microstructure of SBS and EVA polymer modified bitumen. *Constr. Build. Mater.* **2008**, *22*, 1897–1905. [CrossRef]
9. Cai, L.; Shi, X.; Xue, J. Laboratory evaluation of composed modified asphalt binder and mixture containing nano-silica/rock asphalt/SBS. *Constr. Build. Mater.* **2018**, *172*, 204–211. [CrossRef]
10. He, W.; Zhao, Z.; Yuan, J.; Xiao, F. Recent development of ethylene–vinyl acetate modified asphalt. *Constr. Build. Mater.* **2023**, *363*, 129800. [CrossRef]
11. Lee, S.; Akisetty, C.K.; Amirkhanian, S.N. The effect of crumb rubber modifier (CRM) on the performance properties of rubberized binders in HMA pavements. *Constr. Build. Mater.* **2008**, *22*, 1368–1376. [CrossRef]
12. Presti, D.L. Recycled tyre rubber modified bitumens for road asphalt mixtures: A literature review. *Constr. Build. Mater.* **2013**, *49*, 863–881. [CrossRef]
13. Lei, Z.; Golalipour, A.; Tabatabaee, H.A.; Bahia, H.U. Prediction of effect of bio-based and refined waste oil modifiers on rheological properties of asphalt binders. In Proceedings of the Transportation Research Board 93rd Annual Meeting, Washington, DC, USA, 12–16 January 2014.
14. Sun, D.; Sun, G.; Du, Y.; Zhu, X.; Lu, T.; Pang, Q.; Shi, S.; Dai, Z. Evaluation of optimized bio-asphalt containing high content waste cooking oil residues. *Fuel* **2017**, *202*, 529–540. [CrossRef]
15. Johnson, C.; Bahia, H.; Wen, H. Practical application of viscoelastic continuum damage theory to asphalt binder fatigue characterization. *Asph. Paving Technol.-Proc.* **2009**, *28*, 597.
16. Clopotel, C.; Mahmoud, E.; Bahia, H. Modification of the Elastic Recovery Test and its Relationship to Performance Related Properties of Modified Asphalt Binders. In Proceedings of the 90th Annual Meeting of the Transportation Research Board, Washington, DC, USA, 23–27 January 2011.
17. D'Angelo, J. Development of a Performance Based Binder Specification for Rutting Using Creep and Recovery Testing. Ph.D. Thesis, University of Calgary, Calgary, AB, Canada, 2009.
18. D'Angelo, J.A. The relationship of the MSCR test to rutting. *Road Mater. Pavement Des.* **2009**, *10* (Suppl. S1), 61–80. [CrossRef]
19. Wang, Y.; Yi, H.; Liang, P.; Chai, C.; Yan, C.; Zhou, S. Investigation on Preparation Method of SBS-Modified Asphalt Based on MSCR, LAS, and Fluorescence Microscopy. *Appl. Sci.* **2022**, *12*, 7304. [CrossRef]
20. Katiyar, V.; Gupta, A.; Tounsi, A. Microstructural/geometric imperfection sensitivity on the vibration response of geometrically discontinuous bi-directional functionally graded plates (2D FGPs) with partial supports by using FEM. *Steel Compos. Struct. Int. J.* **2022**, *45*, 621–640.
21. Mesbah, A.; Belabed, Z.; Amara, K.; Tounsi, A.; Bousahla, A.A.; Bourada, F. Formulation and evaluation a finite element model for free vibration and buckling behaviours of functionally graded porous (FGP) beams. *Struct. Eng. Mech.* **2023**, *86*, 291.
22. Jia, H.; Chen, H.; Sheng, Y.; Meng, J.; Cui, S.; Kim, Y.R.; Huang, S.; Qin, H. Effect of laboratory aging on the stiffness and fatigue cracking of asphalt mixture containing bamboo fiber. *J. Clean. Prod.* **2022**, *333*, 130120. [CrossRef]
23. Lu, X.; Isacsson, U. Chemical and rheological evaluation of ageing properties of SBS polymer modified bitumens. *Fuel* **1998**, *77*, 961–972. [CrossRef]
24. Li, Z.; Zeng, J.; Li, Y.; Zhao, Z.; Cong, P.; Wu, Y. Effect of bitumen composition on micro-structure and rheological properties of styrene–butadiene–styrene modified asphalt before and after aging. *Mater. Struct.* **2022**, *55*, 165. [CrossRef]
25. Cuciniello, G.; Leandri, P.; Filippi, S.; Presti, D.L.; Losa, M.; Airey, G. Effect of ageing on the morphology and creep and recovery of polymer-modified bitumens. *Mater. Struct.* **2018**, *51*, 136. [CrossRef]

26. Yan, C.; Huang, W.; Zheng, M.; Zhang, Y.; Lin, P. Influence of ageing on high content polymer modified asphalt mixture stripping, cracking and rutting performances. *Road Mater. Pavement Des.* **2020**, *22*, 1824–1841. [CrossRef]
27. Sun, L.; Gu, X.; Hu, D.; Zhou, Z.; Wang, G. Anti-aging mechanism and rheological properties of lignin, quercetin, and gallic acid as antioxidants in asphalt. *Constr. Build. Mater.* **2023**, *369*, 130560. [CrossRef]
28. Bahia, U.H.; Hanson, I.D.; Zeng, M.; Zhai, H.; Khatri, A.M.; Anderson, M.R. *Characterization of Modified Asphalt Binders in Superpave Mix Design*; Transportation Research Board: Washington, DC, USA, 2001.
29. Hossain, Z.; Ghosh, D.; Zaman, M.; Hobson, K. Use of the multiple stress creep recovery (MSCR) test method to characterize polymer-modified asphalt binders. *J. Test. Eval.* **2016**, *44*, 507–520. [CrossRef]
30. Huang, W.; Tang, N. Characterizing SBS modified asphalt with sulfur using multiple stress creep recovery test. *Constr. Build. Mater.* **2015**, *93*, 514–521. [CrossRef]
31. Yang, X.; You, Z.; Dai, Q.; Mills-Beale, J. Mechanical performance of asphalt mixtures modified by bio-oils derived from waste wood resources. *Constr. Build. Mater.* **2014**, *51*, 424–431. [CrossRef]
32. Sperling, L.H. *Introduction to Physical Polymer Science*; John Wiley & Sons: Hoboken, NJ, USA, 2005.
33. Cheng, L.; Yu, J.; Zhao, Q.; Wu, J.; Zhang, L. Chemical, rheological and aging characteristic properties of Xinjiang rock asphalt-modified bitumen. *Constr. Build. Mater.* **2020**, *240*, 117908. [CrossRef]
34. Wang, H.; Apostolidis, P.; Zhu, J.; Liu, X.; Skarpas, A.; Erkens, S. The role of thermodynamics and kinetics in rubber–bitumen systems: A theoretical overview. *Int. J. Pavement Eng.* **2020**, *22*, 1785–1800. [CrossRef]
35. Roylance, D. Atomistic basis of elasticity. *ACE* **2000**, *5*, 3.
36. Ionescu, T.C.; Edwards, B.J.; Keffer, D.J.; Mavrantzas, V.G. Energetic and entropic elasticity of nonisothermal flowing polymers: Experiment, theory, and simulation. *J. Rheol.* **2008**, *52*, 105–140. [CrossRef]
37. Wang, H.; Ma, Z.; Chen, X.; Hasan, M.R.M. Preparation process of bio-oil and bio-asphalt, their performance, and the application of bio-asphalt: A comprehensive review. *J. Traffic Transp. Eng. (Engl. Ed.)* **2020**, *7*, 137–151. [CrossRef]
38. Yan, C.; Huang, W.; Ma, J.; Xu, J.; Lv, Q.; Lin, P. Characterizing the SBS polymer degradation within high content polymer modified asphalt using ATR-FTIR. *Constr. Build. Mater.* **2020**, *233*, 117708. [CrossRef]

coatings

MDPI

Article

Evaluation of Fume Suppression, Viscosity-Retarding, and Rheological Properties of Eco-Friendly High-Viscosity Modified Asphalt

Weidong Ning, Guoqiang Sun *, Kexin Qiu, Xulai Jiang, Chunze Wang and Ruiqi Zhao

Beijing Key Laboratory of Traffic Engineering, Beijing University of Technology, Beijing 100124, China
* Correspondence: gqsun@bjut.edu.cn

Abstract: In order to address the issues of high viscosity and excessive fume exhaust associated with high-viscosity modified asphalt (HVMA), the objective of this study was to develop an eco-friendly HVMA by incorporating fume suppressants and viscosity-retarding agents (VRAs). To begin with, desulfurization rubber powder (DRP) was utilized as a modifier, and fume suppressants, including activated carbon, a chemical reaction fume suppressant, and a composite fume suppressant combining activated carbon and chemical reaction fume suppressant were added to the HVMA separately. The fume suppression effect and odor level were observed to determine the optimal fume suppressant composition for this study. Based on these observations, an area integration method was proposed, utilizing rotational viscosity testing and temperature sweeping experiments, evaluating the viscosity-retarding effect and mixing temperature when different amounts of Sasobit VRA, Evotherm3G VRA, and a composite VRA of Sasobit and Evotherm3G were added to the HVMA. This approach aimed to identify the eco-friendly HVMA with the most effective fume suppression and viscosity-retarding abilities. Furthermore, the morphology and rheological properties of the eco-friendly HVMA were examined through fluorescence microscopy, zero shear viscosity test, multiple stress creep recovery analysis, liner amplitude sweep test, and frequency sweep test. The results demonstrated that the HVMA formulation consisting of 15% DRP and 1% composite fume suppressant exhibited a satisfactory fume suppression effect and odor level. Based on this, the HVMA formulation containing 0.6% Evotherm3G and 3% Sasobit VRAs displayed the best viscosity-retarding effect while reducing the mixing temperature. Moreover, when compared to common HVMA, the eco-friendly HVMA exhibited excellent high-temperature resistance, successfully accomplishing the dual objectives of ecological friendliness and superior performance.

Keywords: eco-friendly high-viscosity modified asphalt; viscosity retarding; fume suppression; rheological properties

Citation: Ning, W.; Sun, G.; Qiu, K.; Jiang, X.; Wang, C.; Zhao, R. Evaluation of Fume Suppression, Viscosity-Retarding, and Rheological Properties of Eco-Friendly High-Viscosity Modified Asphalt. *Coatings* **2023**, *13*, 1497. https://doi.org/10.3390/coatings13091497

Academic Editors: Hadj Benkreira and Valeria Vignali

Received: 12 July 2023
Revised: 19 August 2023
Accepted: 23 August 2023
Published: 24 August 2023

1. Introduction

During the process of urban construction, high-viscosity modified asphalt (HVMA) had attracted significant attention from researchers due to its remarkable performance in both high and low temperatures as well as its superior durability. However, the high mixing temperature of HVMA led to the emission of a considerable amount of asphalt smoke during the production process, particularly for CRMA. The production process of CRMA often generated more asphalt smoke compared to ordinary HVMA. This not only posed a serious threat to human health but also resulted in severe pollution to the surrounding environment. Consequently, these issues greatly impeded the extensive adoption and application of HVMA in pavement engineering [1–3].

Asphalt fume predominantly consisted of volatile organic compounds (VOCs), polycyclic aromatic hydrocarbons (PAHs), sulfur oxides, and nitrogen oxides. VOCs, in terms of content and type, represented the majority, with alkanes, olefins, alkynes, benzene series, and hydrocarbon derivatives (such as aldehydes and ketones) being the main constituents [4]. Inhalation of these compounds could lead to respiratory difficulties, organ damage, and even cancer [5]. PAHs, which were a distinct category of VOCs, were typically considered to be separate components of asphalt fume. They contained highly carcinogenic substances like acyclic naphthalene, fluoroanthracene, fluorene, naphthalene, benzene, and pyrene, posing significant threats to human health [6]. Therefore, reducing the release of asphalt fumes and energy consumption of HVMA have nowadays become a prominent research area in road engineering [7].

Fundamentally, variations in production temperature played a crucial role in causing changes in asphalt fume emissions. Elevated temperatures caused the macromolecular organic compounds in asphalt materials to undergo thermolysis, resulting in the formation of smaller molecules. With increasing temperatures, the intensity of Brownian motion was heightened, enabling the escape of small molecules from the organic interface. By disrupting the Van der Waals forces between asphalt molecules, these small molecules subsequently combined with gases, steam, and aerosols external to the organic interface, forming asphalt fumes [7,8]. However, it is important to note that there were significant differences in the physical and chemical properties between waste rubber and asphalt. These differences resulted in the poor stability and compatibility of asphalt when waste rubber was used as an additive. Specifically, at high temperatures, the components of waste rubber were more susceptible to separation, which led to the release of sulfur-containing smoke and volatile organic compounds [2,9].

Currently, fume detection methods primarily consist of indoor and on-site detection techniques, including gravimetric, UV–Vis, and sampling detection techniques [10,11]. The gravimetric method only permits the measurement of asphalt fume mixture concentration, whereas the UV visible method allows for the determination of both the concentration of asphalt fume mixtures and the content of individual polycyclic aromatic hydrocarbons [1,11,12]. Sampling detection technology enables the measurement of the composition and concentration of polycyclic aromatic hydrocarbons (PAHs) in asphalt smoke. Gas chromatography–mass spectrometry (GC-MS) is presently one of the principal methods employed for quantitatively characterizing and qualitatively analyzing different components of asphalt fume [13–15]. Tang et al., conducted an analysis of fume emissions from the production process of CRMA using activated carbon adsorption tubes. The collected samples were further analyzed through GC-MS. The researchers discovered that CRMA contained nine new compounds that were not present in pure asphalt [3]. Moreover, the presence of internal sulfur bonds in CRMA led to the release of more volatile sulfur compounds (VSCs). These VSCs included hydrogen sulfide, benzothiazole, 3-methylthiophene, 2-methylthiophene, and others [2,16]. It is worth noting that these compounds possess strong carcinogenic properties, posing a significant threat to human health [2,4].

In order to mitigate asphalt fume pollution, researchers typically added different types of fume suppressants to HVMA [17–19]. Lv et al., developed a fume suppressant named UiO-66 and evaluated its effectiveness in reducing fume production. Its impact on the concentration of class 1 carcinogens in CRMA was assessed through GC-MS analysis. The results showed that the addition of UiO-66 resulted in the reduction in class 1 carcinogens concentration in CRMA by 70% [18]. Li et al., aimed to address the root cause of malodorous gases in rubber-modified asphalt and devised a composite modifier using desulfurized rubber powder (DRP) and styrene–butadiene–styrene (SBS). Through desulfurization, the number of disulfide bonds in the rubber was reduced, thereby minimizing the release of malodorous gases. The study determined that the optimal proportions were 15% DRP and 4% SBS. Remarkably, the performance of the modified asphalt with this composite modifier surpassed that of the single-component SBS, resulting in a reduction of over 40% in SO_2 emissions, as well as reductions exceeding 50% in nitrogen dioxide (NO_2) and nitrogen oxide (NOx) emissions [19]. This provided a new approach for this study. However, it was worth noting that during this period, most studies only added traditional smoke inhibitors during the process and did not develop efficient target smoke suppressants for asphalt and asphalt fume components, which should be the direction of future research.

Nowadays, people often reduce the viscosity of HVMA by adding warm mixing agents, thereby indirectly reducing the energy consumption of HVMA [20]. Various warm mix additives such as organic waxes, surfactants, and foaming techniques have been utilized [21–24]. In terms of organic waxes, Jamshidi et al., conducted a review on the rheological properties and performance of asphalt-containing Sasobit, along with its potential for energy-saving and reduction in greenhouse gas emissions. They found that the incorporation of Sasobit resulted in a 17.9% reduction in natural gas consumption and a 10% reduction in the use of recycled oil as industrial fuel [21]. Furthermore, Wan et al., investigated the modification of asphalt using a surfactant called Evotherm, along with aluminum hydroxide (ATH). Their findings indicated that the modified asphalt exhibited improved thermal stability [22]. Additionally, researchers such as Valdes-Vidal et al., had utilized natural zeolite in the preparation of WMA. Their study demonstrated that the incorporation of recycled pavement allowed for lower-temperature mixing and construction, with promising road performance [23]. However, despite the benefits they offer, warm mix additives still have some drawbacks as they can potentially negatively impact the performance of asphalt while lowering the mixing temperature. Therefore, further research is needed to investigate the effects of additive types and dosages on the performance of WMA [24].

To enhance the performance of WMA, researchers have proposed a technology called warm-mix-based viscosity-retarding asphalt (WM-VRA) [25]. This innovative approach reduces the mixing temperature by implementing higher-temperature viscosity-retarding techniques, extends the working time through medium-temperature viscosity-retarding methods, and improves the service viscosity within lower temperature ranges. By achieving both viscosity-retarding and energy-saving goals, WM-VRA ensures the performance of HVMA. Furthermore, this paper introduces an area integration method for viscosity–temperature curves. By comparing the integration areas of viscosity–temperature curves between various types of WM-VRA and traditional HVMA at different temperatures, we can evaluate the viscosity-retarding effect of WM-VRA. This research provides a foundation for selecting HVMAs that could maintain their performance while also being energy-saving and viscosity-retarding.

This study aims to address the limitations of traditional fume suppressants on modified asphalt components and the decreased performance of traditional WMA. To achieve this, an eco-friendly HVMA was developed through the optimization of fume suppressants and viscosity-retarding agents (VRAs), and the rheological properties of this eco-friendly asphalt were assessed against standards. The study focused on several key factors. First, indoor fume evaluations and odor level were conducted to compare the fume suppression effects of activated carbon, chemical reaction fume suppressants, and composite fume suppressants on DRP-modified HVMA and rubber-modified HVMA. The optimal combination of fume suppressants was then identified. Building on these findings, different amounts of Sasobit VRA, Evotherm3G VRA, and a composite VRA of Sasobit and Evotherm3G were added to screen for the most effective eco-friendly HVMA in terms of fume suppression and viscosity-retarding effects. The mixing temperature and viscosity-retarding properties of the HVMAs were evaluated using rotating viscosity, temperature sweep, and viscosity–temperature line integral area tests. Finally, the morphological characteristics and rheological properties of the eco-friendly HVMA were confirmed through fluorescence microscopy and dynamic shear rheological tests, demonstrating the feasibility of its performance. This study presented a research and development framework for eco-friendly HVMA, which aims to reduce asphalt fume emissions and decrease energy consumption during asphalt pavement construction, thereby promoting a green and low-carbon approach to asphalt pavement development.

2. Materials and Methods

2.1. Raw Materials

This study utilized Shell 70# asphalt as the base matrix asphalt. Based on a series of preliminary experiments and the relevant literature, the following additives were selected to prepare the high-viscosity modified asphalt (HVMA): 0.15% sulfur, 4% styrene–butadiene–styrene (SBS), 0.75% polyphosphoric acid (PPA), 0.5% antioxidant, and 15% desulfurization rubber powder (DRP) [17,19]. To determine the most effective fume suppressant, 1% activated carbon, 1% chemical reaction fume suppressant, and 1% composite fume suppressant were added to the HVMA separately. Comparative samples included rubber-powder HVMA with the same components, blank rubber-powder HVMA, and blank DRP-modified asphalt. Indoor fume and odor-level evaluations were conducted to assess the feasibility of reducing fume emissions using DRP and to identify the optimal fume suppressant formulation (as shown in Figure 1).

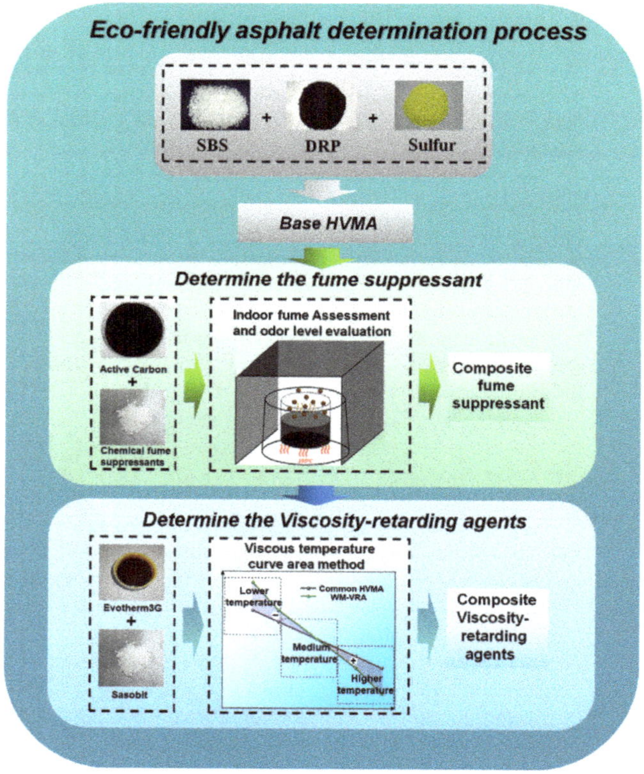

Figure 1. Eco-friendly asphalt determination process.

Subsequently, in the previously optimized fume-suppressing HVMA, different percentages of Sasobit were incorporated, namely 2% Sasobit, 3% Sasobit, 0.5% Sasobit, 0.6% Sasobit, 0.5% Evotherm3G + 2% Sasobit, and 0.6% Evotherm3G + 3% Sasobit [19,20]. Through FM analysis, rotational viscosity testing, and rheological performance testing, eco-friendly HVMA with a remarkable viscosity-retarding effect and exceptional rheological properties was chosen.

In this experiment, the chemical reaction fume suppressant utilized a mixture comprising 75% zinc ricinoleate, 20% sulfurization promoter, 2.5% magnesium sulfate, and 2.5% silane coupling agent. The sulfurization promoter component consisted of stearic acid and zinc oxide. To obtain the composite fume suppressant, the chemical fume suppressant was combined with activated carbon in a 1:1 ratio. The silane coupling agent was prepared by blending KH550 and KH560 in a 1:1 ratio.

2.2. Preparation of the Modified Asphalt Samples

The preparation process of eco-friendly HVMA was as follows:

(1) The base asphalt was heated to a temperature of 180~190 °C, SBS was then added and stirred for 10 min, followed by shearing for 15 min at a speed of 1000~3000 r/min. DRP was added in batches, stirred for 10 min, and sheared for 60 min at a speed of 4000~6000 r/min.
(2) PPA and sulfur were added and sheared for 10 min at a speed of 4000~6000 r/min. Viscosity-retarding agents (VRAs) and antioxidants were then added and sheared for 10 min at a speed of 3000~5000 r/min. Finally, fume suppressant was added and sheared for 10 min at a speed of 4000~6000 r/min.
(3) The temperature was lowered to a range of 170~180 °C, and a four-blade mixer operating at a speed of 600 r/min was used for low-speed stirring for 30 min to eliminate foaming. Subsequently, the mixture was taken out and placed in an oven at 170~180 °C for 30 min to undergo swelling and development.
(4) Once the product was inspected and deemed qualified, it was certified as an eco-friendly HVMA and stored for insulation.

2.3. Experimental

2.3.1. Fluorescence Microscopy (FM) Observation

The FM sample of HVMA was prepared through the hot casting method. The FM micrographs were captured at room temperature using the Olympus Fluorescence microscope system, and a series of images with magnifications of 10 and 20 were obtained. This method was used to observe the crosslinking of HVMA and the distribution of modifiers.

2.3.2. Indoor Fume Assessment

In this study, the fume suppression effect was evaluated by observing the fume release of different fume-reducing asphalt materials. Firstly, in a fume hood, the 10 g of asphalt was heated to 185 °C for ten minutes. A video recording was then started with a duration of 1 min to monitor the fume release of different HVMAs. Finally, images of HVMA fume release at 30 s were collected for comparative analysis.

2.3.3. Odor-Level Evaluation

To assess the intensity of unpleasant odor emitted by high viscosity asphalt, this study employed the Japan Odor Discomfort Scale (JODS) developed by the Ministry of Land, Infrastructure, Transport, and Tourism (MLIT) in Japan. JODS categorized odor levels on a scale ranging from 0 to 5, denoting the absence of discomfort (level 0) to the maximum extent of discomfort (level 5) (Table 1).

Table 1. Japanese odor-level evaluation.

Strength	Grade
0	No odor
1	Can barely feel the odor
2	The odor is weak but can distinguish properties
3	Easy to feel the odor
4	Strong odor
5	An unbearable strong odor

The experiment recruited 10 participants, including 5 males and 5 females, aged between 25 and 50 years old. They were provided with standardized guidelines for scoring the odorous experience. During the evaluation process, participants described and evaluated the odor based on their personal perception and experiences, taking into account factors such as the intensity, discomfort level, color, and duration of the odor stimulus.

The scores provided by the participants, which ranged from 0 to 5, reflected their individual perceptions of the intensity and level of discomfort associated with the odor. Statistical analysis was employed to summarize and report the evaluation results. The interpretation of these results allowed for a comprehensive description and understanding of the odor levels, with higher scores indicating heightened levels of odor perception and discomfort.

2.3.4. Viscosity Test Performed by Rotational Viscometer

In this study, the rotational viscosity test was conducted using the Brookfield viscometer method [26]. The test was mainly performed using a rotational viscometer to measure the apparent viscosity of road asphalt within the temperature range above 155 °C. This measurement was used to determine the construction temperature of various asphalt mixtures and provides a data basis for viscosity-retarding evaluation. The test results might be influenced by external factors, so it was common practice to take the average of three readings as the result during the experiment.

2.3.5. Temperature Sweep (TS) Test

In order to monitor the rheological properties of asphalt as it ages over a wide temperature range, temperature scanning tests were conducted from low temperature to high temperature (30 °C to 100 °C, intervals of 10 °C) at a constant frequency of 10 Hz and a strain of 1%. Throughout the process, a plate with a geometry shape of 8 mm and a gap of 1 mm was used for oscillation testing. The complex shear modulus (G^*) and phase angle (δ) at different temperatures were obtained. Three repeated averages were used, and the coefficient of variation was kept within 5%. This method was used to measure the complex viscosity of HVMA from 30 °C to 100 °C, providing a data basis for viscosity-retarding evaluation.

2.3.6. Shearing Rate Sweep Test

The zero shear viscosity test was conducted using a dynamic shear rheometer with a shear rate scanning experiment. Combined with rheological modeling and fitting analysis, the zero shear viscosity could be obtained. The shear rate scanning experiment measured the rotation speed and torque of the circular plate to obtain viscosity. By continuously applying different torques, the viscosity at different shear rates could be obtained, thus obtaining the flow curve of the asphalt sample [27]. The specific calculation procedures could be found elsewhere [28]. This method was used to evaluate the rheological properties of HVMA.

2.3.7. Multiple Stress Creep Recovery (MSCR) Analysis

The multiple stress creep recovery (MSCR) test using a dynamic shear rheometer (DSR) was conducted to evaluate the elastic recovery of HVMA at temperatures ranging from 58 °C to 82 °C, under creep stresses of 0.1 kPa and 3.2 kPa with intervals of 6 °C. The MSCR test involved applying a 1 s creep load followed by a 9 s recovery period for each cycle. The strain recovery and nonrecoverable creep compliance were computed in accordance with AASHTO 70-1 [29]. In this study, the recovery percentage at 3.2 kPa ($R_{3.2}$) and the non-recoverable creep compliance at 3.2 kPa ($J_{3.2}$) were used as indicators, and each sample was tested twice with duplicates, with the average value taken. The specific calculation procedures could be found elsewhere [30].

2.3.8. Liner Amplitude Sweep (LAS) Test

The LAS test was utilized to evaluate the intermediate-temperature performance based on AASHTO TP101 [31]. The standard loading procedure for the LAS test involved applying controlled strain loadings, with the sinusoidal load amplitude increasing linearly from 0.1% to 30% within 5 min at a frequency of 10 Hz in, and a test temperature of 20 °C. In this study, the LAS test conducted over a duration of 5 min was referred to as LAS-5. This method was used to measure the fatigue life of HVMA.

2.3.9. Frequency Sweep (FS) Test

In this study, FS tests were conducted at temperatures of 85 °C, 65 °C, 45 °C, 25 °C, and 5 °C to provide sufficient data basis for constructing rheological master curves. Sweeping from 5 °C to 35 °C used an 8 mm rotor, while scanning from 40 °C to 100 °C used a 25 mm rotor. The frequency range of the scanning was 0.1 to 100 Hz. The strain level employed was 1%. More detailed information could be found elsewhere in references. It should be noted that each asphalt binder was tested twice, and the average values were used in this study [32,33]. This method was used to evaluate the complex rheological properties of HVMA.

2.4. Evaluation Method

This study employed an evaluation method to assess the effectiveness of viscosity-retarding by comparing the area enclosed by the viscosity–temperature curve and the coordinate axis (as shown in Figure 2). The figure below illustrates the ideal viscosity–temperature curve for a warm-mixing-based viscosity-retarding asphalt (WM-VRA). It could be observed that the viscosity–temperature curve of the WM-VRA intersected with that of a typical HVMA. Within the lower temperature range of 50–100 °C, the viscosity of the WM-VRA was higher. If the area S_1* enclosed by the viscosity–temperature curve of the WM-VRA and the coordinate axis was larger than the area S_1 enclosed by the viscosity–temperature curve of the common HVMA and the coordinate axis, then it indicated that the WM-VRA demonstrated superior performance in viscosity-retarding:

$$S_D = S_1 - S_1^* \quad (1)$$

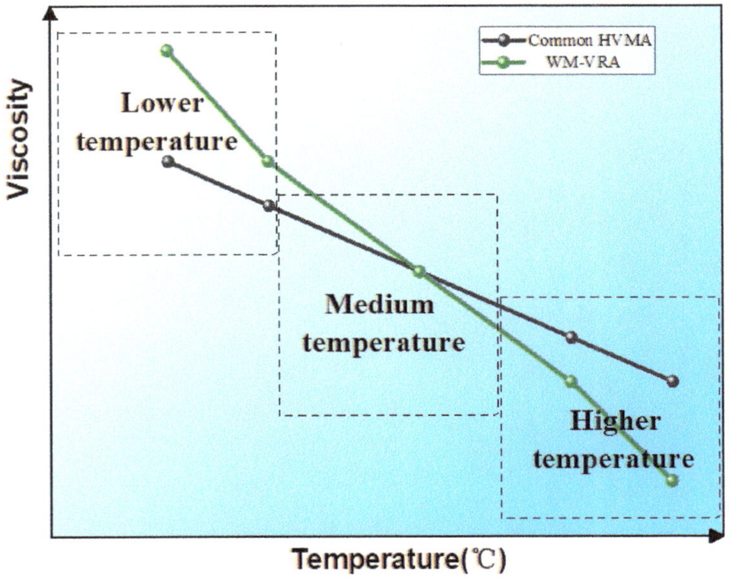

Figure 2. Mechanism diagram of viscosity-retarding effect.

When S_D was less than 1, HVMA was considered to exhibit a lower-temperature viscosity increase effect. A smaller S_D value indicated that the lower-temperature viscosity of the WM-VRA was higher compared to the original sample, suggesting a better lower-temperature viscosity increase effect.

In the higher temperature range (155–200 °C), the viscosity of the HVMA was lower, and the area S_2* enclosed by its viscosity–temperature curve and the coordinate axis was smaller than the area S_2 enclosed by the viscosity–temperature curve of common HVMA.

$$S_G = S_2 - S_2^* \tag{2}$$

When S_G was greater than 1, it was considered that HVMA exhibited a higher-temperature viscosity-retarding effect. As the S_G value increases, the high-temperature viscosity of HVMA decreased compared to the original sample, suggesting a more pronounced improvement in higher-temperature viscosity reduction.

When both conditions were met simultaneously (i.e., $S_D < 1$ and $S_G > 1$), HVMA was considered to have viscosity-retarding properties, and its viscosity-retarding effect was evaluated using a formula. As the S_V increased, the viscosity-retarding effect of HVMA improved:

$$S_V = S_G - S_D \tag{3}$$

2.5. Test Process

The experimental procedure framework is showed in Figure 3. Firstly, through indoor fume suppression assessment and odor level evaluation, optimal fume suppressants and their respective dosages were determined. Subsequently, various types and dosages of VRAs were added to the fume-suppressed modified asphalt. The HVMA viscosity–temperature relationship was determined through the utilization of the Brookfield rotational viscometer (RV) and temperature sweeping experiments, and the effectiveness of viscosity-retarding was evaluated using the area method. Furthermore, the internal network structure of the HVMA was examined using fluorescence microscopy, and the rheological properties of the HVMA were assessed using a dynamic shear rheometer (DSR).

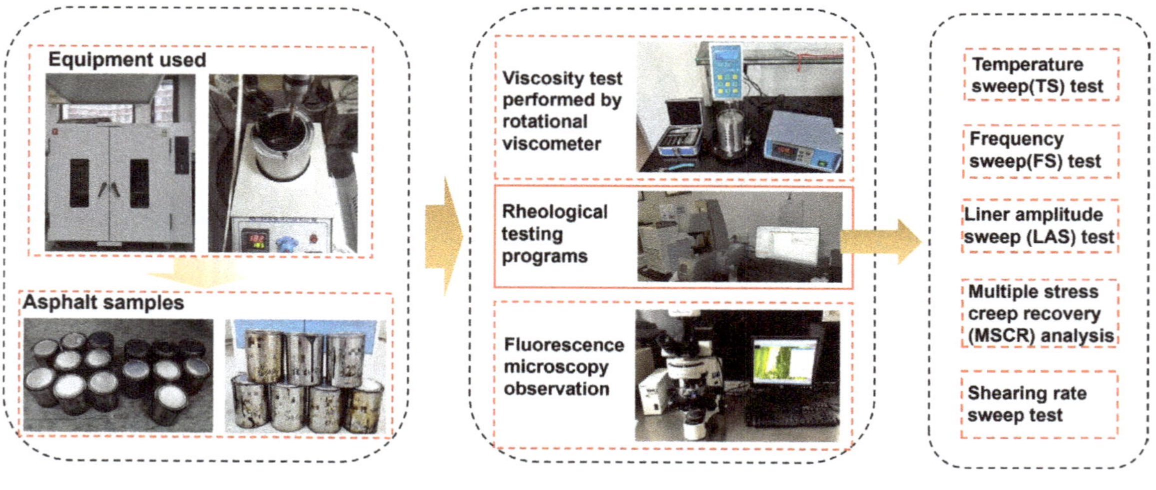

Figure 3. The framework of the experimental procedure.

The modification process of asphalt binder lasted for a period of two weeks, during which, potential thermal and oxidative aging could occur. In order to prevent these aging issues, it was ensured that all test quantities for each tank of HVMA samples were poured in one attempt. This approach was adopted to avoid any thermal aging that could be caused by subjecting the binder to multiple heating cycles. Simultaneously, after the pouring and cooling process, each sample was individually wrapped with silicone oil paper. This precaution was taken to prevent any oxidation and aging problems that may arise during storage.

3. Results and Discussion

3.1. Morphological Analysis

The FM images in Figure 4 depict the morphological features of HVMA modified with varying types and concentrations of viscosity-retarding agents (VRAs) at a magnification of 100×. In Figure 4a,b, the polymer phase is uniformly dispersed within the asphalt phase. The addition of Evotherm3G, in comparison to the control group, enhanced the network structure of high-viscosity modified asphalt (HVMA). With an increasing concentration of Evotherm3G, the network structure became more pronounced. On the other hand, in Figure 4c,d, the addition of Sasobit and fume suppressants did not impact the network structure of SBS in the asphalt, and both Sasobit and fume suppressant particles were evenly disseminated within the asphalt phase. Turning to Figure 4e,f, the presence of both Sasobit and Evotherm3G led to a complex structure in HVMA, where the network structure was less apparent compared to HVMA with a single component added. Nevertheless, a continuous spatial network structure still existed. This observation might arise from a series of unknown reactions that occur when both additives are simultaneously introduced, potentially influencing the subsequent generation of a crosslinked network. Further comprehensive investigation is required to unravel the underlying mechanisms.

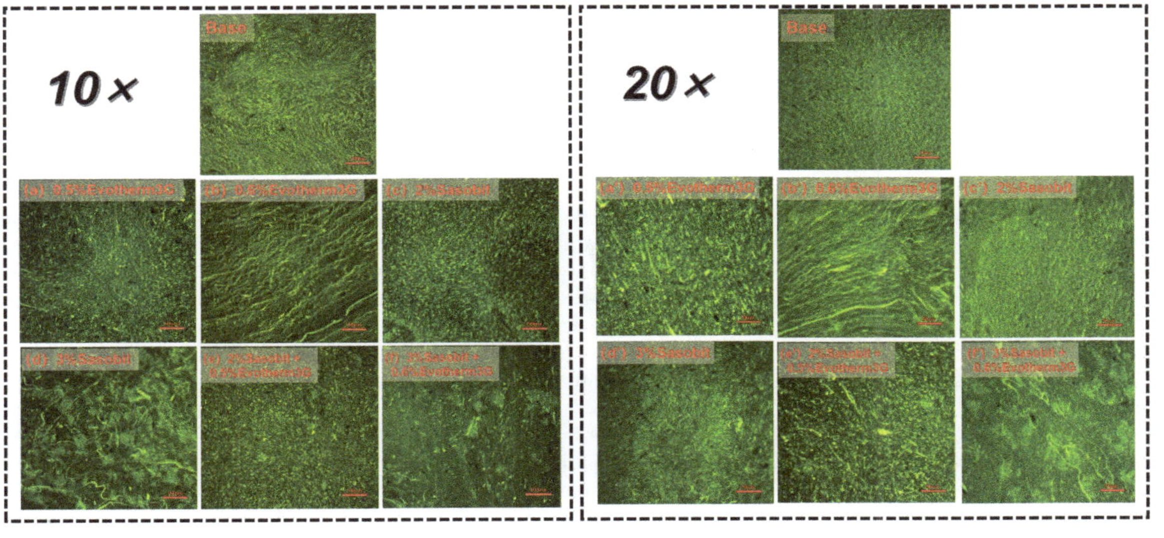

Figure 4. Fluorescence micrograph of different HVMAs.

3.2. Fume-Suppressing Effect Evaluation

Figure 5 shows the fume release and odor levels of different HVMA based on indoor fume assessment and odor level assessment described in Sections 2.3.2 and 2.3.3. According to Figure 5, it is evident that the fume suppression effect and odor level of HVMA using different types of fume suppressants (activated carbon; chemical reaction fume suppressants) were usually better than unmodified asphalt without fume suppressants. Compared with the use of chemical reaction fume suppressants, the addition of activated carbon had a slightly better fume suppression effect, but had a higher odor level. The odor level of HVMA chemical reaction fume suppressants with 15% DRP added was one level higher than that of activated carbon. This difference might be attributed to the composition of chemical reaction inhibitors—zinc castor oleate and vulcanization accelerators. The zinc oxide presented in the vulcanization promoter forms a zinc salt with the help of stearic acid.

This zinc salt was soluble in asphalt, improving the vulcanization efficiency of HVMA, thereby enhancing its high-temperature stability, and indirectly reducing the emission of foul gases such as H_2S and SO_2 during the vulcanization process of the stabilizer. In addition, zinc castor oil effectively eliminated odors through chemical reactions between internal active zinc atoms and the generated sulfur and nitrogen substances. Activated carbon exhibited significant adsorption capacity for asphalt fume through its inherent adsorption principle, visually reducing fume emissions.

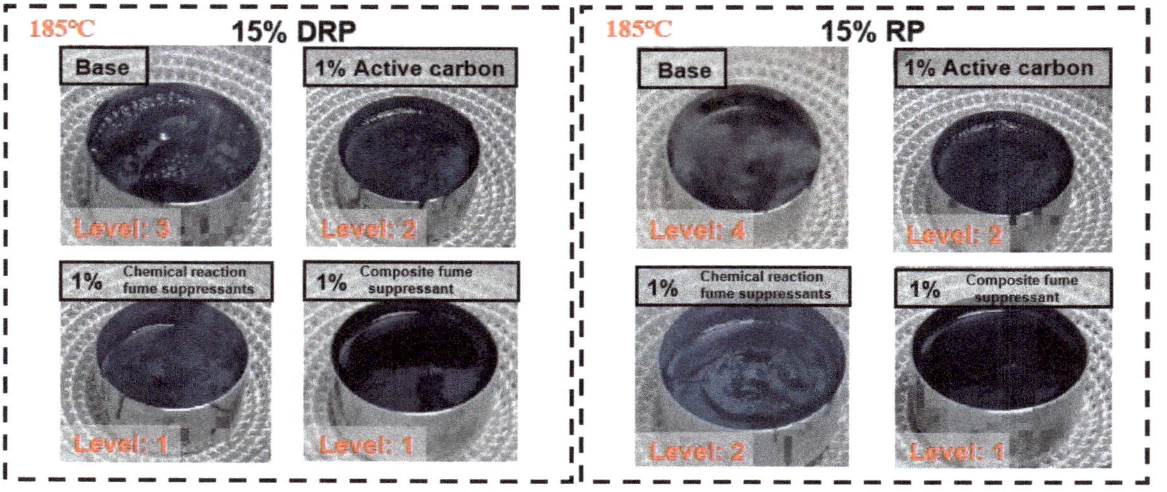

Figure 5. Emission of HVMA fume gas under different fume suppressants.

The emission of fumes from HVMA was significantly improved when composite odor-removing agents were added, compared to modified asphalt with single-component additives. And the odor levels were all level 1, which is superior to the odor level of adding chemical reaction fume suppressants and activated carbon separately. This suggested a synergistic effect between activated carbon and chemical reaction fume suppressant. Therefore, the 1% composite fume suppressant could be considered the optimal choice in this study. Additionally, the fume emission and odor level from high-viscosity modified asphalt (HVMA) with the addition of desulfurization rubber powder (DRP) were lower than ordinary rubber-powder HVMA. This indicated that 15% DRP could be deemed as an effective asphalt addictive in this study. This was because the desulfurized rubber powder was a byproduct of ordinary rubber powder that underwent a desulfurization process, resulting in a significant reduction in sulfur content compared to regular rubber powder. As a result, it helped to prevent the generation of foul gases such as hydrogen sulfide, which arose from the breaking of sulfur bonds in rubber.

3.3. Energy-Saving Effect Evaluation

3.3.1. Rotational Viscosity Test Analysis

As depicted in Figure 6a, the viscosity of all seven types of HVMAs showed a decreasing trend as the temperature increases. HVMA conditioned with VRAs exhibited a more pronounced decrease in viscosity compared to common HVMA. Among these, the most significant viscosity reduction was achieved when adding 2% Sasobit, 2% Sasobit, and 0.6% Evotherm3G + 3% Sasobit to the asphalt separately. This could be attributed to the fact that, at elevated temperatures, the solid particles of Sasobit dissolved completely in the asphalt, leading to the formation of a liquid structural wax film, ultimately resulting in a reduction in asphalt viscosity [34].

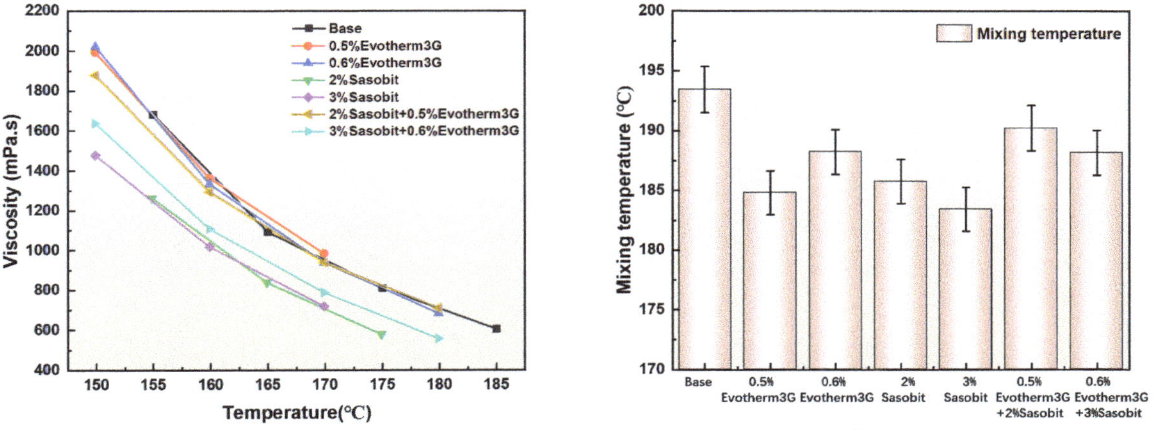

Figure 6. The viscosity temperature curve (**a**) and mixing temperature (**b**) of HVMAs under different VRAs.

Compared to regular HVMA, the addition of Evotherm3G led to a reduction in viscosity, although it still remained higher than that of Sasobit additives. The mixing temperature of different HVMA formulations could be determined by analyzing the viscosity–temperature curve, as depicted in Figure 6b. Notably, the mixing temperature of regular HVMA was the highest, reaching 193 °C, while the mixing temperature of single-component VRAs asphalt was the lowest. The rotational viscosity of HVMA after incorporating Evotherm3G did not differ significantly from that of regular HVMA, but the mixing temperature was lower. This suggested that although the initial addition of Evotherm3G might result in an increase in HVMA viscosity, its viscosity decreased more rapidly as the temperature rises. In the end, there was no substantial difference in the final mixing temperature when compared to HVMA supplemented solely with Sasobit as an additive. Furthermore, when compared to single-component additives, the mixing temperature of the two HVMA groups with composite VRAs was slightly higher. However, the HVMA formulation with 0.6% Evotherm3G + 3% Sasobit did not significantly differ from the HVMA formulation with 0.6% Evotherm3G alone. Hence, this specific combination could be deemed as an appropriate dosage for achieving the desired viscosity reduction in this eco-friendly HVMA formulation.

3.3.2. Evaluation of the Viscosity-Retarding Effect

Based on the previously obtained viscosity–temperature curve, a polynomial was employed to accurately fit the data. Moreover, temperature scanning experiments were conducted to measure the complex viscosity of different high-viscosity asphalt samples within the temperature range of 30 °C to 100 °C (as depicted in Figure 7a). The viscosity-retarding performance of HVMA was evaluated using the on the integral area method (as shown in Table 1). The areas enclosed by each HVMA sample and the coordinate axes conditions were separately calculated under higher temperatures and lower temperatures. The outcomes of the integral area calculations for seven groups of HVMA samples are presented in Figure 7. The results were accurate to one decimal place with Scientific notation (Table 2).

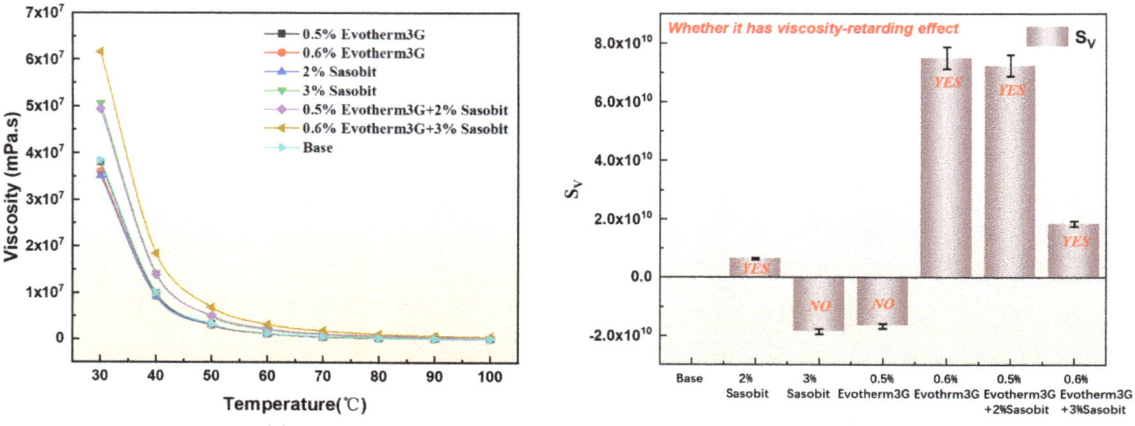

Figure 7. The low temperature complex viscosity curve (**a**) and viscosity-retarding effect results of HVMAs under different VRAs (**b**).

Table 2. Results of the viscosity-retarding effect.

Addictive	S_1	S_2	S_G	S_D	Retarding Viscosity Effect Presence (Yes/No)	Sv
Base	5.9×10^{10}	4.9×10^4	0	0	--	--
0.5% Evotherm3G	4.2×10^{10}	6.5×10^4	1.7×10^{10}	-1.6×10^4	No	--
0.6% Evotherm3G	1.3×10^{11}	4.2×10^4	-7.5×10^{10}	7.0×10^3	Yes	7.5×10^{10}
2% Sasobit	6.5×10^{10}	3.9×10^4	-6.4×10^9	1.0×10^4	Yes	6.4×10^9
3% Sasobit	4.0×10^{10}	4.2×10^4	1.9×10^{10}	6.6×10^3	No	--
2% Sasobit + 0.5% Evotherm3G	1.3×10^{11}	4.2×10^4	-7.2×10^{10}	7.1×10^3	Yes	7.2×10^{10}
3% Sasobit + 0.6% Evotherm3G	7.7×10^{10}	2.4×10^4	-1.8×10^{10}	2.5×10^4	Yes	1.8×10^{10}

According to the data in Table 1, it was evident that the S_G value of HVMA containing 0.5% Evotherm3G and 3% Sasobit was greater than 0, indicating that these two HVMAs did not achieve any low-temperature viscosity increasing effect. It could be seen that the addition of single-component Sasobit and Evotherm3G not only reduced high-temperature viscosity, but also led to a decrease in low-temperature viscosity. This was because when Sasobit and Evotherm3G were added separately, the long-chain hydrocarbons in Sasobit melted and formed an organic wax film inside HVMA, while Evotherm3G generated a structural water film inside the asphalt during the addition process, thereby reducing friction between asphalt molecules and further reducing the viscosity of the mixture [34].

The composition of HVMA with 0.6% Evotherm3G, 2% Sasobit, 0.5% Evotherm3G + 2% Sasobit, and 0.6% Evotherm3G + 3% Sasobit satisfied the criteria of S_D values less than 0 and S_G values greater than 0. By comparing the Sv values of these three methods, the ranking of the viscous effects could be established as follows: 0.6% Evotherm3G > 0.5% Evotherm3G + 2% Sasobit > 0.6% Evotherm3G + 3% Sasobit > 2% Sasobit (as indicated in Figure 7). This analysis revealed that Evotherm3G played a primary role in promoting slow adhesion in HVMA, with 2% Sasobit being the next most effective additive.

3.4. Rheological Performance Characterization

3.4.1. Shearing Rate Sweep Test Analysis

The flow curve of the eco-friendly asphalt is shown in Figure 8. When the shear stress surpassed the yield stress, the asphalt underwent gradual orientation, extension, deformation, and dispersion in the direction of flow, which exhibited a Newtonian flow state known as the Newtonian flow region. Within this region, the viscosity approaches a constant value, which was referred to as zero shear viscosity (ZSV). A higher ZSV for asphalt materials signified a greater resistance to long-term deformation under load, improved asphalt film on aggregate surfaces, enhanced adhesion, increased water stability, and heightened fatigue resistance.

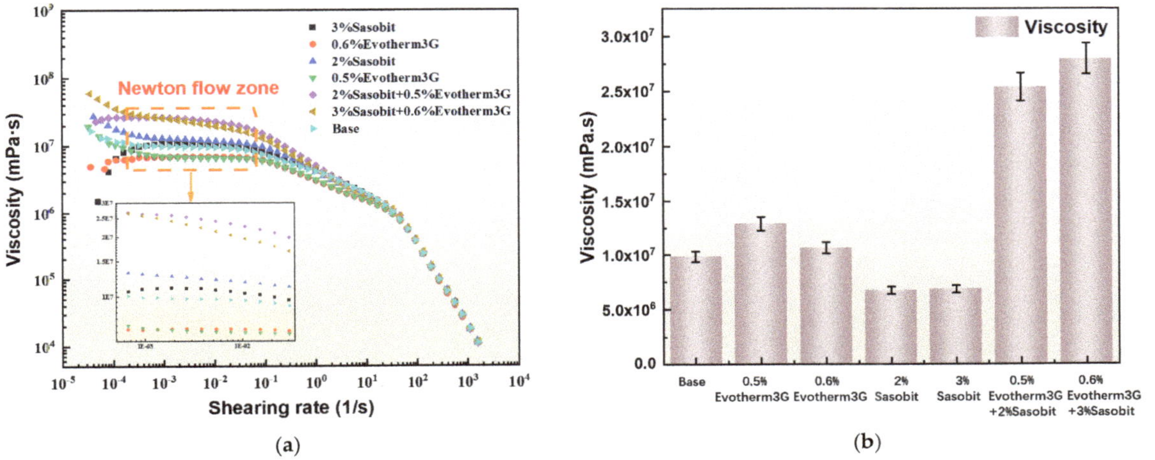

Figure 8. Scatter plots (**a**) and bar plots (**b**) of zero shear viscosity of HVMAs under different VRAs.

Figure 8 depicts that the ZSV of the modified asphalt, containing varying amounts of single-component Sasobit, surpassed that of the blank control group. This observation suggested that the incorporation of Sasobit enhanced the long-term load deformation resistance of the modified asphalt to some extent. It is worth noting that compared to HVMA with 2% Sasobit added, HVMA with 3% Sasobit added had a slightly lower zero shear viscosity. This could be attributed to the organic wax-based nature of Sasobit, which upon increasing dosage, formed an organic wax film within the asphalt. Consequently, this film acted as a lubricant, thereby slightly reducing the viscosity of the HVMA [34]. Furthermore, the ZSV of the HVMA, incorporating single-component Evotherm3G, was lower than that of the blank control group. This discrepancy might be attributed to the formation of a structural water film due to the presence of the surfactant, which subsequently led to a decrease in asphalt viscosity [35]. While the decrease in ZSV resulted in a reduced deformation resistance of the HVMA after adding Evotherm3G, it also implied that the HVMA could more easily attain the required viscosity for mixing, effectively reducing the mixing temperature. The difference in ZSV between high and low contents of Evotherm3G was not significant, indicating that the viscosity decrease in the HVMA with increasing content of Evotherm3G was not pronounced, and it still maintained the deformation resistance at low content.

The ZSV of the composite VRAs Sasobit/Evotherm3G at high and low contents was significantly higher compared to the addition of either of the two single-component additives or the blank control group. This indicated that the combination of these two additives could significantly enhance the deformation resistance of the HVMA by possibly operating through different mechanisms. These distinct mechanisms of action between Sasobit and Evotherm3G resulted in varying effects of the two additives on the viscosity of the mixture, consequently leading to an increase in solute concentration within the mixture system [21,24]. As a result, the ZSV was elevated. Although the ZSV of the high-content composite additive was slightly lower than that of the low-content additive, the overall difference between the two was minimal.

3.4.2. Multiple Stress Creep Recovery (MSCR) Analysis

The Multiple Stress Creep and Recovery (MSCR) test was widely considered as the preferred method for characterizing the high-temperature performance of asphalt binders. This test results accurately reflected the elastic recovery properties of HVMA, as shown in Figure 9. Specifically, the parameter $R_{3.2}$ provided a direct reflection of the binder's elastic recovery, allowing for the evaluation of its elastic response and dependency. Furthermore, $J_{3.2}$ represented the non-recoverable creep compliance of the binder, serving as an indicator of the asphalt binder's resistance to permanent deformation under repeated loading at high temperatures. In this study, the recovery rates $R_{0.1}$, $R_{3.2}$, and non-recoverable creep compliance $J_{nr0.1}$, $J_{nr3.2}$ of various HVMA types were plotted at stress levels of 1.0 kPa and 3.2 kPa, while considering a range of temperatures, respectively.

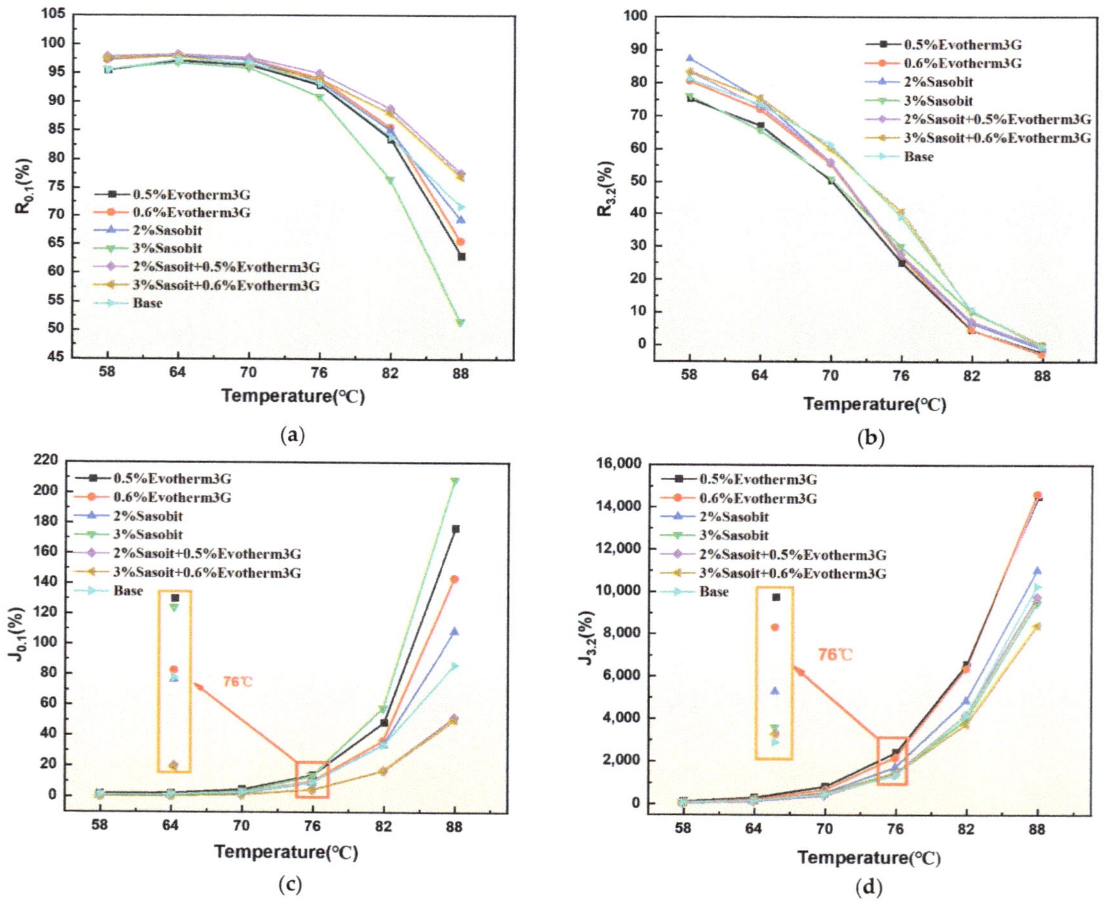

Figure 9. The $R_{0.1}$ (**a**), $R_{3.2}$ (**b**), and $J_{0.1}$ (**c**), $J_{3.2}$ (**d**) of HVMAs under different VRAs.

Figure 9a illustrates the relationship between $R_{0.1}$ and experimental temperature for seven asphalt samples that were subjected to a stress of 0.1 kPa. The horizontal axis represents the experimental temperature, while the vertical axis represents the elastic recovery rate. Initially, the creep recovery rate of the blank control group decreased slightly as the temperature rose. However, when the temperature reached around 75 °C, the creep recovery rate sharply decreased. The $R_{0.1}$ of the other six modified HVMA samples also decreased with increasing temperature. Among them, the sample with only one type of VRA showed a much larger decrease compared to the blank control group, especially the sample with 3% Sasobit, which exhibited the greatest decrease. This decrease could be attributed to the degradation of polymers in HVMA at high temperatures, leading to a significant decrease in creep recovery rate. On the other hand, samples with a combination of two rejuvenators showed higher creep recovery rates than the blank control group, and the rate of decrease was slower compared to the blank control group. This was because the addition of both can improve the cohesion between asphalt molecules, forming a more compact structure, thereby improving the elastic recovery performance of asphalt.

Figure 9b depicts the relationship between the $R_{3.2}$ and experimental temperature of seven asphalt samples under a stress of 3.2 kPa. The creep recovery rate of all seven groups of samples continuously decreased with the increase in temperature. In the initial stages, the decline rate of HVMA was relatively slow as the temperature rises. When the temperature reached around 75 °C, the creep recovery rate sharply decreased. Except for the HVMA with 0.5% Sasobit and 3% Sasobit added, the $R_{3.2}$ of the six groups of samples with VRA was not significantly different from the blank control group. Among them, the $R_{3.2}$ of HVMA after 0.6% Evotherm3G + 3% Sasobit was slightly higher than that of ordinary HVMA. This indicated that the elastic recovery performance of HVMA had been enhanced to a certain extent after the addition of composite VRA. Figure 9c illustrates the relationship between $J_{nr0.1}$ and experimental temperature for seven asphalt samples subjected to a stress of 0.1 kPa. Prior to reaching a temperature of 76 °C, the increase in $J_{nr0.1}$ was relatively gradual compared to the control group. The differences in $J_{nr0.1}$ among the seven groups of samples were not statistically significant, suggesting that HVMA demonstrated similar resistance to rutting within this temperature range. However, after the temperature surpassed 76 °C, $J_{nr0.1}$ displayed a sudden increase, and the samples with a single VRA exhibited higher $J_{nr0.1}$ values compared to those with common HVMA. This observed variance could be attributed to the degradation of polymers in HVMA at high temperatures, which correlated with the trend observed in $R_{0.1}$. It is noteworthy that the introduction of combination VRAs demonstrated robust resistance to rutting. Specifically, under high-temperature conditions, the upward trend of $J_{nr0.1}$ was noticeably slower and considerably lower compared to common HVMA. These findings indicated that HVMA with composite VRA had excellent resistance to rutting under a stress of 0.1 kPa in high-temperature conditions.

Figure 9d depicts the relationship between $J_{nr3.2}$ and experimental temperature for seven asphalt samples subjected to a stress of 3.2 kPa. Prior to reaching a temperature of 76 °C, the $J_{nr3.2}$ values of the seven HVMAs were found to be higher than that of common HVMA. This suggested that HVMAs exhibited weaker resistance to rutting than common HVMA within the mid-temperature range, although the difference was not statistically significant. However, as the temperature exceeded 76 °C, the $J_{nr3.2}$ values of the three groups with 3%Sasobit, 0.5% Evotherm3G + 2% Sasobit, and 0.6% Evotherm3G + 3% Sasobit VRAs decreased compared to that of common HVMA. Among these, the reduction was most pronounced for the 0.6% Evotherm3G + 3% Sasobit group. These findings indicated that under a stress of 3.2 kPa, high-dosage combination VRAs exhibited excellent resistance to rutting in high-temperature conditions for HVMA.

3.4.3. Analysis of Fatigue Life Prediction from LAS Test

Figure 10 illustrates the fatigue life values of different HVMAs under different strain conditions (2.5%, 5%). For a strain condition of 2.5%, except for 0.6% Evotherm3G + 3% Sasobit, the fatigue life of HVMA with other additives exhibited higher values compared to ordinary HVMA. It is worth noting that HVMA with the addition of 2% Sasobit,

3% Sasobit, and 0.5% Evotherm3G + 2% Sasobit, respectively, had a higher fatigue life, with an Nf value exceeding 40,000. The reason for this notable improvement could be attributed to the addition of Sasobit, which increased the zero shear viscosity of HVMA. This, in turn, ensured that the asphalt maintains high stability, effectively resisting road reactions and preventing oxidation damage even under high temperature conditions. As a result, the durability and fatigue life of the asphalt were substantially improved. While an increase in the dosage of single-component Sasobit marginally improved the fatigue life of HVMA, the enhancement was not particularly significant. Remarkably, the N_f value of 0.5% Evotherm3G + 2% Sasobit reached as high as 45,300, which stood as the highest fatigue life value among the seven groups of HVMA.

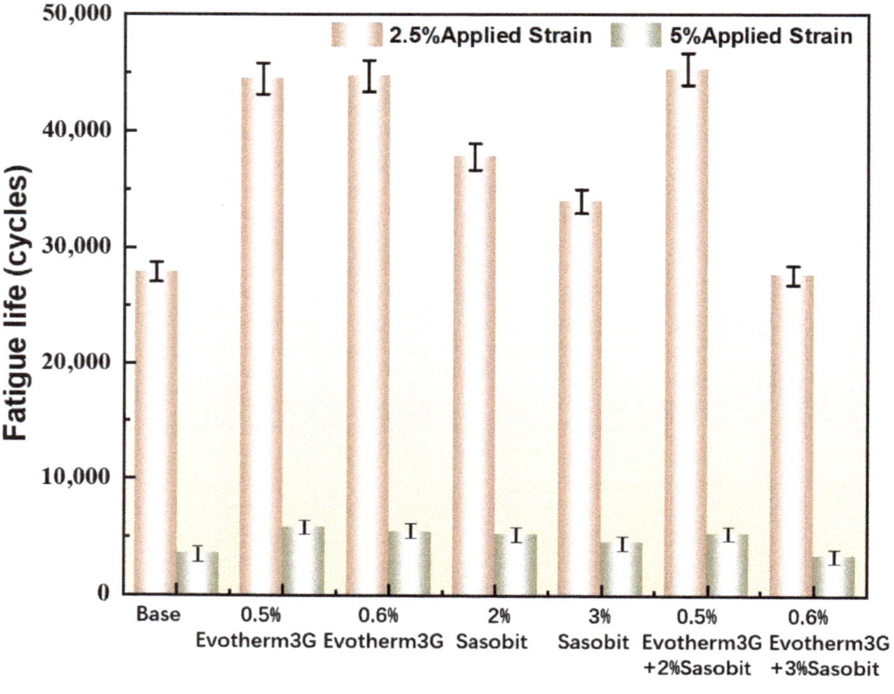

Figure 10. Fatigue life for tested HVMAs.

The fatigue life of HVMA with the addition of the VRA Evotherm3G was slightly lower compared to Sasobit, but still higher than that of common HVMA. As the dosage of Evotherm3G increased, the fatigue life began to decrease slightly. Among them, the fatigue life of HVMA after adding 0.6% Evotherm3G + 3% Sasobit was 27,658, which had the lowest fatigue life among the seven groups of HVMA, but still met the requirements for road performance. It could be inferred that increasing the dosage of Sasobit further enhanced the fatigue life of the HVMA, while increasing the dosage of Evotherm3G slightly decreased the fatigue life of the HVMA. This could be attributed to a series of unknown reactions between Evotherm3G and Sasobit, resulting in a decrease in fatigue life. However, further research is needed to provide additional evidence. Considering various performance indicators of the HVMA, adding 0.5% Evotherm3G + 2% Sasobit and 0.6% Evotherm3G + 3% Sasobit remained the optimal choice for eco-friendly HVMA in this screening process.

3.4.4. Frequency Sweep Test Results

Figure 11 illustrates the relationship between the complex modulus and phase angle of different HVMA samples with varying frequencies. A comparison of the seven images revealed that, during the high-temperature stage at 85 °C, the complex modulus of the blended HVMA with Sasobit/Evotherm3G demonstrated a significantly higher value compared to the other five groups. Specifically, the main curve of the complex modulus for the sample containing 0.6% Evotherm3G + 3% Sasobit exhibited a particularly noticeable increase when compared to the remaining six groups. This finding suggested that the addition of both Saosbit and Evotherm3G enhanced the elasticity of HVMA, thereby improving its high-temperature performance for road applications. This improvement could be attributed to the ability of Sasobit and Evotherm3G to enhance the fluidity and plasticity of asphalt by altering its viscosity and flow characteristics, resulting in a higher deformation capacity and improved crack resistance. Consequently, the high-temperature stability of HVMA was significantly enhanced [21].

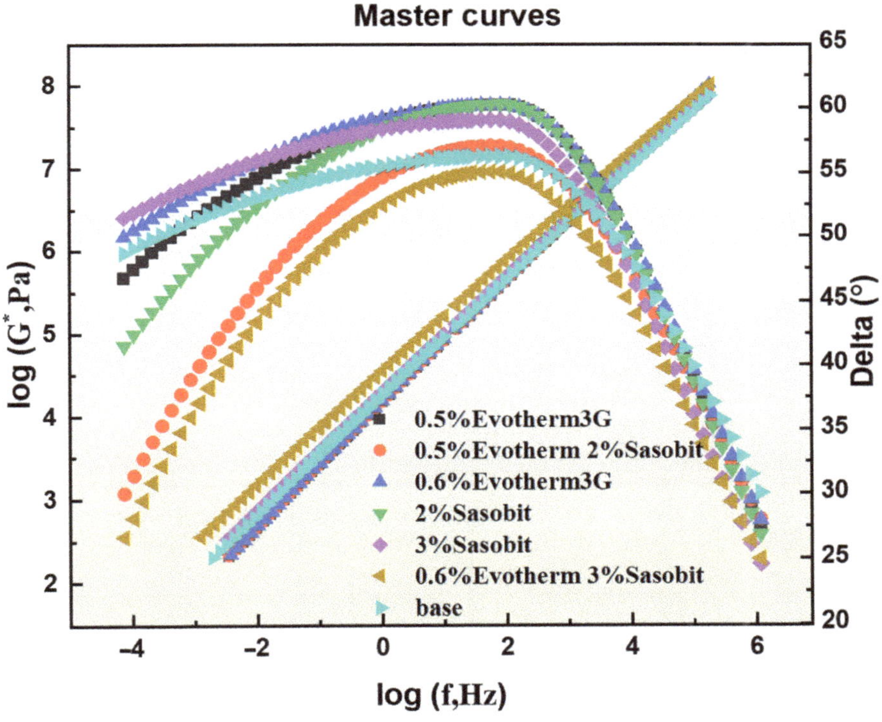

Figure 11. Moduli and phase angles of HVMAs under different VRAs.

Figure 11 illustrates the relationship between the complex modulus and phase angle main curves as a function of frequency. Upon observation, it was evident that the main curves of the HVMA samples displayed an increasing trend in the complex modulus with a higher frequency. Conversely, the main curves of the phase angle initially increased and then decreased. The low-frequency region represented the high-temperature characteristics of HVMA, whereas the high-frequency region represented its low-temperature behavior. Consequently, an elevation in temperature corresponded to a decrease in the complex modulus of HVMA. When comparing the main curves of the complex modulus for various VRAs within the HVMA, it became apparent that, with the exception of the sample containing 0.6% Evotherm3G + 3% Sasobit, the differences in complex modulus in the low-frequency region were minimal; this indicated that the remaining HVMA groups exhibited similar properties to common HVMA. However, due to the limitations imposed by high frequencies (low temperatures) on the movement of asphalt molecules in HVMA, the impact of VRA type and dosage on the complex modulus gradually weakened. Consequently, all the modulus curves converged to a single intersection point at high frequencies.

Upon analyzing the main curves of the phase angle, it was discerned that all seven groups of HVMA exhibited an increase followed by a subsequent decrease in the phase angle. This decrease in the high-frequency region was primarily attributed to the volatilization and transformation of light components within the asphalt. These transformations influenced the elastic response of the asphalt, ultimately leading to an enhancement in its overall elasticity. Conversely, the dominant factor affecting the phase angle was the degradation behavior of the polymer network within the HVMA. The degraded network structure was unable to provide additional elasticity to the HVMA, resulting in an increase in the phase angle. Notably, it was observed that as the dosage of the VRA increased, the polymer content within the HVMA similarly rose, thereby causing a more significant increase in the phase angle.

In high-temperature environments (specifically in the low-frequency region), the inclusion of a higher dosage of Sasobit/Evotherm3G resulted in an amplified degree of polymer crosslinking. Consequently, this led to a more noticeable decrease in the main curve of the phase angle. Conversely, the phase angle experienced minimal changes in the high-frequency region. This could be attributed to the fact that the viscoelastic properties within the high-frequency region were predominantly governed by the asphalt phase itself. Although the VRAs underwent reactions with certain functional groups within the asphalt, these reactions were relatively limited in comparison to their interactions with the polymers. As a result, the addition of the VRA did not exert a significantly impact the viscoelastic proportion of the asphalt phase.

4. Conclusions and Outlook

This study evaluated the fume suppression effect and odor level of different fume suppressants on high-viscosity modified asphalt (HVMA) through indoor fume assessment firstly. Based on this, the influence of different viscosity-retarding agents (VRAs) on the mixing temperature, viscosity-retarding, and rheological properties of HVMA were investigated. The main conclusions of this research are as follows:

(1) According to the fume-suppressing effect and odor-level evaluation results, the fume suppression effect and odor level effect of the DRP-modified HVMA was superior to that of rubber-modified HVMA when different types of fume suppressants were added. Furthermore, the ranking in terms of fume suppression effects was: composite fume agent > activated carbon > chemical reaction fume suppressant. And the odor level of adding composite fume suppressants sharply decreased to level one, far lower than the level three of ordinary base HVMA. Ultimately, it was determined that the 1% composite fume suppressants was selected as the optimal fume suppression plan.

(2) The mixing temperatures of different HVMAs were determined based on their viscosity–temperature curves. It was found that the addition of VRA could reduce the mixing temperature of HVMA to 4–10 °C. A novel method based on the integrated area of viscosity–temperature curve was proposed in this research to evaluate the viscosity-retarding effect of different VRAs. Finally, the HVMA containing with 0.6% Evotherm3G + 3% Sasobit VRAs had the optimal viscosity-retarding effect and lower mixing temperature, and was determined as the eco-friendly HVMA.

(3) The morphological characteristics and rheological properties of various HVMAs were evaluated. After Sasobit/Evotherm3G was added, the network structure of the HVMA was not as apparent as that of the HVMA with a single component added, but a continuous spatial network structure still existed. The research results of ZSV, MSCR, and LAS indicated that the selected eco-friendly HVMA exhibited better high-temperature resistance compared to common HVMAs, and the fatigue life also met the performance requirements.

Overall, the focus of this study was to study an eco-friendly HVMA and propose a new evaluation method to contribute to the sustainable development of road materials.

However, this study solely illustrated the inhibitory effects of composite fume suppressants on visible fume and foul odor gases in asphalt fumes. In future research, our attention would shift toward exploring the impact of composite fume suppressants on the invisible components of asphalt fume, as well as on enhancing the evaluation method for the viscosity-retarding effect. Furthermore, we intended to broaden the scope of our investigation to encompass the fume release of asphalt mixtures during the paving process, providing technical support for the development of high-performance and eco-friendly asphalt. In addition, the material combinations selected in this study and the evaluation methods for fume suppression still needed to be further optimized.

Author Contributions: Methodology, G.S.; Validation, X.J.; Formal analysis, R.Z.; Investigation, W.N. and C.W.; Resources, R.Z.; Data curation, K.Q.; Writing—original draft, W.N.; Writing—review & editing, G.S. and X.J.; Visualization, K.Q.; Supervision, G.S. All authors have read and agreed to the published version of the manuscript.

Funding: The work described in this paper was supported by the National Natural Science Foundation of China, China (No. 52108390); Beijing Municipal Education Commission, China (No. KM202210005002); Key Laboratory of Road and Traffic Engineering of the Ministry of Education, Tongji University, China (No. K202205); Open Funding of Engineering Research Center of Ministry of Education for Traffic Pavement Materials, Chang'an University, China (No. 300102312503).

Institutional Review Board Statement: Not applicable.

Informed Consent Statement: Not applicable.

Data Availability Statement: Not applicable.

Conflicts of Interest: The authors declare no conflict of interest.

References

1. Yang, X.; Wang, G.; Rong, H.; Meng, Y.; Liu, X.; Liu, Y.; Peng, C. Review of fume-generation mechanism, test methods, and fume suppressants of asphalt materials. *J. Clean. Prod.* **2022**, *347*, 131240. [CrossRef]
2. Cao, L.; Yang, C.; Li, A.; Wang, P.; Zhang, Y.; Dong, Z. Flue gas composition of waste rubber modified asphalt (WRMA) and effect of deodorants on hazardous constituents and WRMA. *J. Hazard. Mater.* **2021**, *403*, 123814. [CrossRef] [PubMed]
3. Tang, N.; Zhang, Z.; Dong, R.; Zhu, H.; Huang, W. Emission behavior of crumb rubber modified asphalt in the production process. *J. Clean. Prod.* **2022**, *340*, 130850. [CrossRef]
4. Li, H.; Jia, M.; Zhang, X.; Wang, Z.; Liu, Y.; Yang, J.; Yang, B.; Sun, Y.; Wang, H.; Ma, H. Laboratory investigation on fumes generated by different modified asphalt binders. *Transp. Res. Part D Transp. Environ.* **2023**, *121*, 103828. [CrossRef]
5. Cui, P.; Schito, G.; Cui, Q. VOC emissions from asphalt pavement and health risks to construction workers. *J. Clean. Prod.* **2020**, *244*, 118757. [CrossRef]
6. Xiu, M.; Wang, X.; Morawska, L.; Pass, D.; Beecroft, A.; Mueller, J.F.; Thai, P. Emissions of particulate matters, volatile organic compounds and polycyclic aromatic hydrocarbons from warm and hot asphalt mixes. *J. Clean. Prod.* **2020**, *275*, 123094. [CrossRef]
7. Clark, C.R.; Burnett, D.M.; Parker, C.M.; Arp, E.W.; Swanson, M.S.; Minsavage, G.D.; Kriech, A.J.; Osborn, L.V.; Freeman, J.J.; Barter, R.A.; et al. Asphalt fume dermal carcinogenicity potential: I. dermal carcinogenicity evaluation of asphalt (bitumen) fume condensates. *Regul. Toxicol. Pharmacol.* **2011**, *61*, 9–16. [CrossRef]
8. Bao, J.; Wang, H.; Cong, Y.; Wang, X.; Liu, S.; Wang, J.; Chen, M. Research progress of road asphalt smoke suppressant. *Contemp. Chem. Ind.* **2020**, *49*, 988–992. (In Chinese)
9. Zhang, F.; Yu, J.; Wu, S. Effect of ageing on rheological properties of storage-stable SBS/sulfur-modified asphalts. *J. Hazard. Mater.* **2010**, *182*, 507–517. [CrossRef]
10. Autelitano, F.; Garilli, E.; Pinalli, R.; Montepara, A.; Giuliani, F. The odour fingerprint of bitumen. *Road Mater. Pavement Des.* **2017**, *18*, 178–188. [CrossRef]
11. Wang, M.; Li, P.; Nian, T.; Mao, Y. An overview of studies on the hazards, component analysis and suppression of fumes in asphalt and asphalt mixtures. *Constr. Build. Mater.* **2021**, *289*, 123185. [CrossRef]
12. Mo, S.; Wang, Y.; Xiong, F.; Ai, C. Effects of asphalt source and mixing temperature on the generated asphalt fumes. *J. Hazard. Mater.* **2019**, *371*, 342–351. [CrossRef] [PubMed]
13. Yang, X.; You, Z.; Perram, D.; Hand, D.; Ahmed, Z.; Wei, W.; Luo, S. Emission analysis of recycled tire rubber Modified Asphalt in hot and warm mix conditions. *J. Hazard Mater.* **2019**, *365*, 942–951. [CrossRef] [PubMed]
14. Liang, M.; Ren, S.; Fan, W.; Wang, H.; Cui, W.; Zhao, P. Characterization of fume composition and rheological properties of asphalt with crumb rubber activated by microwave and TOR. *Constr. Build. Mater.* **2017**, *154*, 310–322. [CrossRef]

15. Wang, F.; Li, N.; Hoff, I.; Wu, S.; Li, J.; Barbieri, D.M.; Zhang, L. Characteristics of VOCs generated during production and construction of an asphalt pavement. *Transp. Res. Part D Transp. Environ.* **2020**, *87*, 102517. [CrossRef]
16. Gągol, M.; Boczkaj, G.; Haponiuk, J.; Formela, K. Investigation of volatile low molecular weight compounds formed during continuous reclaiming of ground tire rubber. *Polym. Degrad. Stabil.* **2015**, *119*, 113–120. [CrossRef]
17. Cui, P.Q.; Wu, S.P.; Xiao, Y.; Zhang, H.H. Experimental study on the reduction of fumes emissions in asphalt by different additives. *Mater. Res. Innov.* **2015**, *19*, S158–S161. [CrossRef]
18. Lv, Y.; Wu, S.; Li, N.; Cui, P.; Wang, H.; Amirkhanian, S.; Zhao, Z. Performance and VOCs emission inhibition of environmentally friendly rubber modified asphalt with UiO-66 MOFs. *J. Clean. Prod.* **2023**, *385*, 135633. [CrossRef]
19. Li, H.; Feng, Z.; Liu, H.; Ahmed, A.T.; Zhang, M.; Zhao, G.; Guo, P.; Sheng, Y. Performance and inorganic fume emission reduction of desulfurized rubber powder/styrene–butadiene–styrene composite modified asphalt and its mixture. *J. Clean. Prod.* **2022**, *364*, 132690. [CrossRef]
20. Cheraghian, G.; Falchetto, A.C.; You, Z.; Chen, S.; Kim, Y.S.; Westerhoff, J.; Moon, K.H.; Wistuba, M.P. Warm mix asphalt technology: An up to date review. *J. Clean. Prod.* **2020**, *268*, 122128. [CrossRef]
21. Jamshidi, A.; Hamzah, M.O.; You, Z. Performance of Warm Mix Asphalt containing Sasobit®: State-of-the-art. *Constr. Build. Mater.* **2013**, *38*, 530–553. [CrossRef]
22. Wan, Z.; Zheng, B.; Xie, X.; Yang, J.; Zhou, H.; Wan, R.S.; Wang, J.; Wang, J.; Luo, W.; Xue, R.; et al. Preparation method and performance test of Evotherm pre-wet treatment aluminum hydroxide type warm-mixed flame-retardant asphalt. *Constr. Build. Mater.* **2020**, *262*, 120618. [CrossRef]
23. Valdes-Vidal, G.; Calabi-Floody, A.; Sanchez-Alonso, E. Performance evaluation of warm mix asphalt involving natural zeolite and reclaimed asphalt pavement (RAP) for sustainable pavement construction. *Constr. Build. Mater.* **2018**, *174*, 576–585. [CrossRef]
24. Sukhija, M.; Saboo, N. A comprehensive review of warm mix asphalt mixtures-laboratory to field. *Constr. Build. Mater.* **2021**, *274*, 121781. [CrossRef]
25. Duan, Y. Research of the Features and Mechanism of Retardable Viscosity's Asphalt in Cold Region. Master's Thesis, Chang'an University, Xi'an, China, 2013. (In Chinese).
26. *AASHTO T316*; Standard Method of Test for Viscosity Determination of Asphalt Binder Using Rotational Viscometer. AASHTO: Washington, DC, USA, 2010.
27. Sun, G.; Zhu, X.; Zhang, Q.; Yan, C.; Ning, W.; Wang, T. Oxidation and polymer degradation characteristics of high viscosity modified asphalts under various aging environments. *Sci. Total Environ.* **2022**, *813*, 152601. [CrossRef] [PubMed]
28. Biro, S.; Gandhi, T.; Amirkhanian, S. Determination of zero shear viscosity of warm asphalt binders. *Constr. Build. Mater.* **2009**, *23*, 2080–2086. [CrossRef]
29. *AASHTO TP 70-1*; Multiple Stress Creep Recovery (MSCR) Test of Asphalt Binder Using a Dynamic Shear Rheometer (DSR). AASHTO: Washington, DC, USA, 2009.
30. Huang, W.; Tang, N. Characterizing SBS modified asphalt with sulfur using multiple stress creep recovery test. *Constr. Build. Mater.* **2015**, *93*, 514–521. [CrossRef]
31. *AASHTO TP101-12*; Standard Method of Test for Estimating Fatigue Tolerance of Asphalt Binders Using the Linear Amplitude Sweep. AASHTO: Washington, DC, USA, 2012.
32. Asgharzadeh, S.; Tabatabaee, N.; Naderi, K.; Partl, M. An empirical model for modified bituminous binder master curves. *Mater. Struct.* **2012**, *46*, 1459–1471. [CrossRef]
33. Asgharzadeh, S.; Tabatabaee, N.; Naderi, K.; Partl, M. Evaluation of rheological mastercurve models for bituminous binders. *Mater. Struct.* **2013**, *48*, 393–406. [CrossRef]
34. Ingrassia, L.P.; Lu, X.; Canestrari, F.; Ferrotti, G. Tribological characterization of bituminous binders with Warm Mix Asphalt additives. *Constr. Build. Mater.* **2018**, *172*, 309–318. [CrossRef]
35. Kassem, E.; Cucalon, L.G.; Masad, E.; Little, D. Effect of Warm Mix Additives on the Interfacial Bonding Characteristics of Asphalt Binders. *Int. J. Pavement. Eng.* **2018**, *19*, 1111–1124. [CrossRef]

coatings

MDPI

Article

Study on the Performance and Modification Mechanism of Polyphosphoric Acid (PPA)/Styrene–Butadiene–Styrene (SBS) Composite Modified Asphalt

Xiangjie Niu [1], Yuanzhao Chen [1,2,*], Zhenxia Li [1], Tengteng Guo [1], Jing Wang [1] and Lihui Jin [1]

1 School of Civil Engineering and Communication, North China University of Water Resources and Electric Power, Zhengzhou 450045, China; xiangjieniu@163.com (X.N.); zhenxiali2009@ncwu.edu.cn (Z.L.); guotth@ncwu.edu.cn (T.G.); wj@ncwu.edu.cn (J.W.); lihuijin2023@163.com (L.J.)

2 Henan Province Engineering Technology Research Center of Environment Friendly and High-Performance Pavement Materials, Zhengzhou 450045, China

* Correspondence: cyz740513@ncwu.edu.cn

Abstract: In order to address the high preparation cost of styrene–butadiene–styrene block copolymer (SBS) modified asphalt, four kinds of polyphosphoric acid (PPA) content (0%, 0.5%, 0.75%, and 1% PPA by weight of the matrix asphalt) were selected to prepare composite modified asphalt with better high-temperature performance. The physical properties of composite modified asphalt were evaluated by conventional performance tests. The rheological properties of composite modified asphalt were evaluated by dynamic shear rheometer (DSR) test and bending beam rheometer (BBR) test. The synergistic modification mechanism of PPA and SBS was revealed by the Fourier transform infrared spectroscopy test. The results show that with the increase of PPA content, the penetration of PPA/SBS composite modified asphalt is reduced by 20.92%, 25.07% and 28.94%, respectively, compared with matrix asphalt, and the softening point is increased by 5.46%, 22.69% and 34.03%, respectively. In addition, PPA can improve the thermal oxidative aging resistance of asphalt. PPA can improve the shear resistance, high-temperature performance and temperature sensitivity of asphalt. At 82 °C, compared with SBS modified asphalt, the phase angle of PPA/SBS composite modified asphalt can be decreased by 8.63%, 13.23% and 19.24%, respectively, and $G^*/\sin\delta$ can be increased by 41.97%, 67.62% and 70.97%, respectively. SBS mainly exists in asphalt in the form of physical blending, and PPA has a new chemical reaction with asphalt, which increases the macromolecules and chain hydrocarbon components in asphalt, and the macroscopic performance is the improvement of high-temperature performance of asphalt. However, PPA has a negative effect on the low-temperature performance of the SBS modified asphalt.

Keywords: polyphosphoric acid; modified asphalt; rheological properties; zero shear viscosity; temperature sensitivity; microscopic mechanism

Citation: Niu, X.; Chen, Y.; Li, Z.; Guo, T.; Wang, J.; Jin, L. Study on the Performance and Modification Mechanism of Polyphosphoric Acid (PPA)/Styrene–Butadiene–Styrene (SBS) Composite Modified Asphalt. *Coatings* **2023**, *13*, 2003. https://doi.org/10.3390/coatings13122003

Academic Editor: Andrea Simone

Received: 26 October 2023
Revised: 14 November 2023
Accepted: 22 November 2023
Published: 25 November 2023

1. Introduction

Asphalt pavement is widely used in road construction because of its comfortable driving performance and low noise. However, in the service process of asphalt pavement, it is susceptible to rutting, cracks and potholes due to the coupling of climatic change, load, light and other factors, which seriously affect the service life of asphalt pavement [1–3]. As a by-product of crude oil processing, the extensive use of asphalt will inevitably lead to frequent exploitation of crude oil, resulting in environmental pollution and energy shortage, which is contrary to current global green initiatives and concepts related to environmental protection and sustainable development [4–6].

Researchers usually use styrene–butadiene–styrene block copolymer (SBS), styrene butadiene rubber (SBR), polyethylene (PE) and other polymers to modify asphalt, which is a common method to solve the deterioration of pavement materials [7–10]. Among these modifiers, SBS modifier is widely used because it has the advantage of significantly improving the high- and low-temperature performance of matrix asphalt, but its compatibility with matrix asphalt is poor, and it is prone to segregation during transportation. In the actual use process, there are problems such as road performance degradation and weak anti-aging ability, which in turn lead to the reduction of pavement service life and the increase of life cycle construction cost [11–13]. Polyphosphoric acid (PPA) is an inorganic acid modifier formed by the polymerization of various phosphoric acids, but its acidity is less than that of sulfuric acid and nitric acid. It is a colorless, transparent and viscous liquid at room temperature, and the viscosity is positively correlated with the content of phosphoric acid (H_3PO_4). It plays a corresponding chemical role as a different functional admixture in organic synthesis [14]. PPA has the characteristics of easy access and low cost, so it has quickly attracted the attention of researchers. The physical properties, high-temperature properties and anti-aging properties of PPA modified matrix asphalt were studied [15–17]. For example, Ramayya et al. [18] found that PPA significantly increased the proportion of asphaltene to improve the high-temperature performance of asphalt by studying the rheological properties of PPA modified asphalt in the middle and high temperature range. Xiao F et al. [19] analyzed the rotational viscosity, anti-rutting factor, phase angle and failure temperature of polyphosphate modified asphalt and found that the rotational viscosity and anti-rutting factor of polyphosphate modified asphalt showed a gradually increasing trend with the increase of PPA content. Peng Liang et al. [20] found that polyphosphoric acid increased the viscosity and elasticity of modified asphalt at high-temperatures by gelation. Zhang et al. [21] used chemical composition analysis and atomic force microscopy to observe that the proportion and dispersion of asphaltenes are directly related to the content of PPA. With the continuous increase of its content, the proportion of asphaltenes increases and the dispersion is more uniform, so the physical properties and rheological properties of asphalt can be improved.

Although polyphosphoric acid can improve the high-temperature performance and aging resistance of asphalt, it will also weaken the low-temperature performance of asphalt to a certain extent [22,23]. For example, Jun Liu et al. [24] found that the low-temperature crack resistance of warm mix modified asphalt decreased after adding polyphosphoric acid by means of BBR test. Cui [25] studied the low-temperature performance of PPA modified asphalt and its mixture by low-temperature ductility, bending creep stiffness test and low-temperature bending test. It was found that the feasibility of improving the low-temperature crack resistance of asphalt and its mixture by PPA alone was low, and it needed to work together with other polymers or rubber modifiers. Therefore, in recent years, researchers have gradually turned to the study of multiphase composite modified asphalt [26–29]. For example, Yajin Han et al. [30] found that an appropriate amount of PPA can reduce the phase separation phenomenon of SBR modified asphalt, and its low-temperature performance degradation is not significant. Liu et al. [31] found that the incorporation of SBR modifier significantly made up for the deficiency of PPA modified asphalt in low-temperature performance through a semi-circular bending test.

Based on the analysis of the findings, the research object is mainly based on PPA single modified asphalt, and the research on PPA/polymer composite modified asphalt is insufficient. Therefore, PPA modifier was selected to partially replace the SBS modifier, and PPA/SBS composite modified asphalt was prepared by double doping of PPA and SBS. The road performance of composite modified asphalt was evaluated by conventional asphalt performance tests. The rheological properties of composite modified asphalt were evaluated by dynamic shear rheometer (DSR) test and bending beam rheometer (BBR) test. The synergistic modification mechanisms of PPA and SBS were revealed by Fourier transform infrared spectroscopy tests.

2. Raw Materials

2.1. Asphalt

The grade A 70# road petroleum asphalt provided by Zhengfa Municipal Construction Co., Ltd. in Zhengzhou City, Henan Province, China was used. The technical indicators are shown in Table 1, which meets the requirements of the specification.

Table 1. Technical properties of matrix asphalt.

Item		Unit	Result	Specification
Penetration (25 °C, 100 g, 5 s)		0.1 mm	69.8	60~80
5 °C ductility (5 cm/min)		cm	10.3	——
10 °C ductility (5 cm/min)		cm	51	≥20
15 °C ductility (5 cm/min)		cm	>150	≥100
Softening point		°C	47.6	≥46
Penetration index PI		——	−0.768	−1.5~1.0
Rolling thin-film oven test (RTFOT)	Mass variation	%	−0.264	−0.8~0.8
	Residual penetration ratio (25 °C)	%	62.6	≥61
	Residual ductility (10 °C)	cm	7.1	≥6

2.2. Polyphosphoris Acid (PPA)

The industrial-grade polyphosphoric acid 115% H_3PO_4 base provided by Anhui Longhua Chemical Co., Ltd. (Chizhou, China) was used. It is a transparent viscous liquid, and its specific technical indicators are shown in Table 2.

Table 2. Technical properties of polyphosphoric acid.

Item	Unit	Result
P_2O_5 Concentration	%	84.7
25 °C Vapor pressure	Pa	2.64×10^{-6}
Boiling point	°C	552
Chloride (Cl) content	%	0.0002
Iron (Fe) content	%	0.0013
Arsenic (As) content	%	0.0068
Heavy metal (Pb) content	%	0.0017

2.3. SBS Modifier

SBS1401E-modified material produced by Baling Petrochemical Branch of Sinopec Asset Management Co., Ltd. (Baling, China) was used. It is a white solid particle. Its basic technical indicators are shown in Table 3, which meet the requirements of the specification.

Table 3. Technical properties of the SBS modifier.

Item	Unit	Result	Specification
Molecular structure	——	line style	line style
Ash	%	0.09	≤0.20
300% Stress at definite elongation	MPa	4.1	≥3.5
Tensile strength	MPa	27.8	≥24.0
Tensile elongation	%	736	≥730
Volatile	%	0.57	≤0.70
S/B mass ratio	——	20/80	20/80

3. Test Method

3.1. Preparation of Composite Modified Asphalt

SBS modified asphalt and PPA/SBS composite modified asphalt were prepared by high-speed shear method. The preparation process is shown in Figure 1. The specific steps are as follows:

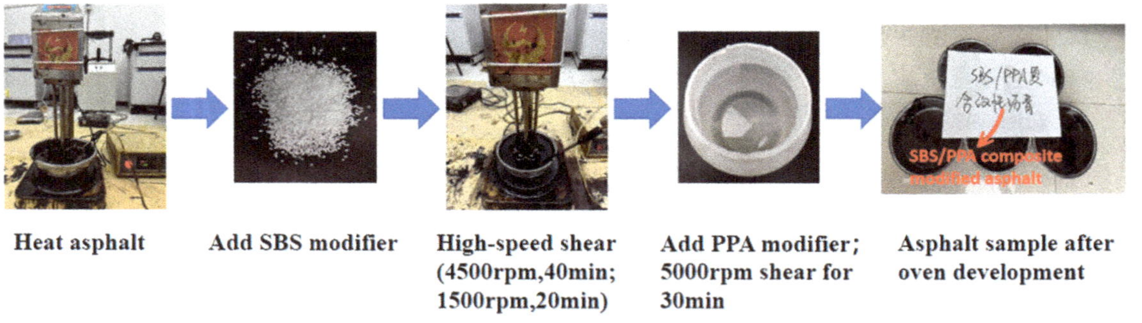

Figure 1. Preparation process of composite modified asphalt.

(1) The matrix asphalt was placed in an electric blast drying oven at 160 °C and heated to full melting and dehydration.

(2) The asphalt was taken out and placed on a constant temperature heating table for heat preservation. The SBS modifier was slowly and uniformly added at 3000 rpm, and sheared at a speed of 4500 rpm for 40 min. Finally, the constant temperature was maintained, and the SBS modified asphalt was fully swelled at a low speed for about 20 min.

(3) The temperature was raised to 170 °C, the speed was increased to 4500 rpm and certain amounts of PPA (0.5%, 0.75% and 1% PPA by weight of the matrix asphalt) were added at a constant speed. After complete addition, the shear was continued at 5000 rpm for 30 min.

(4) The sheared SBS modified asphalt and PPA/SBS composite modified asphalt were placed in an oven at 180 °C for 1 h to fully swell and develop.

3.2. Conventional Performance Test

Three index tests (penetration test, softening point test and ductility test) and Brinell rotational viscosity test were carried out to evaluate the physical properties of asphalt according to the test method of JTG E20-2011 (referred to as the Specification). The rolling thin-film oven test (RTFOT) was carried out on the asphalt according to the specification [32]. The quality, penetration and softening point of the aged asphalt were tested, and the quality change, residual penetration ratio and softening point increment of asphalt before and after aging were calculated to evaluate the anti-aging performance of asphalt.

3.3. DSR Test

The DSR test can better reflect the viscoelastic properties of asphalt materials. The principle is to apply sinusoidal strain to the sample. The sample generates corresponding stress with the strain. The applied sinusoidal strain and the sinusoidal stress generated by the sample will generate two amplitudes with a time difference. The complex modulus G^* can be obtained by amplitude, and the phase angle δ of the test sample can be obtained by calculating the time difference before and after the two signals. Through these two indexes, more evaluation indexes of rheological properties can be obtained to evaluate the high-temperature rheological properties of asphalt.

According to the test method of T 0628-2011 in the Specification [32], the sample size was 25 mm in diameter and 1 mm in thickness, and the DHR-1 dynamic shear rheometer was used, as shown in Figure 2. The frequency sweep test and temperature sweep test were carried out, respectively. Among them, the test temperature of the frequency sweep test was 40~88 °C, the test interval was 12 °C, the angular frequency changes from 0.1 rad/s to 100 rad/s, and the strain level was controlled to be 1% during the test. The temperature range of the temperature sweep test was 46~82 °C, the test interval was 6 °C, the angular frequency was adjusted to 10 rad/s, and the strain level was controlled to 10% during the test.

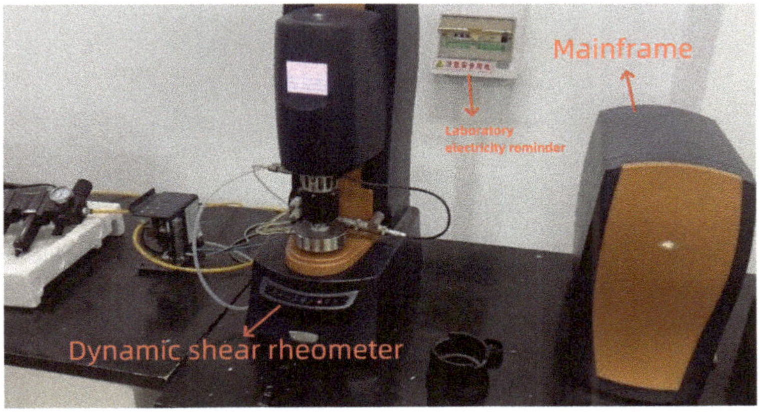

Figure 2. DHR-1 dynamic shear rheometer.

3.4. BBR Test

The low-temperature crack resistance of asphalt was studied with BBR test. The test applies the theory of a simply supported beam. The bending beam rheometer is used to apply creep load to the beam specimen to simulate the mechanical response of asphalt under the action of pavement temperature stress. The ATS low-temperature bending rheometer is shown in Figure 3. The 127 mm × 12.7 mm × 6.35 mm trabecular specimens were formed according to the test method of AASHTO T313 [33], and the test temperatures were −12 °C, −18 °C and −24 °C. The creep stiffness modulus (S) and creep rate (m) under different loading times can be obtained by experiment. According to the specification requirements of the American Highway Strategy Research Program (SHRP), $S \leq 300$ MPa, $m \geq 0.3$ at 60 s [34].

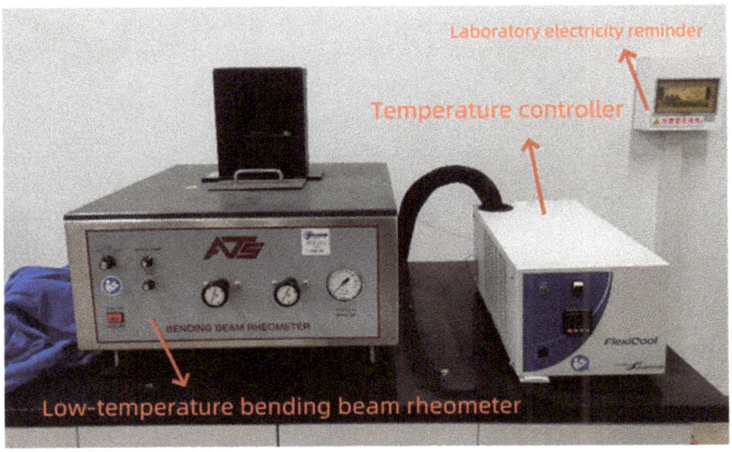

Figure 3. Low-temperature bending beam rheometer.

3.5. Fourier Transform Infrared Spectroscopy Test

With the help of the Nicolet iS10 Fourier transform infrared spectrometer produced by Thermo Fisher Scientific(Waltham, MA, USA), as shown in Figure 4, the evolution behavior of characteristic groups in the spectra after the addition of the modifier was analyzed. Firstly, the KBr wafer was prepared with the tabletting method, and the asphalt was evenly coated on it. Then, the absorption infrared spectra of each asphalt sample were collected by transmission light with a wavelength range of 500~4000 cm^{-1}, and the scanning times were 64 times.

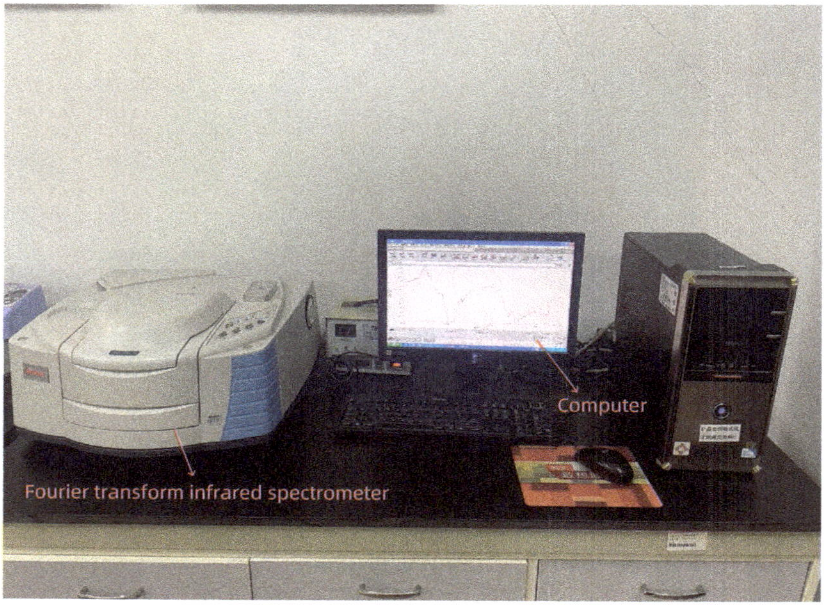

Figure 4. Nicolet iS10 Fourier transform infrared spectrometer.

4. Results and Analysis

4.1. Conventional Performance Test Analysis

The purpose of this section is to study the influence of the PPA modifier on the road performance of the SBS modified asphalt. According to the above test method, the conventional performance test of four kinds of asphalt was carried out. The test results are shown in Table 4.

Table 4. Conventional performance test results.

Item	Unit	Matrix Asphalt	5%SBS	0.5%PPA/3.5%SBS	0.75%PPA/3.5%SBS	1%PPA/3.5% SBS
Penetration (25 °C, 100 g, 5 s)	0.1 mm	69.8	52.4	55.2	52.3	49.6
Softening point	°C	47.6	67.0	50.2	58.4	63.8
Ductility (5 cm/min, 5 °C)	cm	10.3	30.1	26.9	20.5	17.0
Brookfield viscosity (135 °C)	Pa·s	0.47	1.16	1.56	2.02	2.72
Mass variation	%	0.264	0.220	0.206	0.194	0.169
RTFOT Residual penetration ratio	——	62.6	65.9	62.9	64.1	67.0
Softening point increment	——	6.6	8.5	6.8	6.2	3.9

It can be seen from Table 4:

(1) Compared with the matrix asphalt, the penetration of the 5%SBS modified asphalt decreased by 24.9%, and the softening point increased by 40.8%. The 5 °C ductility of the 5%SBS modified asphalt is 2.9 times that of matrix asphalt, which further verifies the improvement effect of SBS on the high- and low temperature properties of asphalt. Compared with 5%SBS modified asphalt, the penetration of the 0.5%PPA/3.5%SBS composite modified asphalt increased by 5.3%, and the softening point decreased by 25.1%. This is mainly because the high-temperature performance of asphalt is weakened after reducing

the content of the SBS modifier, and the content of the 0.5%PPA modifier alone cannot make up for the deficiency of high-temperature performance. However, the high-temperature performance of the 0.5%PPA/3.5%SBS composite modified asphalt is still better than that of matrix asphalt.

The content of the PPA modifier changes from 0.5% to 1%, the penetration of PPA/SBS composite modified asphalt decreases from 55.2 to 49.6, while the softening point increases from 50.2 to 63.8. This may be because the PPA modifier increases the viscous component of the asphalt, thereby improving the high-temperature performance of the asphalt. The ductility index of asphalt decreases with the increase of the content of the PPA modifier. It can be found that although the content of the PPA modifier is very small, the ductility value shows a significant change, indicating that the SBS modified asphalt is gradually hardened due to the addition of PPA, and the low temperature ductility deformation ability of asphalt is weakened. The 135 °C Brookfield rotational viscosity of the SBS modified asphalt was 1.16 Pa·s. Compared with it, the 135 °C Brookfield rotational viscosity of the three groups of composite modified asphalt increased by 34.5%, 74.1% and 134.5%, respectively, and all of them met the requirements of no more than 3 Pa·s. This may be that polyphosphoric acid reacts with asphalt while changing the composition of asphalt, to improve the viscosity of asphalt.

(2) The five groups of asphalt after RTFOT all suffered mass loss, which was caused by the loss of light components in the asphalt during heating and aging. However, PPA reduced the mass change of asphalt, indicating that PPA played a positive role in anti-aging performance. Aged asphalt generally shows increased consistency and hardness, which will reduce its penetration and increase its softening point. With the increase of PPA content, the residual penetration ratio gradually increases, and the softening point increment gradually decreases, which also shows that PPA can play a beneficial role in the anti-aging performance of the SBS modified asphalt, so that it can meet the performance requirements during production mixing and construction.

4.2. Frequency Sweep Test Analysis

4.2.1. High-Temperature Rheological Properties Analysis

Due to the complex chemical composition of asphalt, conventional performance tests cannot fully characterize the performance of asphalt. Therefore, in this section, the frequency sweep test of asphalt was carried out to obtain the complex shear modulus of asphalt in the temperature range of 40~88 °C (interval 12 °C) and frequency change (0.1~100 rad/s) to analyze the variation of complex shear modulus of four modified asphalts with angular frequency under different temperature conditions. The test results are shown in Figure 5.

By observing the complex modulus change curves of four kinds of asphalt in Figure 5, it can be concluded that at any experimental temperature, the complex modulus of the four modified asphalt shows an increasing trend with increasing angular frequency. The increase in angular frequency in the experiment can be used to characterize the increase in road load frequency in practical applications. Therefore, the more vibration generated by the road surface per unit time, the increase in vibration leads to a decrease in strain generated by asphalt, and thus the complex modulus shows an increasing trend. It can also be understood that when the vehicle load is constant, when the speed of the road vehicle is accelerated, the impact on the road surface is instantaneous, the deformation is small, the stiffness is large, and the complex modulus shows an increasing trend. Especially under high-frequency and low-temperature conditions, the complex modulus is the largest and has little effect on the pavement. Under high-temperature conditions, when the vehicle runs slowly, brakes sharply or stops, the load frequency is small, resulting in a low complex modulus, and the road surface is prone to permanent deformation.

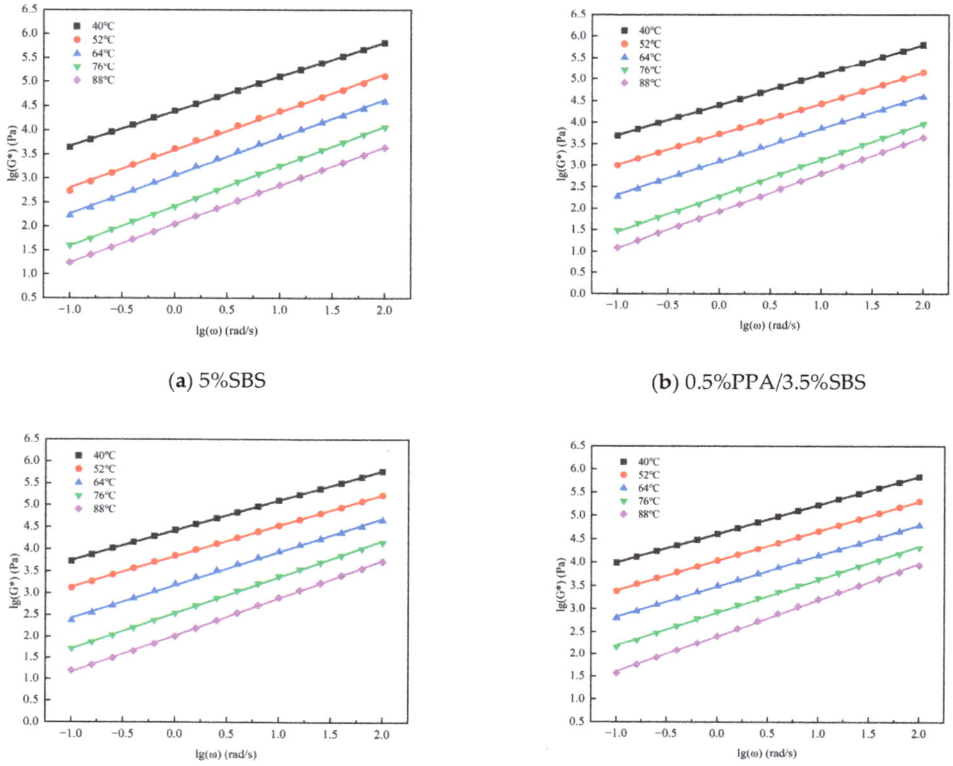

(**a**) 5%SBS (**b**) 0.5%PPA/3.5%SBS

(**c**) 0.75%PPA/3.5%SBS (**d**) 1%PPA/3.5%SBS

Figure 5. The variation trend of complex shear modulus of asphalt with frequency.

In the test temperature range of 40~88 °C, there is almost no difference in the change trend of G* of the same kind of asphalt under five temperature conditions, and the curves are generally parallel. The G* values of the four kinds of asphalt reduce with the rise in temperature, which indicates that the rise in temperature leads to the softening of asphalt, and the asphalt gradually turns from elastomer to viscoelastic body, its G* value degrades gradually.

4.2.2. Viscoelastic Characteristic Analysis of Principal Curve

The frequency sweep test is to study the viscoelastic properties of asphalt binder at different loading frequencies under the proposed temperature state. When quantitatively measuring the mechanical properties of asphalt, the test frequency cannot be infinitely expanded based on the test. Therefore, the mechanical response data in a wider frequency range can be obtained by using the time-temperature equivalence principle (TTSP) to observe the change of material properties in a wider frequency or temperature range.

(1) Determination of displacement factor

Through the displacement factor, the complex modulus curves at different temperatures are translated to obtain the complex modulus master curve [35,36]. Taking 5%SBS modified asphalt as an example, the complex modulus change curves at 40~88 °C in Figure 5a were linearly fitted one by one to obtain the double logarithmic fitting equations at different temperatures. The relationship between lgG* and lgω was clarified by the fitting equation. The fitting results are shown in Table 5.

Table 5. SBS modified asphalt double logarithmic fitting curve equation table.

Temperature/°C	Curve-Fitting Equation	R^2
40	y = 4.3875 + 0.71555x	0.99982
52	y = 3.58965 + 0.78339x	0.99798
64	y = 3.05983 + 0.78662x	0.99849
76	y = 2.42496 + 0.82765x	0.99985
88	y = 2.04866 + 0.80207x	0.99993

The angular frequency logarithm lgω (G* = 1 kP, rad/s) was obtained by bringing G* = 1 kPa into the fitting equation at each temperature in Table 5, and the displacement factor was obtained by taking 40 °C as the reference temperature. The results are shown in Table 6.

Table 6. Displacement factor of the 5%SBS modified asphalt.

Temperature/°C	lgω(G* = 1 kP, rad/s)	Displacement Factor
40	−1.9391	0
52	−0.7527	1.1864
64	−0.0761	1.8630
76	0.6948	2.6339
88	1.1861	3.1252

Through this idea and method, the linear fitting equations and displacement factors of the other three modified asphalts were obtained, respectively. The results are shown in Tables 7 and 8.

According to the calculated results in Tables 5–8, the complex modulus of four kinds of asphalt at different temperatures was translated, and the complex modulus master curve at 40 °C was obtained, as shown in Figure 6.

Table 7. PPA/SBS composite modified asphalt double logarithmic fitting curve equation summary table.

Scheme	Temperature/°C	Curve-Fitting Equation	R^2
0.5%PPA/3.5%SBS	40	y = 4.401 + 0.69712x	0.99997
	52	y = 3.72489 + 0.70901x	0.99989
	64	y = 3.08663 + 0.76396x	0.99898
	76	y = 2.29102 + 0.83766x	0.99957
	88	y = 1.93556 + 0.85827x	0.99993
0.75%PPA/3.5%SBS	40	y = 4.42343 + 0.67596x	0.99992
	52	y = 3.83727 + 0.69351x	0.9997
	64	y = 3.18477 + 0.75266x	0.99835
	76	y = 2.5458 + 0.82381x	0.99951
	88	y = 2.02573 + 0.86171x	0.9996
1%PPA/3.5%SBS	40	y = 4.60696 + 0.61414x	0.99998
	52	y = 4.03362 + 0.63192x	0.99986
	64	y = 3.48315 + 0.65615x	0.99972
	76	y = 2.91344 + 0.71499x	0.99875
	88	y = 2.3976 + 0.78802x	0.99915

Table 8. Summary of displacement factors of PPA/SBS composite modified asphalt.

Scheme	Temperature/°C	lgω(G* = 1 kP, rad/s)	Displacement Factor
0.5%PPA/3.5%SBS	40	−2.0097	0
	52	−1.0224	0.9873
	64	−0.1134	1.8963
	76	0.8464	2.8561
	88	1.2402	3.2499
0.75%PPA/3.5%SBS	40	−2.1058	0
	52	−1.2073	0.8985
	64	−0.2455	1.8603
	76	0.5513	2.6571
	88	1.1306	3.2364
1%PPA/3.5%SBS	40	−2.6166	0
	52	−1.6357	0.9809
	64	−0.7363	1.8803
	76	0.1211	2.7377
	88	0.7644	3.3810

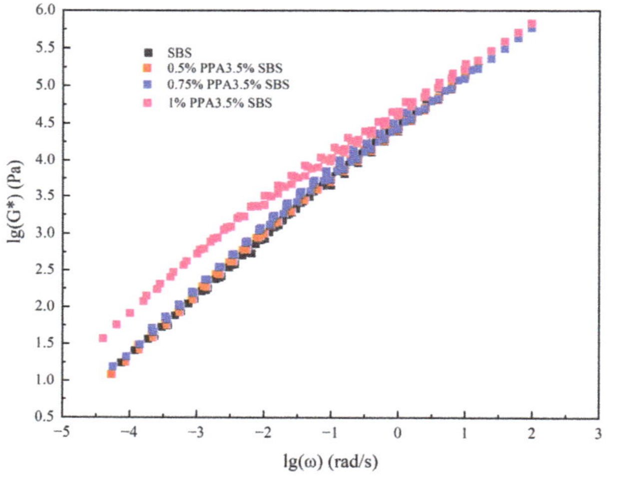

Figure 6. Principal curves of complex shear modulus of four asphalts.

It can be observed from Figure 6 that at the same frequency, due to the addition of the PPA modifier, the master curve of asphalt complex modulus shows an obvious change trend, and the more the content, the more obvious the change trend. When the PPA content increases from 0.75% to 1%, the modulus of PPA/SBS asphalt increases the most, indicating that the addition of PPA is the main factor for the modulus growth of PPA/SBS asphalt. In the low-frequency region, when the PPA content increases from 0.75% to 1%, the complex modulus difference of the composite modified asphalt is the most obvious, indicating that the addition of the PPA modifier under low-frequency and high-temperature conditions makes the asphalt have better high-temperature deformation resistance. It is assumed that the addition of PPA makes the asphalt sticky, so as to improve the high-temperature deformation resistance of the asphalt. In addition, the 0.5%PPA/3.5%SBS composite modified asphalt and the 5%SBS modified asphalt are at the same level. In the high-frequency region, the main curves of the complex shear modulus of the four kinds of asphalt tend to be concentrated, and the PPA/SBS composite modified asphalt is at the same level as the 5%SBS. This shows that the G* of PPA/SBS composite modified asphalt performs better at a lower driving speed. At a faster driving speed, the G* discrimination of PPA/SBS modified asphalt is not large.

4.3. Zero Shear Viscosity Analysis

The viscosity of asphalt material will gradually decrease with the increase of load shear rate in practical pavement application. The related research of DSR test using the limit shear rate level also shows that the viscosity of asphalt material tends to be stable under the condition of too high or too low shear rate [37]. At this time, the increase or decrease of shear rate has no obvious effect on the viscosity of asphalt material. When the shear rate is extremely low, the stable viscosity value is the zero shear viscosity (ZSV) of the asphalt material. The viscosity value at an extremely high shear rate usually means that the asphalt material is in the second Newtonian fluid state, and the asphalt material is already in the shear thinning state [38].

Under the action of an external load, the matrix asphalt is mainly flow deformation, and its deformation increases linearly with the load. However, due to its nonlinear viscoelasticity and delayed elasticity, the modified asphalt contains delayed elastic deformation in addition to flow deformation, so it is more accurate to characterize the high-temperature viscosity of modified asphalt by zero shear viscosity (ZSV) [39,40].

The frequency sweep of each group of asphalt samples was carried out at 64 °C, and the four-parameter Cross model and Carreau model were used to fit the complex modulus shear rate non-curve. The Cross and Carreau models are shown in Equations (1) and (2).

$$\frac{\eta - \eta_\infty}{\eta_0 - \eta_\infty} = \frac{1}{1 + (k\omega)^m} \tag{1}$$

$$\frac{\eta - \eta_\infty}{\eta_0 - \eta_\infty} = \frac{1}{\left(1 + (k\omega)^2\right)^{m/2}} \tag{2}$$

where η is the complex viscosity, Pa·s, ω is the angular frequency, rad/s, k and m are the characteristic constants of the material, η_0 is First Newtonian viscosity, i.e., ZSV, Pa·s, η_∞ is viscosity of the Second Newtonian region of the flow curve, Pa·s.

In the frequency sweep test, the set shear frequency is generally 0.1~100 rad/s; in this range, $\eta_0 \geq \eta \geq \eta_\infty$ can be considered. In the actual fitting process, when the test shear frequency is large enough, the asphalt is completely elastic, and the viscosity is basically zero; that is, $\eta_\infty = 0$, so the equation can be simplified as a parameter expression, as shown in Equations (3) and (4).

$$\frac{\eta}{\eta_0} = \frac{1}{1 + (k\omega)^m} \tag{3}$$

$$\frac{\eta}{\eta_0} = \frac{1}{\left(1 + (k\omega)^2\right)^{m/2}} \tag{4}$$

where η is the complex viscosity, Pa·s, ω is the angular frequency, rad/s, k and m are the characteristic constants of the material, η_0 is First Newtonian viscosity, i.e., ZSV, Pa·s.

The complex shear viscosity values of four kinds of asphalt with the change of angular frequency are shown in Figure 7. The fitting results of the Cross model and Carreau model of different asphalts are shown in Table 9.

Table 9. Cross model and Carreau model fitting results.

Scheme	ZSV		R^2	
	Cross Model	**Carreau Model**	**Cross Model**	**Carreau Model**
5%SBS	2347.69435	1662.84815	0.99959	0.99316
0.5%PPA/3.5%SBS	3145.37399	1937.65434	0.99996	0.99699
0.75%PPA/3.5%SBS	3516.56269	2385.98834	0.9999	0.99657
1%PPA/3.5%SBS	20,935.59206	7371.37261	0.99989	0.99982

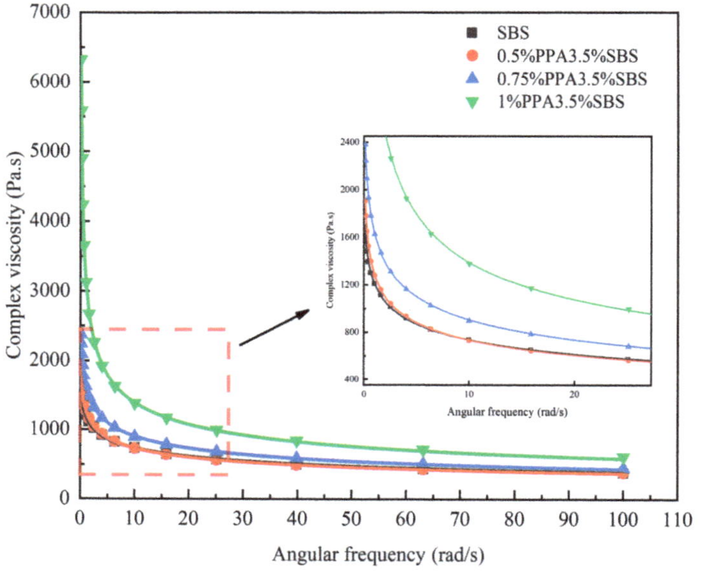

Figure 7. Curve of complex shear viscosity changing with angular frequency.

From Figure 7 and Table 9, it can be concluded that the complex viscosity of PPA/SBS composite modified asphalt gradually reduces with increasing frequency at different dosages. The complex viscosity curve of the 0.5%PPA/3.5%SBS composite modified asphalt almost coincides with the complex viscosity curve of the 5%SBS modified asphalt, and with the continuous increase of PPA content, the complex viscosity curve of PPA/SBS composite modified asphalt is obviously superior to that of the 5%SBS modified asphalt, indicating that the shear performance of PPA/SBS composite modified asphalt is superior to that of the SBS modified asphalt.

The fitting correlation between three types of PPA/SBS composite modified asphalt and 5%SBS modified asphalt is good, with R^2 above 0.99. The fitting effect of the Cross model is better than that of the Carreau model, but the ZSV result obtained after the Cross model fitting is significantly higher than that of the Carreau model fitting. The results of accelerated loading tests in the United States showed that the ZSV value obtained by the Carreau model was better than the ZSV result fitted by the Cross model. In addition, the study of Wu et al. [41] also found that the ZSV value obtained by the Cross model has the characteristics of a virtual high. However, both the Cross model fitting results and the Carreau model fitting results show that the zero shear viscosity (ZSV) of asphalt shows a significant growth trend with the increase of PPA content. In the Carreau model fitting results, the PPA content changed from 0.5% to 1%, and the ZSV value of the asphalt increased by 16.53%, 23.14% and 208.94%, indicating that the PPA modifier has a remarkable improvement effect on the high-temperature shear resistance of the asphalt, and the more the content, the more obvious the improvement effect.

4.4. Temperature Sweep Test Analysis

4.4.1. Complex Modulus and Phase Angle Analysis

The complex modulus and phase angle of four kinds of modified asphalt at 46~82 °C were obtained by temperature sweep test, and the dependence of complex modulus and phase angle of four kinds of modified asphalt on temperature was analyzed. The test results are shown in Figure 8.

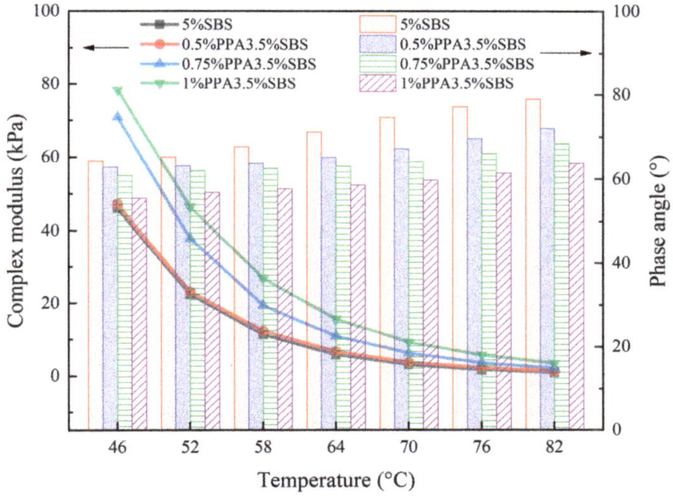

Figure 8. The curves of complex shear modulus and phase angle of asphalt with temperature change.

It can be obviously seen from Figure 8 that the complex shear modulus G* values of the four kinds of asphalt show a decreasing trend with the increase in temperature. Therefore, whether it is SBS modified asphalt or PPA/SBS composite modified asphalt, its high-temperature deformation resistance is weakening. Among them, in the temperature range of 52~64 °C, the decrease rate of complex shear modulus of asphalt is faster, while in the temperature range of 70~82 °C, the decrease rate is slower. This shows that the asphalt is greatly affected at the initial stage of temperature rise, and the degree of influence gradually decreases with the accumulation of temperature. The reason is that in the process of temperature rise, the molecular chain movement inside the asphalt becomes active, and the correlation between molecules decreases.

When the temperature changes from 46 °C to 82 °C, the phase angle δ of the four kinds of asphalt gradually increases, and the phase angle of the SBS modified asphalt is the largest. Due to the fact that the phase angle represents the ratio of elasticity to viscosity performance of asphalt, the larger phase angle means a larger proportion of viscosity, and it also means that asphalt begins to change from elasticity to viscosity, making it harder to recover its deformation after being loaded. Overall, the change in asphalt phase angle is SBS modified asphalt > 0.5PPA/SBS > 0.75PPA/SBS > 1PPA/SBS composite modified asphalt, indicating that the addition of PPA is beneficial for improving the ability of asphalt to resist deformation at high temperatures.

4.4.2. Anti-Rutting Factor Analysis

The anti-rutting factor of asphalt under the condition of 46~82 °C (interval 6 °C) was obtained by temperature sweep test, as shown in Figure 9. And its variation with temperature was studied. The effect of the PPA modifier on the anti-rutting performance of the SBS modified asphalt was analyzed.

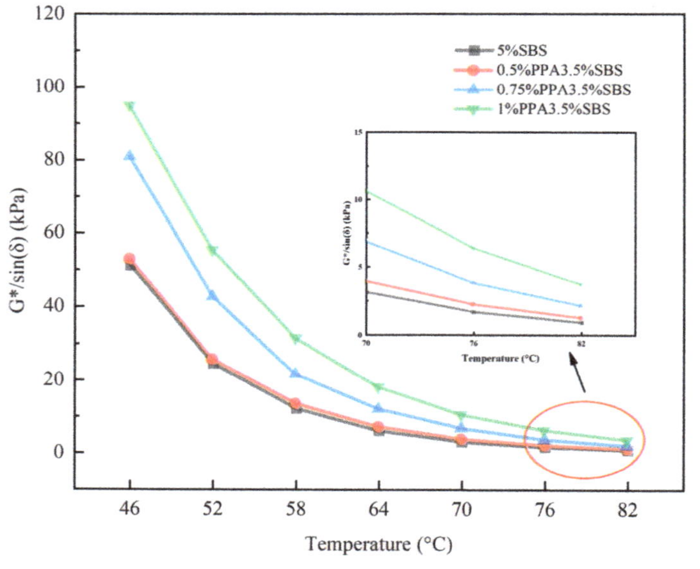

Figure 9. Rutting factor curve of asphalt.

It can be seen from Figure 9 that the rutting factors of the four kinds of asphalt show an evolution law that gradually decreases with increasing temperature, indicating that at a high temperature, the ability of asphalt to resist plastic deformation is weak, and it cannot better resist the driving load and increase the rutting deformation. However, in the high-temperature region, with the increase of PPA content, the G*/sinδ of asphalt increases. The G*/sinδ of asphalt with 0.5%PPA content is almost at the same level as that of the SBS modified asphalt, and the G*/sinδ of asphalt with 1%PPA content is the highest. Under the condition of 82 °C, the G*/sinδ increased by 41.97%, 67.62% and 70.97% when the content of PPA changed from 0.5% to 1%. This shows that the addition of the PPA modifier can improve the ability of asphalt to resist plastic deformation and improve the high temperature performance of asphalt.

4.5. Temperature Sensitivity Analysis

During the service process of asphalt pavement, due to the change in temperature, the asphalt material will change accordingly. If the penetration index in the conventional physical test is adopted, it has certain disadvantages and great differences. VTS involves a wide temperature range, and the characterization of the temperature sensitivity of asphalt is more accurate [42].

This index was used to evaluate the temperature sensitivity of four kinds of asphalt. The complex shear modulus G^* and angular frequency ω of asphalt can be obtained by temperature sweep. The test temperature range was 46~82 °C (interval is 6 °C), the angular frequency was 10 rad/s, and the strain level was controlled to be 10% during the test.

Furthermore, the complex viscosity η^* can be obtained by complex shear modulus G^* and angular frequency ω, as shown in Equation (5). Because asphalt is a non-Newtonian fluid, in order to accurately calculate the equivalent viscosity η' of various types of temperature-regulated asphalt, the equivalent viscosity η' is calculated by the conversion method proposed by Cox–Merz, as shown in Equation (6). The research results of many scholars show that it is more accurate to calculate the *VTS* of modified asphalt by Kelvin temperature [43,44]. After calculating the equivalent viscosity η' of all kinds of asphalt, the viscosity temperature index VTS of all kinds of asphalt can be calculated by Equation (7). The calculation results of four kinds of asphalt are shown in Tables 10–13.

$$\eta^* = \frac{G^*}{\omega} \tag{5}$$

where G^* is the complex shear modulus and Pa, ω is the angular frequency, rad/s.

$$\eta\prime = \frac{(\sin\delta)^{-4.8628}|G^*|}{\omega} \tag{6}$$

where δ is the phase angle, G^* is the complex shear modulus and ω is the angular frequency.

$$VTS = \frac{\lg(\lg\eta_1 - \lg\eta_2)}{\lg_{T_{K,1}} - \lg_{T_{K,2}}} \tag{7}$$

where η is viscosity, ω is the angular frequency, η_1 and η_2 are the viscosity corresponding to the adjacent temperature and T_K is the Kelvin temperature.

Table 10. SBS modified asphalt equivalent viscosity.

Temperature/°C	Kelvin Temperature/K	G*/Pa	sinδ	η′/Pa·s	lg(lg(η′))
46	318.92	46,115.5	0.9011	7652.8754	0.5893
52	325.10	22,352.8	0.9079	3576.1537	0.5506
58	331.11	11,348.9	0.9247	1661.0299	0.5079
64	337.12	5797.77	0.9463	758.1339	0.4594
70	343.10	3024.18	0.9639	361.6789	0.4080
76	349.12	1643.94	0.9747	186.1669	0.3560
82	355.10	908.523	0.9812	99.6240	0.3007

Table 11. Equivalent viscosity of the 0.5% PPA3.5% SBS composite modified asphalt.

Temperature/°C	Kelvin Temperature/K	G*/Pa	sinδ	η′/Pa·s	lg(lg(η′))
46	318.92	47,193.5	0.8897	8331.7158	0.5934
52	325.10	23,119.1	0.8921	4027.5924	0.5569
58	331.11	12,314.7	0.8965	2094.8806	0.5213
64	337.12	6645.97	0.9069	1068.9688	0.4813
70	343.10	3667.66	0.9217	545.1113	0.4372
76	349.12	2138.89	0.9373	293.0784	0.3922
82	355.10	1250.69	0.9514	159.3204	0.3429

Table 12. Equivalent viscosity of the 0.75% PPA3.5% SBS composite modified asphalt.

Temperature/°C	Kelvin Temperature/K	G*/Pa	sinδ	η′/Pa·s	lg(lg(η′))
46	318.92	70,805.5	0.8727	13,729.9659	0.6168
52	325.10	37,805.3	0.8823	6949.2353	0.5846
58	331.11	19,368.2	0.8868	3473.2014	0.5491
64	337.12	10,950.1	0.8916	1913.1437	0.5161
70	343.10	6248.01	0.9001	1042.2974	0.4797
76	349.12	3530.76	0.9139	547.0778	0.4374
82	355.10	2049.36	0.9301	291.5504	0.3918

Table 13. Equivalent viscosity of 1% PPA 3.5% SBS composite modified asphalt.

Temperature/°C	Kelvin Temperature/K	G*/Pa	sinδ	η′/Pa·s	lg(lg(η′))
46	318.92	78,196.9	0.8231	20,153.8209	0.6339
52	325.10	46,442.8	0.8366	11,058.2239	0.6068
58	331.11	26,802.3	0.8443	6102.9137	0.5781
64	337.12	15,639	0.8532	3385.2959	0.5477
70	343.10	9250.52	0.8637	1886.0519	0.5153
76	349.12	5656.84	0.8775	1067.7738	0.4812
82	355.10	3377.48	0.8965	574.4235	0.4408

The above calculation results were linearly fitted, and the fitting results are shown in Figure 10.

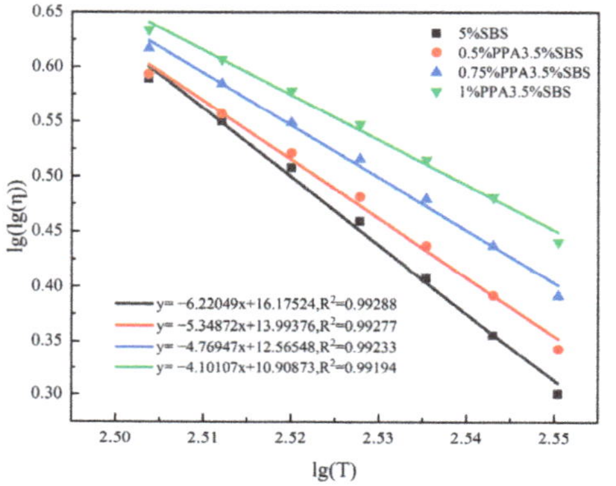

Figure 10. Linear fitting of equivalent viscosity of four kinds of asphalt.

It can be seen from Figure 10 that there is a great linear relationship between the double logarithm of the four asphalt viscosities and the logarithm of the absolute temperature T. The absolute values of VTS of the 5%SBS modified asphalt, 0.5%PPA/3.5% SBS composite modified asphalt, 0.75%PPA/3.5%SBS composite modified asphalt and 1%PPA/3.5%SBS composite modified asphalt are 6.22049, 5.34872, 4.76947 and 4.10107, respectively. In addition, with the incorporation of the PPA modifier, it showed a significant reduction in the absolute value of VTS. The smaller the absolute value of VTS of asphalt, the lower the temperature sensitivity. Therefore, PPA modifier can reduce the sensitivity of asphalt to ambient temperature, and with the increase of PPA content, the improvement of temperature sensitivity is more obvious.

4.6. BBR Test Analysis

The stiffness modulus S and creep rate m of four kinds of modified asphalt at −12 °C, −18 °C and −24 °C were obtained by BBR test. According to the requirements of American Highway Strategy Research Program (SHRP) specification, the stiffness modulus S and creep rate m at 60 s were selected to evaluate the low-temperature crack resistance of asphalt. The test results are shown in Figures 11 and 12.

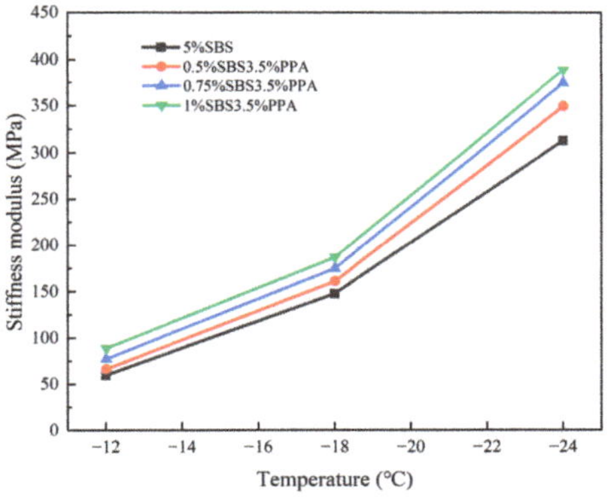

Figure 11. Variation of stiffness modulus with temperature.

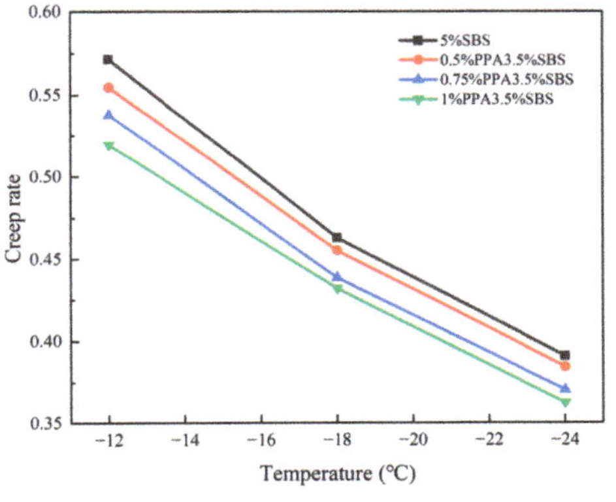

Figure 12. Creep rate changing with temperature.

From Figures 11 and 12, it can be seen that the slope of the stiffness modulus S value of the four kinds of asphalt in the range of −18~−24 °C becomes significantly larger, and the value increases the fastest, indicating that as the temperature continues to decrease, the low-temperature flexibility of the asphalt is worse, and the low-temperature anti-deformation ability is weakened. From the change rule of creep rate m value, the creep rate of the four kinds of asphalt decreases with the decrease in temperature from −12 °C to −24 °C, which indicates that the decrease in temperature leads to the decrease of stress relaxation performance of asphalt, which makes it more prone to brittle fracture. The stiffness modulus of the three groups of PPA/SBS composite modified asphalt is higher than that of the SBS modified asphalt, and the creep rate is lower than that of the SBS modified asphalt. At −12 °C, compared with SBS modified asphalt, its stiffness modulus increased by 11.17%, 29.78% and 48.86%, respectively, and the creep rate decreased by 3.00%, 5.93% and 9.18%, respectively. It shows that after PPA modifier partially replaces SBS modifier, its low-temperature deformation resistance and stress relaxation ability have different degrees of weakening effect.

4.7. Fourier Transform Infrared Spectroscopy Test Analysis

The infrared spectra of matrix asphalt, 5%SBS modified asphalt and 1%PPA/3.5%SBS composite modified asphalt were obtained by the Fourier transform infrared spectroscopy test. The results are shown in Figure 13, and then the synergistic modification mechanism of PPA and SBS was analyzed.

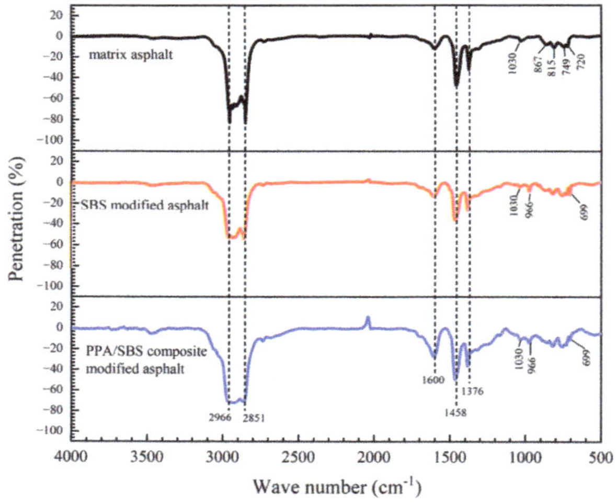

Figure 13. Infrared spectra of three kinds of asphalt.

From Figure 13, it can be observed that in the matrix asphalt map, the absorption peak near 3500 cm^{-1} is the stretching vibration of -OH, and there is an obvious absorption peak at 2800–3000 cm^{-1}, which is mainly due to the formation of-CH_2-stretching vibration. The vibration of the C=C double bond skeleton of toluene results in an absorption peak at 1600 cm^{-1}. The 1376 cm^{-1} is due to the symmetric stretching vibration of C-H bond in methyl-CH_3-, and the 1458 cm^{-1} is due to the characteristic peak caused by the antisymmetric stretching vibration, in which the antisymmetric stretching vibration amplitude is stronger. The stretching vibration of sulfoxide group S=O in asphalt causes energy fluctuation, which is reflected at 1030 cm^{-1}. A small functional group absorption peak within 1000 cm^{-1} is due to the presence of aromatics in the asphalt, and the benzene ring in the aromatics has massive C-H bonds, so the absorption peak here is generated by the bending vibration of the C-H bond. It can be seen that the composition of matrix asphalt is complex, and it is a hydrocarbon containing a variety of hydrocarbons.

Because SBS is an immiscible system, the infrared spectrum of the SBS is only a simple superposition of the infrared spectrum of polystyrene and polybutadiene. The position and intensity of the absorption peak are basically unchanged, and no new absorption peak appears. The C=C double bond bending vibration of the butadiene block in the SBS modifier causes an absorption peak at 966 cm^{-1}, and the C-H bond out-of-plane bending vibration in the benzene ring of the polystyrene block in the SBS modifier causes an absorption peak at 699 cm^{-1}.

After adding PPA, the asphalt formed a new mixed absorption peak in the 800–1300 cm^{-1} band, indicating that PPA had a chemical reaction with asphalt and changed the chemical structure unit of the original SBS modified asphalt molecule. It is a chemical modifier. The strong absorption peak at 2800–3000cm^{-1} is mainly due to the formation of-CH_2-stretching vibration. In addition, a convex peak appeared in the PPA/SBS composite modified asphalt map near 2030 cm^{-1}, indicating that PPA as a chemical modifier, its addition and asphalt formed a new compound. It is assumed that PPA reacts with alcohols in asphalt, and -OH in alcohols is neutralized by phosphoric acid to form phosphate ester. The increase of esterification degree is manifested by the increase of macromolecules and chain hydrocarbon components in asphalt, which leads to its thickening, and the macroscopic performance is the reinforcement of the high-temperature performance of asphalt.

5. Conclusions

(1) Compared with matrix asphalt, when the content of the PPA modifier is increased from 0.5% to 1%, the penetration of PPA/SBS composite modified asphalt can be decreased by 20.92%, 25.07% and 28.94% respectively, and the softening point can be increased by 5.46%, 22.69% and 34.03%, respectively. However, even a low content of PPA will reduce the ductility of asphalt at 5 °C. In addition, PPA has the ability to inhibit the thermal oxidative aging of asphalt, and the more the content, the more obvious the effect.

(2) With the increase of PPA modifier content, the complex modulus of PPA/SBS composite modified asphalt in low-frequency and high-frequency regions is significantly improved. Compared with SBS modified asphalt, the ZSV (Carreau model fitting results) of PPA/SBS composite modified asphalt can be increased by 16.53%, 23.14% and 208.94%, which significantly improves the shear resistance of asphalt. At the same time, the temperature sweep results show that at 82 °C, the phase angle of PPA/SBS composite modified asphalt can be decreased by 8.63%, 13.23% and 19.24%, and the anti-rutting factor can be increased by 41.97%, 67.62% and 70.97%, indicating that PPA could improve the ability of asphalt to resist plastic deformation. The high-temperature performance of asphalt is greatly improved, and the absolute value of VTS of PPA/SBS composite modified asphalt is reduced by 14.01%, 23.33% and 34.07%, respectively.

(3) At −12 °C, compared with SBS modified asphalt, the stiffness modulus of PPA/SBS composite modified asphalt increased by 11.17%, 29.78% and 48.86%, respectively, and the creep rate decreased by 3.00%, 5.93% and 9.18%, respectively, indicating that after PPA modifier partially replaces SBS modifier, its low-temperature deformation resistance and stress relaxation ability have different degrees of a weakening effect. This is consistent with the conclusion that PPA has a negative impact on the low-temperature performance of asphalt in the survey results.

(4) After SBS modification, the molecular chemical structure and properties of asphalt remain unchanged. The modification process is mainly physical blending, while PPA and SBS modified asphalt undergo a new chemical reaction, manifested by the increase of macromolecules and chain hydrocarbon components in the asphalt, leading to the thickening of the asphalt, and the macroscopic manifestation is the improvement of the high-temperature performance of the asphalt.

Author Contributions: X.N.: Writing—original draft, Writing—review & editing, Visualization, Validation, Investigation. Y.C.: Conceptualization, Project administration, Supervision. Z.L.: Data curation, Funding acquisition. T.G.: Resources, Methodology. J.W.: Investigation, Validation. L.J.: Investigation, Validation. All authors have read and agreed to the published version of the manuscript.

Funding: This work was supported by Key R&D and Promotion of Special Scientific and Technological Research Projects of Henan Province: [Grant Number 182102210061, 212102310089]; Key Scientific Research Projects of Colleges and Universities in Henan Province in 2021: Study on pavement performance of cotton straw cellulose modified asphalt [Grant Number 21A580004].

Institutional Review Board Statement: Not applicable.

Informed Consent Statement: Not applicable.

Data Availability Statement: Data are contained within the article.

Conflicts of Interest: The authors declare no conflict of interest.

References

1. Qu, X.; Ding, H.; Wang, H. Research progress on aging evaluation methods of road asphalt. *China J. Highw.* **2022**, *35*, 205–220.
2. Meng, G.; Xin, R.; Yubo, J.; Meichen, L. Review on aging and anti-aging of asphalt and asphalt mixture. *China J. Highw.* **2022**, *35*, 41–59.
3. Decky, M.; Papanova, Z.; Juhas, M.; Kudelcikova, M. Evaluation of the Effect of Average Annual Temperatures in Slovakia between 1971 and 2020 on Stresses in Rigid Pavements. *Land* **2022**, *11*, 764. [CrossRef]

4. Yaro, N.S.A.; Sutanto, M.H.; Baloo, L.; Habib, N.Z.; Usman, A.; Yousafzai, A.K.; Ahmad, A.; Birniwa, A.H.; Jagaba, A.H.; Noor, A. A comprehensive overview of the utilization of recycled waste materials and technologies in asphalt pavements: Towards environmental and sustainable low-carbon roads. *Processes* **2023**, *11*, 2095. [CrossRef]
5. Aryan, Y.; Dikshit, A.K.; Shinde, A.M. A critical review of the life cycle assessment studies on road pavements and road infrastructures. *J. Environ. Manag.* **2023**, *336*, 117697. [CrossRef]
6. Liu, Y.; Li, H.; Wang, H.; Wang, Y.; Han, S. Integrated Life Cycle Analysis of Cost and CO_2 Emissions from Vehicles and Construction Work Activities in Highway Pavement Service Life. *Atmosphere* **2023**, *14*, 194. [CrossRef]
7. Behnood, A.; Gharehveran, M.M. Morphology, rheology, and physical properties of polymer-modified asphalt binders. *Eur. Polym. J.* **2019**, *112*, 766–791. [CrossRef]
8. Qian, J.; Dong, F.; Chen, X.; Gao, Y.; Zhang, L.; Li, M.; Guan, N. Research progress on high temperature rheological properties of polymer modified asphalt. *Mater. Introd.* **2023**, 1–11. Available online: http://kns.cnki.net/kcms/detail/50.1078.TB.20230628.1149.002.html (accessed on 1 November 2023).
9. Li, C.; Wang, L.; Feng, L. Study on low temperature rheological properties of polymer modified asphalt binder before and after aging. *Funct. Mater.* **2016**, *47*, 2206–2211.
10. Chen, Q.; Wang, C.; Li, Y.; Huang, S. Performance development of polyurethane elastomer composites in different construction and curing environments. *Constr. Build. Mater.* **2023**, *365*, 130047. [CrossRef]
11. Wang, S.; Dong, F.; Yu, X.; Zu, Y.; Jiang, Y. Effect of compatibilizer composition on anti-aging properties of SBS modified asphalt. *J. Build. Mater.* **2023**, 1–15.
12. Cao, X.; Gao, H.; Li, Z.; Wu, B.; Li, X. Study on aging characteristics of SBS modified asphalt. *Appl. Chem. Eng.* **2021**, *50*, 2739–2744.
13. Ran, L. Study on Aging Mechanism of SBS Modified Asphalt and Development of High Performance Regenerant under the Condition of Heat, Light and Water Coupling. Ph.D. Thesis, Chongqing Jiaotong University, Chongqing, China, 2017.
14. Liu, C.; Li, Z.; Liu, S.; Fu, L. Technical analysis and prospect of polyphosphoric acid modified asphalt. *China Adhes.* **2020**, *29*, 63–67.
15. Zhang, L.; Huang, W.; Wei, M.; Ouyang, X.; Wang, Y. Analysis of rheological properties of polyphosphoric acid modified asphalt. *J. Mater. Sci. Eng.* **2020**, *38*, 638–642.
16. Liu, Y.; Pang, S.; Ma, Q. Research progress and prospect of polyphosphoric acid modified asphalt technology. *Transp. Res.* **2022**, *8*, 153–164.
17. Jafari, M.; Akbari Nasrekani, A.; Nakhaei, M.; Babazadeh, A. Evaluation of rutting resistance of asphalt binders and asphalt mixtures modified with polyphosphoric acid. *Pet. Sci. Technol.* **2017**, *35*, 141–147. [CrossRef]
18. Ramayya, V.V.; Ram, V.V.; Krishnaiah, S.; Sandra, A.K. Performance of VG30 paving grade bitumen modified with polyphosphoric acid at medium and high temperature regimes. *Constr. Build. Mater.* **2016**, *105*, 157–164. [CrossRef]
19. Xiao, F.; Amirkhanian, S.; Wang, H.; Hao, P. Rheological property investigations for polymer and polyphosphoric acid modified asphalt binders at high temperatures. *Constr. Build. Mater.* **2014**, *64*, 316–323. [CrossRef]
20. Liang, P.; Liang, M.; Fan, W.; Zhang, Y.; Qian, C.; Ren, S. Improving thermo-rheological behavior and compatibility of SBR modified asphalt by addition of polyphosphoric acid (PPA). *Constr. Build. Mater.* **2017**, *139*, 183–192. [CrossRef]
21. Zhang, H.; Shi, C.; Yu, J.; Shen, J. Study on the modification and modification mechanism of polyphosphoric acid on different asphalt. *J. Build. Mater.* **2013**, *16*, 255–260.
22. Zhang, Z.; Jia, M.; Wei, L.; Ban, X.; Han, J. Research Progress of Polyphosphoric Acid Modified Asphalt. *Zhongwai Highw.* **2016**, *36*, 230–233.
23. Zhou, Y.; Huang, W.; Fu, X. Low temperature performance of polyphosphoric acid composite modified asphalt. *J. Build. Mater.* **2017**, *20*, 996–1000.
24. Liu, J.; Yan, K.; You, L.; Ge, D.; Wang, Z. Laboratory performance of warm mix asphalt binder containing polyphosphoric acid. *Constr. Build. Mater.* **2016**, *106*, 218–227. [CrossRef]
25. Cui, L. Anti-cracking performance evaluation of polyphosphoric acid and its composite modified asphalt and asphalt mixture. *Highw. Eng.* **2016**, *41*, 178–187.
26. Wang, W.; Li, J.; Wang, D.; Liu, P.; Li, X. The Synergistic Effect of Polyphosphates Acid and Different Compounds of Waste Cooking Oil on Conventional and Rheological Properties of Modified Bitumen. *Materials* **2022**, *15*, 8681. [CrossRef]
27. Amir, T.; Mohammad, R.; Pouria, H.; Behzad, R.; Yousef, Y.; Ali, K. Investigation on the Effect of Ethylene Bis(Stearamide) and Polyphosphoric Acid Modification of Bitumen for Paving Applications. *J. Mater. Civ. Eng.* **2022**, *34*, 04022165.
28. Wang, L.; Pei, K.; Li, C. Study on low temperature rheological properties and constitutive relationship of polyphosphoric acid-SBS composite modified asphalt mixture. *J. Build. Mater.* **2021**, *24*, 842–850.
29. Zhou, Y.; Wei, J.; Shi, S.; Gao, J.; Duan, X.; Chen, Y. The performance of polyphosphoric acid and rubber powder composite modified asphalt. *J. Chang. Univ. Nat. Sci. Ed.* **2018**, *38*, 9–17.
30. Han, Y.; Tian, J.; Ding, J.; Shu, L.; Ni, F. Evaluating the storage stability of SBR-modified asphalt binder containing polyphosphoric acid (PPA). *Case Stud. Constr. Mater.* **2022**, *17*, e01214. [CrossRef]
31. Liu, H.; Chang, R.; Wang, C.; Cao, X.; Hao, P. Road performance of polyphosphoric acid composite modified asphalt mixture. *J. Build. Mater.* **2017**, *20*, 293–299.
32. *JTG E20-2011*; Test Procedures for Asphalt and Asphalt Mixtures for Highway Engineering. People's Transportation Publishing House: Beijing, China, 2011.

33. *AASHTO T313-2012;* Standard Method of Test for Determining the Flexural Creep Stiffness of Asphalt Binder Using the Bending Beam Rheometer (BBR). American Association of State Highway and Transportation Officials: Washington, DC, USA, 2012.
34. Lesueur, D.; Elwardany, M.D.; Planche, J.P.; Christensen, D.; King, G.N. Impact of the asphalt binder rheological behavior on the value of the ΔTc parameter. *Constr. Build. Mater.* **2021**, *293*, 123464. [CrossRef]
35. Chen, Y.; Wang, Q.; Li, Z.; Ding, S. Rhysiological properties of graphene nanoplatelets/rubber crowd composite modified asphalt. *Constr. Build. Mater.* **2020**, *261*, 120505. [CrossRef]
36. Li, Z.; Guo, T.; Chen, Y.; Dong, L.; Chen, Q.; Hao, M.; Zhao, X.; Liu, J. Study on Rheological Properties of Graphene Oxide/Rubber Crowd Composite-Modified Asphalt. *Materials* **2022**, *15*, 6185. [CrossRef]
37. Zeiada, W.; Liu, H.; Al-Khateeb, G.G.; Shanableh, A.; Samarai, M. Evaluation of test methods for measurement of zero shear viscosity (ZSV) of asphalt binders. *Constr. Build. Mater.* **2022**, *325*, 126794. [CrossRef]
38. Li, G.; Tan, Y.; Fu, Y.; Liu, P.; Fu, C.; Oeser, M. Density, zero shear viscosity and microstructure analysis of asphalt binder using molecular dynamics simulation. *Constr. Build. Mater.* **2022**, *345*, 128332. [CrossRef]
39. Wang, W.; Chen, B.; Shen, J. Zero Shear Viscosity of Hybrid Modified Asphalts and Its Gray Correlation with Other Properties. *Materials* **2022**, *15*, 7056. [CrossRef]
40. Ou, L.; Zhu, H.; Xu, Y.; Chen, R.; Yang, X. Gray correlation entropy analysis of zero shear viscosity and high-temperature rheological parameters of phosphogypsum-modified asphalt. *Case Stud. Constr. Mater.* **2022**, *17*, e01448. [CrossRef]
41. Wu, J.; Liu, Q. Application of zero shear viscosity in the evaluation of anti-rutting performance of agricultural film modified asphalt. *Zhongwai Highw.* **2015**, *35*, 296–299.
42. Abed, Y.H.; Al-Haddad, A.H.A. Temperature susceptibility of modified asphalt binders. In Proceedings of the IOP Conference Series: Materials Science and Engineering, Kerbala, Iraq, 4–6 November 2019; Volume 671, p. 012121.
43. Mirzaiyan, D.; Ameri, M.; Amini, A.; Sabouri, M.; Norouzi, A. Evaluation of the performance and temperature susceptibility of gilsonite-and SBS-modified asphalt binders. *Constr. Build. Mater.* **2019**, *207*, 679–692. [CrossRef]
44. Zhou, Z.; Chen, G. Temperature Sensitivity Analysis of High Viscosity-Modified Asphalt. *J. Highw. Transp. Res. Dev. Engl. Ed.* **2023**, *17*, 1–8. [CrossRef]

coatings

MDPI

Article

Analyzing Pore Evolution Characteristics in Cementitious Materials Using a Plane Distribution Model

Yezhen Yuan [1,2], Zhe He [3], Kaimin Niu [1,3], Bo Tian [1,3], Liangliang Chen [4], Wei Bai [5], Shaopeng Zheng [4,*] and Guoman Yu [6]

1 School of Civil Engineering, Chongqing Jiaotong University, Chongqing 400074, China; jd.xie@rioh.cn (Y.Y.); km.niu@rioh.cn (K.N.); b.tian@rioh.cn (B.T.)
2 Guangxi Nanbai Expressway Co., Ltd., Nanning 530029, China
3 Research Institute of Highway, Ministry of Transport, Beijing 100088, China; z.he@rioh.cn
4 Yunnan Provincial Transportation Planning and Design Research Institute Co., Ltd., Kunming 650000, China; chenll@stwip.com
5 Guangxi Nantian Expressway Co., Ltd., Nanning 530022, China; hf198507@gmail.com
6 Construction Headquarters of Nujiang Beautiful Highway Greenway, Nujiang 673200, China; ynjtgh@126.com
* Correspondence: zhengshaopeng@stwip.com

Abstract: This research aims to analyze the distribution and evolution of pores within the planar structure of cement-based materials. Utilizing digital imaging methods, a model for pore plane distribution was established, and the evolutionary patterns of both total pore numbers and varying pore sizes in cement-based materials were investigated. The research introduced an innovative experimental method for analyzing pore distribution within cement-based planar structures. Additionally, a hybrid method was proposed, combining automated image binarization thresholding with manual comparative analysis, thereby enhancing the feasibility of comparative research. Pores were categorized into four distinct sizes: tiny pores (5–200 μm), small pores (200–500 μm), medium pores (500–1000 μm), and large pores (>1000 μm). Areas with apertures <5 μm were classified as dense areas. The findings indicated that the overall number of pores in cement-based materials increased due to the influence of styrene butadiene latex additives. However, at a 15% dosage, the rate of pore formation reached an inflection point, confirming that various factors, such as styrene butadiene latex, air bubbles, and the cement-based material itself, collectively influenced pore formation. The research also demonstrated that styrene butadiene latex affected the four categorized pore sizes differently. Importantly, a higher latex dosage did not necessarily lead to a proportional increase in pore content. Pore content was influenced by multiple factors and exhibited different distribution patterns. The number of micropores, although relatively small, gradually increased with higher latex dosages, while small and medium pores generally showed an upward trend. At a 10% latex dosage, both small and medium pores reached a turning point in their rate of increase. Large pores also exhibited a general increase, peaking at a latex dosage of 10%. It was confirmed that both the total pore volume and the content of micropores were critical factors in determining the mechanical properties of cementitious materials. Higher porosity and micropore content generally weakened mechanical performance. However, at a small latex dosage, there was an improvement in flexural strength. When the latex dosage reached 15%, the total pore and micropore content declined, resulting in a balanced increase in flexural strength and a mitigated decline in compressive strength. This study offers valuable insights into the evolution of total pore volume and the content of pores of various sizes, providing a theoretical basis for the meticulous selection of additive types and dosages from a microscopic perspective.

Keywords: cementitious materials; stomatal plane distribution model; digital images; pore evolution characteristics

Citation: Yuan, Y.; He, Z.; Niu, K.; Tian, B.; Chen, L.; Bai, W.; Zheng, S.; Yu, G. Analyzing Pore Evolution Characteristics in Cementitious Materials Using a Plane Distribution Model. *Coatings* **2023**, *13*, 2023. https://doi.org/10.3390/coatings13122023

Academic Editor: Bohayra Mortazavi

Received: 26 October 2023
Revised: 27 November 2023
Accepted: 28 November 2023
Published: 29 November 2023

1. Introduction

Cementitious materials are of paramount importance in road engineering due to their multiphase porous systems. Understanding the distribution patterns of pore structures is crucial for ensuring both durability and sustainable performance. Factors such as pore number, size, and distribution patterns within these materials directly influence the mechanical properties and durability [1]. The internal pore structure of cementitious materials indeed plays a fundamental role in defining their overall performance. To enhance specific aspects of cementitious materials, additives are commonly employed. For example, butylbenzene latex polymers are added to cement concrete to improve its impermeability [2,3]. Styrene butadiene latex, a block copolymer, has been the subject of extensive research regarding its impact on the microscopic pore structure of cementitious materials. Li et al. discovered that the incorporation of styrene butadiene emulsion shifted the pore size distribution in cementitious materials toward micropores, which also had an effect on their macroscopic mechanical properties [4]. Zhang et al. demonstrated that polymer emulsions influence the pore structure of cement mortar, resulting in an increased proportion of smaller pores [5]. Yang emphasized that a decrease in the proportion of smaller pores has a negative impact on the mechanical properties and frost resistance of cement concrete [6]. Sakai et al. conducted a study on the effects of water-reducing agents on the pore structure of hardened cement, outlining the variations in different pore sizes [7]. Kondraivendhan et al. established a relationship between the compressive strength of cement slurry and the average pore distribution radius, affirming the practical value of such a relationship [8].

Zhang et al. conducted a study on the impact of polymer emulsion on concrete's pore structure. They observed that an increased admixture of polymer emulsion led to a reduction in smaller pores (within 100 μm) and an increase in larger pores (above 400 μm), thereby having a detrimental effect on the material's mechanical properties [9]. Wu et al. categorized pores in cementitious materials into four size-based classes, including benign pores (below 200 nm). Their argument highlighted that an increase in the number of pores smaller than 50 nm and a decrease in those larger than 100 nm contribute to the material's durability [10,11]. Jinteng et al. differentiated pores in cementitious materials into gel microcrystalline endospores (1.2 nm), gel pores (0.6–1.6 nm), capillary pores (3.2–200 nm), and macropores (above 200 nm). They emphasized that pores larger than 10 nm have a significant impact on frost resistance [12]. Jawed et al. also investigated pore structures, classifying them into macropores (above 5 μm) and capillary pores. Capillary pores were further subdivided into large capillary pores (50–5000 nm), interstitial pores (2.6–50 nm), and micropores (2.6 nm) [13]. This classification closely aligns with that of Jinteng et al. [12]. Macropores were identified as the primary contributors to the mechanical and durability properties of cementitious materials. The European Concrete Board's CEB Guide for the design of durable concrete structures specified that pore sizes between 0.1 and 10 nm are considered micropores, 10–10,000 nm are capillary pores, and pores larger than 10,000 nm are macropores. Pores exceeding 1000 μm in size were explicitly mentioned as introducing air holes, which significantly deteriorate the performance of cementitious materials [14,15]. F.H. introduced a pore classification system that included micro, delicate, and macro scales while studying cementitious materials [16]. IO.C. et al. classified cementitious materials by pore size into gel pores (below 1000 nm) and highlighted that pores larger than 10 μm were considered macropores [17]. Yan and Yang et al. referred to pores in cementitious materials larger than 10 μm as stomata. Their observations indicated that general air-entraining agents introduce more stomata ranging from 10 μm to 200 μm, followed by 200 μm to 500 μm stomata, with the fewest stomata being larger than 500 μm [18,19].

Several studies have provided valuable insights into the internal pore structure of cementitious materials and the influence of external factors [20–25]. Most of these studies have focused on the distribution of air bubbles in freshly mixed cementitious materials and the rules governing pore changes within the three-dimensional internal structure. The primary tools used for these analyses are often costly instruments such as mercury pressure instruments, and the scale of study is mainly in the nanometer range or smaller. However, there has been limited research on the evolution of pore structures within the two-dimensional cross-section of cementitious materials after hardening and the impact of these two-dimensional pore distribution characteristics on the mechanical properties of cementitious materials.

Building upon the existing research, this study aims to address specific test constraints, such as the minimum particle size of the white powder used to fill pores, which is 5 μm. Recognizing that smaller pore structures have a less adverse impact on the performance of cementitious materials and contribute to greater internal compactness, a 5 μm threshold is established as the boundary between compact and porous areas. Specifically, the study focuses on pores larger than 5 μm; regions with pore sizes smaller than 5 μm are designated as compact areas. Furthermore, given the substantial influence of pores larger than 1000 μm on the properties of cement mortar, these are classified as areas of large porosity. For pore sizes ranging between 5 and 1000 μm, pores are further categorized into tiny (5–200 μm), small (200–500 μm), and medium (500–1000 μm) groups. The planar distribution model, as depicted in Figure 1, serves as the foundational tool for the analysis of pore distribution in cementitious materials. This model characterizes various pore distribution ratios based on two-dimensional planar area sizes and classifies them according to pore aperture size. This framework offers an intuitive comprehension of the proportion and attributes of different pore sizes within the cementitious material. By adjusting the ratio within each category, the model can depict the overall pore distribution and the specific sizes of individual pores, providing a theoretical framework for future investigations into the evolution of internal pore distribution in cementitious materials.

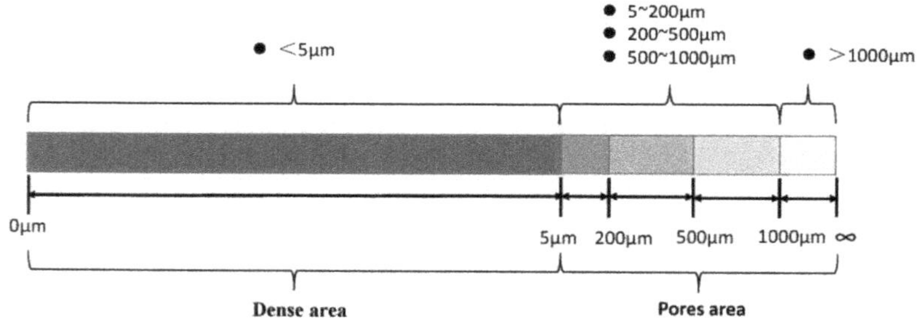

Figure 1. Pore plane distribution model. (<5 μm, 5~200 μm, 200~500 μm, 500~1000 μm, >1000 μm).

To address this gap, the study establishes a planar distribution model for the pore structure in cementitious materials. A low-cost, easy-to-operate digital imaging method is introduced to analyze the pore structure evolution in hardened cementitious materials. By utilizing styrene butadiene latex polymer as the external influencing factor and focusing on the micron-sized pore scale, the development and evolution patterns of cementitious materials' pore structures are examined in the study. The objective is to gain insights into the effects of two-dimensional planar microscopic pore distribution on the mechanical properties of cementitious materials. This study provides a theoretical foundation for enhancing the properties of cementitious materials and offers a simple and accurate experimental evaluation method to ensure their sustainable development.

2. Materials and Methods

2.1. Raw Materials

To investigate the characteristics of pores in cementitious materials, styrene butadiene latex polymer is used as a modifying factor for the internal pore structure. The concentration of the polymer is adjusted to analyze alterations in the internal pore structure of the cementitious materials. These adjustments are expressed as a percentage of the total cement mass and are detailed in Table 1.

Table 1. Mix proportions for test.

Serial Number	Water (g)	Cement (g)	Standard Sand (g)	Latex (g)	Defoamer (g)	Latex Dosage (%)
1	225	450	1350	0	0	0
2	213.53	450	1350	22.5	4.12	5
3	202.05	450	1350	45.0	4.18	10
4	190.58	450	1350	67.5	4.25	15
5	179.10	450	1350	90.0	4.31	20

For the experimental setup, ordinary Portland cement (P.O 42.5) and sand were used in accordance with ISO standard specifications. Test water meeting all relevant criteria was used. Throughout the experiment, a consistent water–cement ratio of 0.5 was maintained, with unaltered cement and standard sand dosages. To mitigate the adverse effects of harmful pores produced by styrene butadiene latex in cementitious materials [26–28], a defoamer was added to the mixture. The defoamer constituted 6% of the total mass of water, cement, and latex, aligning with real-world engineering standards. Experimental mix proportions for the test are detailed in Table 1, while the key properties of the selected styrene butadiene latex polymer are presented in Table 2.

Table 2. Basic performance of styrene butadiene latex.

Solid Content (%)	PH	Viscosity (20 °C) (mPa.s)	Char Yield (450 °C) (%)	Glass Transition Temperature (°C)
47.0	8.25	48.02	15.4	10.2

2.2. Methods

Cement mortar specimens were molded into dimensions of 150 mm × 150 mm × 60 mm. After curing for 28 days, they were cut and smoothed into planar samples measuring 150 mm × 150 mm × 30 mm for testing. From these samples, a 100 mm × 100 mm section was selected and coated with a non-fading black oil-based pen as a primer. This coating was allowed to dry at room temperature for a minimum of two hours. Subsequently, white powder with a particle size of 5 μm was applied to fill the surface pores of the sample [29]. Excess powder was removed, ensuring that only the pores were filled while leaving the black-coated background exposed. The specimen preparation primarily involved three stages: molding, cutting and smoothing, and pre-treatment. These processes are illustrated in Figure 2.

An EOS 2000D camera, Canon (Tokyo, Japan), equipped with a CMOS sensor boasting an effective pixel count of 24.1 million and an optical viewfinder, was employed for capturing images of specimen cross-sections under even, shadowless illumination. MATLAB software (Version Number: R2023) was utilized for grayscale conversion and binarization. This method allowed for the analysis of pore distribution on the specimen's surface using digital imaging techniques. Based on the pixel values in the captured images [30–32], "1" was used to represent white powder-filled pores after binarization, while "0" denoted unfilled areas in black, signifying denser regions (as depicted in Figure 3a). Although grayscale binarization is a widely adopted technique, the selection of the threshold is of utmost importance. Various options are typically considered, including using 0.5 as a starting point, computing the average grayscale values, or employing an automated method to determine the threshold. It is worth noting that each of these methods comes with a certain margin of error.

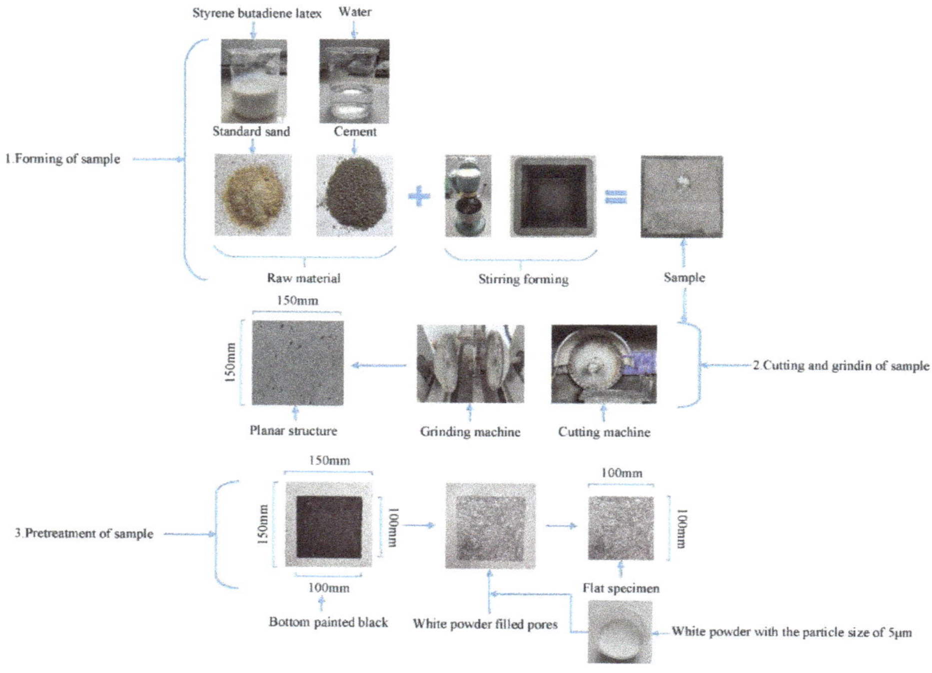

Figure 2. Flow chart of sample preparation process.

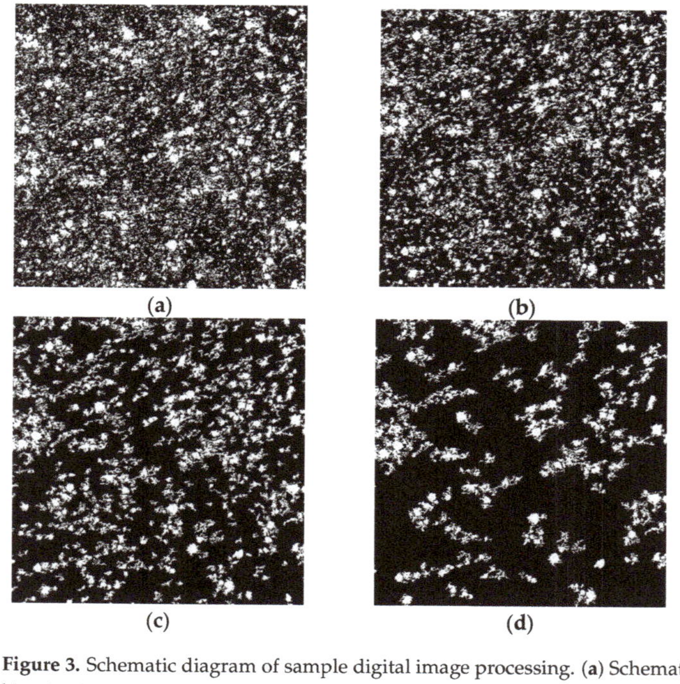

Figure 3. Schematic diagram of sample digital image processing. (**a**) Schematic diagram of sample binarization treatment; (**b**) Schematic diagram of aperture > 200 μm pores; (**c**) Schematic diagram of aperture > 500 μm pores; (**d**) Schematic diagram of aperture > 1000 μm pores.

In this study, MATLAB was initially employed to automatically determine the binarization threshold, which was set at 0.6. Subsequently, manual adjustments were made by incrementally increasing or decreasing the threshold by 0.05 and 0.1, ultimately arriving at the optimal threshold of 0.65. By combining automated threshold confirmation with manual fine-tuning, the binarization threshold was refined to closely align with the actual image characteristics, thereby establishing a dependable foundation for the experimental investigation. While this approach could be influenced by subjective factors, its feasibility and accuracy are generally sufficient for conducting comparative studies under consistent conditions.

Within the experimental framework, the pixel value range was considered in relation to the actual dimensions of the 100 mm × 100 mm planar section. The values "1" and "0" in the binarized image were designated as Z_1 and Z_0, respectively. The sum of Z_1 and Z_0 constituted the total pixel value, represented as Z. The total porosity content, denoted as P_1, was equivalent to Z_1. Therefore, the total content, P_1, is calculated as follows:

$$P_1 = Z_1/Z \tag{1}$$

The actual cross-sectional area is denoted as $S = 0.01\ \text{m}^2$, and the number of pixel points Q per unit area is represented as follows:

$$Q = Z/S \tag{2}$$

In the study, the maximum particle size of the white powder used for pore filling was 5 μm. As a result, pores smaller than this threshold remained unfilled, manifesting as black regions in the digital images. Taking advantage of this scale effect, pixel values corresponding to pores filled with white powder were extracted. Assuming the pores to be approximately circular, the formula πr^2 was utilized to calculate the number of pixel points representing pore sizes of 200 μm, 500 μm, and 1000 μm, denoted as Z_{200}, Z_{500}, and Z_{1000}, respectively.

$$Z_{200} = \left(\pi r_{200}^2\right) * Q$$

$$Z_{500} = \left(\pi r_{500}^2\right) * Q$$

$$Z_{1000} = \left(\pi r_{1000}^2\right) * Q$$

During the digital image processing, the MATLAB function bwareaopen was utilized to remove objects with areas smaller than Z_{200}, Z_{500}, and Z_{1000} from the binary images. Consequently, the number of pixel points exceeding 200 μm, 500 μm, and 1000 μm, denoted as $Z_{>200}$, $Z_{>500}$, and $Z_{>1000}$, were obtained, as depicted in Figure 3b–d. The corresponding contents of stomata larger than 200 μm, 500 μm, and 1000 μm, labeled as P_{200}, P_{500}, and P_{1000}, were calculated as follows:

$$P_{200} = Z_{>200}/Z$$

$$P_{500} = Z_{>500}/Z$$

$$P_{1000} = Z_{>1000}/Z$$

The stomatal contents for <200 μm, 200–500 μm, 500–1000 μm, and >1000 μm were calculated as follows for $P_{<200}$, $P_{200\text{–}500}$, $P_{500\text{–}1000}$, and $P_{>1000}$.

$$P_{<200} = P_1 - P_{200}$$

$$P_{200-500} = P_{200} - P_{500}$$

$$P_{500-1000} = P_{500} - P_{1000}$$

$$P_{>1000} = P_{1000} = Z_{>1000}/Z$$

The methods outlined above yielded insight into the spatial distribution of internal pores in cementitious materials during the continuous hardening process. Evolving trends in pore content across different size ranges were observed, and a planar porosity model supported by extensive experimental data was constructed. This model served as a theoretical foundation for the analysis of the mechanical properties and durability of cementitious materials.

The methods outlined above yielded insights into the spatial distribution of internal pores in cementitious materials during the ongoing hardening process. Evolving trends in pore content across different size ranges were observed, and a planar porosity model, supported by extensive experimental data, was constructed. This model serves as a theoretical foundation for the analysis of the mechanical properties and durability of cementitious materials.

3. Results and Discussion

3.1. Analysis of Evolutionary Pattern of Total Stomatal Volume

In cementitious materials, the blending of various components gave rise to a complex solid–liquid–gas three-phase system, leading to the formation of a porous structure within the cement mortar. The porosity was governed by the intrinsic properties of the material [33–35]. Styrene butadiene latex, as an additive in cementitious materials, also constituted a solid–liquid–gas three-phase mixing system, which inevitably influenced the formation and evolution of internal pores to a certain extent. The study focused on hardened cement mortar, in which internal air bubbles underwent a multi-stage formation process. The first stage was the initiation phase, during which air bubbles were introduced into the fresh mortar mix within the complex solid–liquid–gas three-phase system. The second stage was characterized by disorder, as bubbles rearranged, compressed, merged, and escaped due to external forces such as stirring and compaction applied to the fresh mortar. This was followed by a sub-stabilization stage, where the bubbles gradually transitioned from a disordered state to a more stable one under standard temperature and humidity conditions. However, they were still influenced by factors such as cement hydration and water evaporation during the mortar's hardening process. Finally, in the stabilization stage, which typically occurred after 28 days of curing, the internal pores attained a fixed structural pattern. Specimens for this study were sourced from this last stage.

Figure 4 illustrates the impact of butylbenzene latex on the total number of internal pores in cementitious materials. In comparison to latex-modified mortar, the standard mortar exhibited the fewest pores, highlighting that latex inclusion increased the overall pore content within the hardened mortar plane. As the concentration of styrene butadiene latex increased, there was a general upward trend in the total pore count within the mortar. Interestingly, at a 15% latex dosage, the data revealed a turning point: the total pore count decreased to levels comparable to those observed at a 5% dosage. However, with further increases in latex concentration to 20%, the pore count within the hardened mortar plane continued to rise, albeit at a slower rate of increase—only 5.31% compared to the 10% latex dosage.

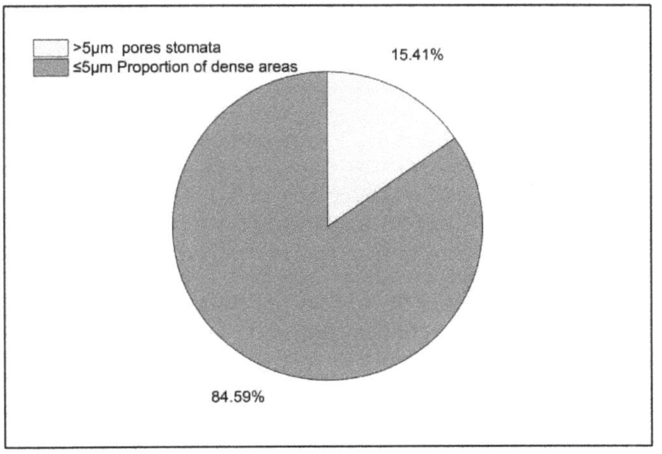

(**a**) Standard mortar (mix ratio 1)

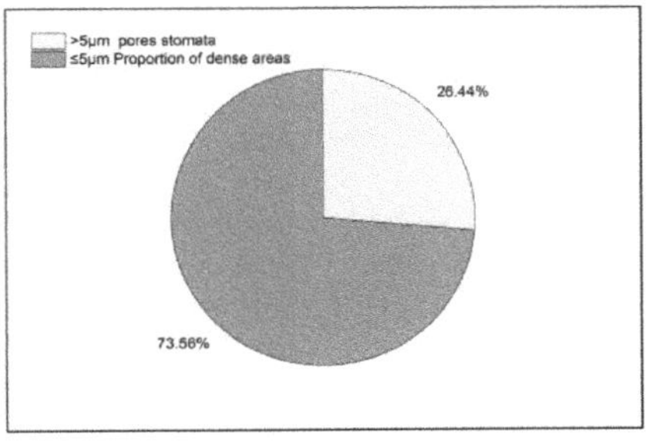

(**b**) Latex dosage of 5% (mix ratio 2)

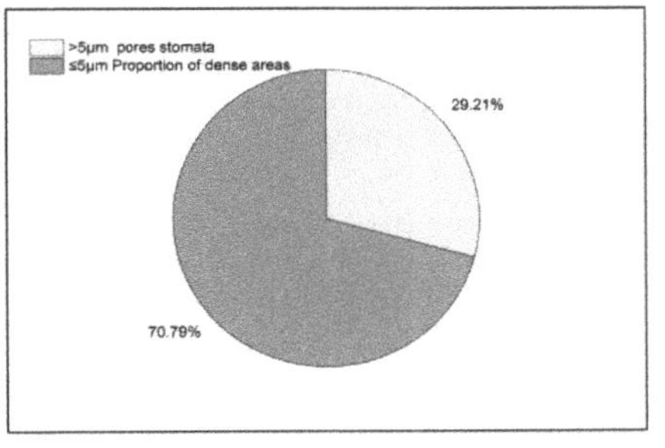

(**c**) Latex dosage of 10% (mix ratio 3)

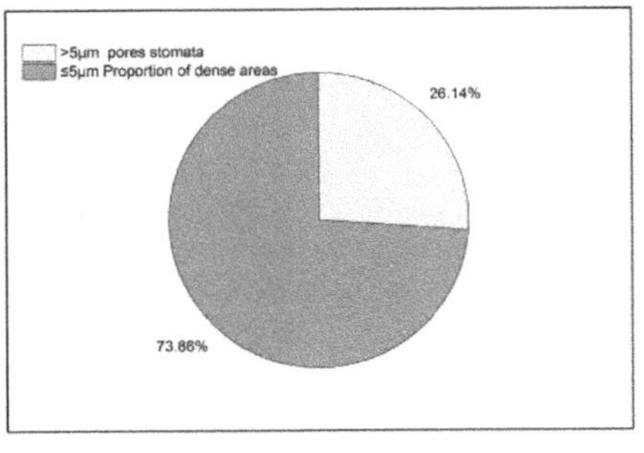

(**d**) Latex dosage of 15% (mix ratio 4)

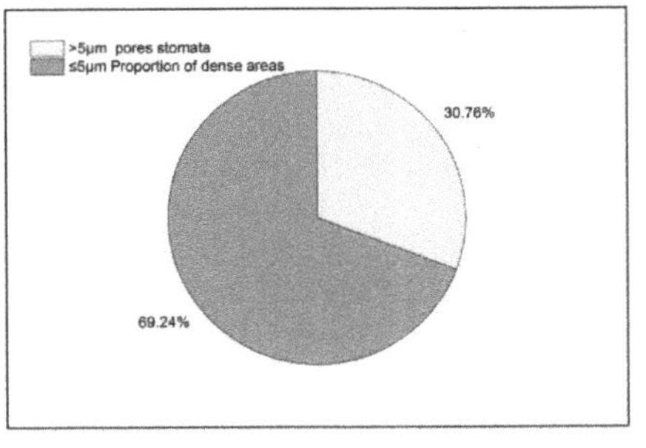

(**e**) Latex dosage of 20% (mix ratio 5)

Figure 4. Evolution law of total porosity in cement-based materials.

Styrene butadiene latex is a material characterized by a three-phase system of solid–liquid–gas. When utilized as an additive in mortar, its adhesive properties enhance the interface between the mortar and air. This facilitates the introduction of air into the interior of the mortar. The adhesive qualities of the latex compound this effect, making it more challenging for internal pores to escape. As a result, the hardened mortar experienced an increase in the total number of planar pores.

At a low latex dosage of 5%, the latex dispersed non-uniformly within the cementitious material. The bubbles also exhibited uneven dispersion, occupying various spatial positions. The mutual compression and merging of these bubbles were relatively limited at this stage. Consequently, the environment favored the formation of bubbles, leading to a doubling of porosity under this lower latex dosage. As the latex dosage increased, the dispersion gradually transitioned from isolated to more uniform within the cementitious material. Simultaneously, the number of bubbles increased, and they began to occupy shared spatial locations more frequently. This heightened the mutual compression, merging, and rupture of bubbles, creating fluctuations in porosity levels.

Upon reaching a latex dosage of 15%, the total number of pores in the mortar began to decrease. This decline was attributed to increased competition for limited spatial positions, leading to some bubbles being compressed and escaping, ultimately reducing the overall pore count in the hardened mortar. With a further increase in latex dosage to 20%, the number of introduced bubbles also increased. However, these additional bubbles began to rupture and reduce each other's numbers. As some bubbles escaped and formed transient channels, the total porosity of the mortar experienced a slight increase. This suggests that, due to interactions among the bubbles, the pore count cannot increase indefinitely, even with higher latex concentrations.

3.2. Analysis of Stomatal Evolution Patterns of Different Pore Sizes

In cementitious materials, pores of varying sizes are distributed throughout. According to the planar distribution model, these pores fell into four main categories: micropores (5–200 μm), tiny pores (200–500 μm), medium pores (500–1000 μm), and large pores (>1000 μm). The addition of styrene butadiene latex polymer as an additive introduced a considerable number of air bubbles into the cement mortar. This not only influenced the total number of pores after the material had hardened but also affected the sizes of these pores to varying degrees. From the perspective of performance enhancement, the goal was to maintain or increase the total number of pores, particularly favoring smaller pores. Smaller pores were more favorable for improving various performance indicators, such as mechanical strength and durability. However, the specific distribution of different pore sizes within cementitious materials, influenced by the addition of latex, and how this pore structure evolved from the fresh mix to the hardened state, needed to be verified through experimental tests.

Using standard cement mortar (test mix ratio 1) as a reference, Figure 5 reflects the planar distribution of pores of various sizes within the standard mortar. Pores smaller than 5 μm were categorized as the dense region, constituting 84.59% of the plane cross-section of the standard mortar. The remaining 15.41% was occupied by pores of different sizes: 5–200 μm pores accounted for 2.46%, 200–500 μm pores represented 4.61%, 500–1000 μm pores made up 5.36%, and pores larger than 1000 μm constituted 2.98%. The distribution exhibited a characteristic pattern: fewer pores at the extremes and more in the middle ranges. Using this standard mortar as a benchmark, the analysis of pore distribution across different size ranges in cementitious materials impacted by external factors formed the basis for understanding changes in material performance at the microstructural level.

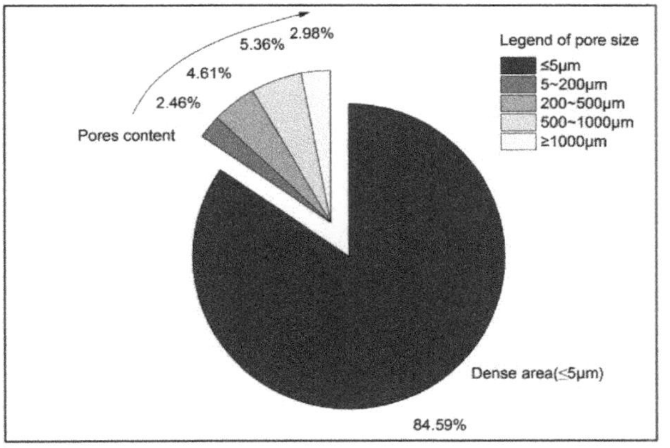

Figure 5. Plane distribution of pores with different pore sizes in standard mortar.

An experimental investigation was undertaken to analyze the distribution characteristics of pores of different sizes in cementitious materials. In this study, the control group consisted of standard mortar without latex, while the primary focus was on latex-modified mortar with varying levels of latex. Figure 6 illustrates the evolution of pore content across different size ranges. Significantly, the control group (standard mortar) exhibited the lowest pore content in all size categories, affirming that the addition of styrene butadiene rubber latex generally increased pore content within each category.

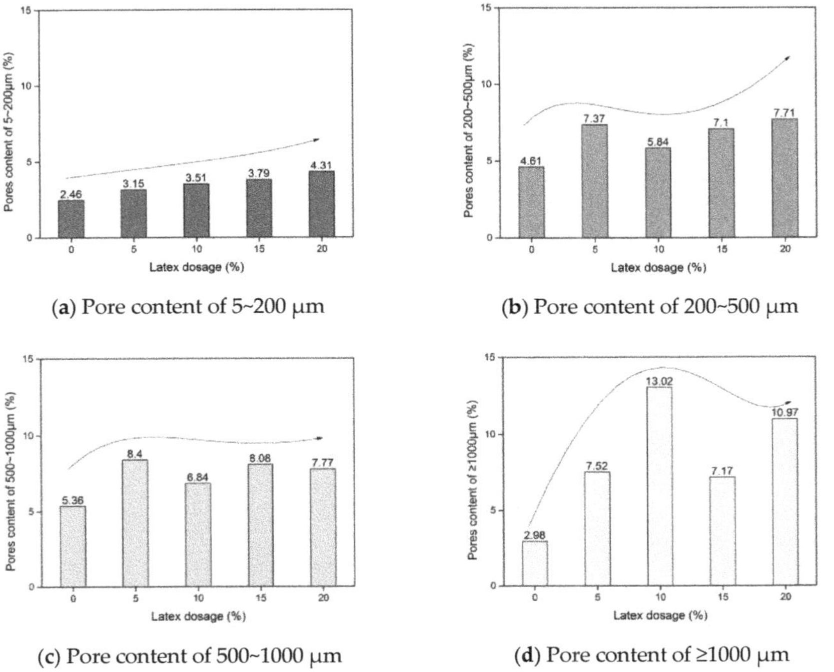

(**a**) Pore content of 5~200 μm

(**b**) Pore content of 200~500 μm

(**c**) Pore content of 500~1000 μm

(**d**) Pore content of ≥1000 μm

Figure 6. Evolution law of pores with different sizes.

Within the 5–200 μm pore size range, the content of pores in latex-modified mortar progressively increased with higher levels of latex doping. The most substantial increase was observed in the data corresponding to a 20% latex doping. In comparison to the standard mortar, the pore content in this range increased by 75.2%. This suggests that the introduction of latex as an admixture led to a shift in the distribution of tiny pores (5–200 μm) in cementitious materials, resulting in increased pore content.

In the pore size ranges of 200–500 μm and 500–1000 μm, a similar pattern of evolution in pore content was observed. With a low latex dosage of 5%, both types of pores exhibited exponential growth. When the latex dosage reached 10%, there was a noticeable change in the growth of pore content. In comparison to the standard mortar, the pore content increased by 26.68% and 27.61% in these size ranges, respectively. As the latex dosage continued to increase, the pore content exhibited a wave-like growth trend. This analysis revealed that at lower latex dosages, the latex was relatively isolated and unevenly distributed within the cementitious material. In this state, there was ample space for the bubbles, weak interactions between them, and limited introduction of bubbles into the mortar, which were conducive to the formation of porosity.

As the latex levels increased, its distribution within the cementitious material became more uniform, as depicted in Figure 7. This led to increased competition for space among the bubbles, more frequent interactions, and intensified processes of bubble extrusion, merging, and rupture. Some bubbles grew into larger sizes, some were crushed due to extrusion, and others escaped the confines of the cementitious material. These intricate internal changes resulted in an oscillatory trend in pore content. However, higher latex dosages altered the solid–liquid–gas three-phase system within the cementitious material, leading to a general increase in the content of both types of pores.

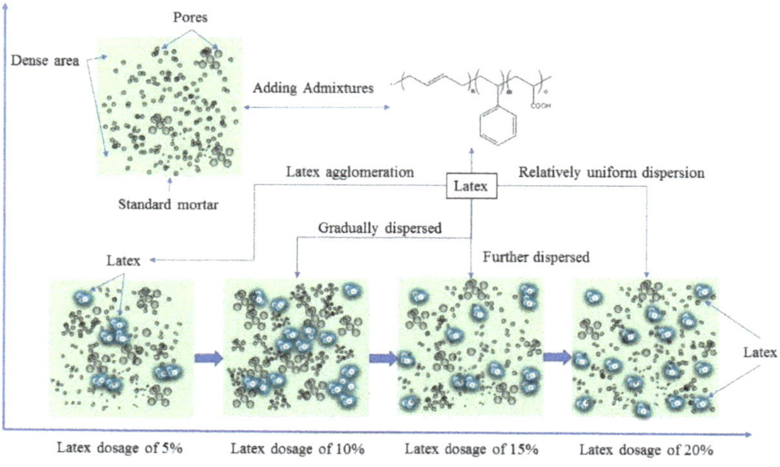

Figure 7. Schematic diagram of dispersion pattern of latex in cement-based materials.

Pore sizes exceeding 1000 μm are categorized as macroporosity, which is known to have detrimental effects on the mechanical properties and durability of cementitious materials, as referenced in [10–21]. Hence, we sought to minimize the presence of these large pores. The test data presented in Figure 6 indicates that standard mortar had the lowest macropore content (>1000 μm) compared to the other pore size ranges, such as 200–500 μm and 500–1000 μm. This low macropore content contributed to the superior performance of the standard mortar.

As the quantity of the latex additive increased, the macropore content in the latex-modified mortar displayed a significant overall upward trend. With a low dosage of 5%, the macropore content essentially doubled. The most substantial increase was observed at a 10% latex dosage, where the macropore content in the latex-modified mortar was 4.37 times that of the standard mortar. When the dosage reached 20%, the macropore content was 3.68 times that of the standard mortar. With increasing latex dosages, the dispersion state of the latex within the cementitious material underwent significant changes. Initially, it was isolated and unevenly distributed, but it eventually became uniformly dispersed within the cementitious matrix. During this transition, the spatial conditions for the bubbles introduced by the latex also changed significantly, shifting from relatively spacious to increasingly crowded. This led to heightened bubble interactions, including rupture, merging, and escape, which in turn drastically increased the content of large pores. This escalation had a particularly adverse impact on the mechanical properties of the cementitious materials.

Figure 8 presents the total pore count and the variations in pore content across different size ranges. When analyzed in conjunction with the planar distribution model of pores in cementitious materials, it became apparent that standard cement mortar featured a relatively low proportion of tiny pores (5–200 μm) and large pores (>1000 μm). In contrast, medium-sized pores (500–1000 μm) and small pores (200–500 μm) constituted a larger proportion. This distribution is characteristic of mortar materials and plays a crucial role in shaping their mechanical properties and durability.

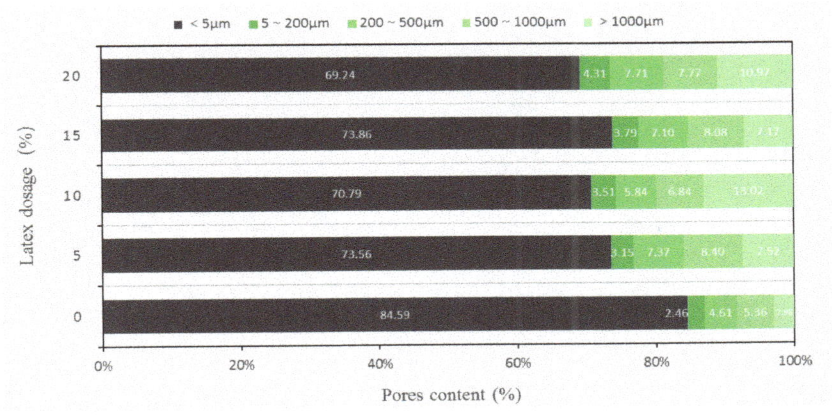

Figure 8. Evolution law of pore size in cement-based materials.

However, the incorporation of styrene butadiene latex generally increased the internal porosity of the cementitious materials. The growth in tiny pores (5–200 μm) was relatively moderate, while the increase in medium-sized (500–1000 μm) and small pores (200–500 μm) followed similar trends. It is worth noting that even with a 10% latex admixture in the modified mortar, the proportion of medium and small pores remained significant. Despite this, the most significant increase was observed in the content of large pores (>1000 μm), which had a detrimental impact on the overall performance of cementitious materials.

3.3. *Influence Law of Porosity on Mechanical Properties of Cementitious Materials*

The creation of porosity exerts a nuanced influence on the mechanical properties of cementitious materials. Specifically, excessive porosity and the presence of larger pores can have an adverse impact on the material's densification, resulting in a deterioration of its mechanical properties. However, the existence of smaller pores can yield beneficial effects on these properties, as demonstrated by various studies [33]. Testing was carried out following the guidelines specified in the Test Procedure for Cement and Cement Concrete in Highway Engineering (JTG 3420-2020) [29]. The experiments encompassed flexural and compressive strength tests conducted on cement mortar specimens, all of which were prepared using an identical proportionate mix. The testing process is depicted in Figure 9.

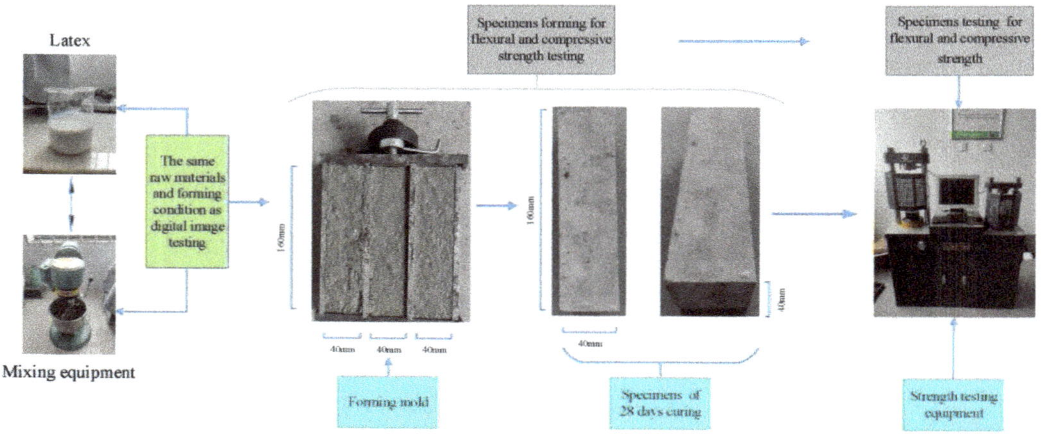

Figure 9. Flow chart of strength test for cementitious materials.

Figure 10 presents the correlation between the total number of pores and the flexural and compressive strength of cementitious materials. Generally, an inverse relationship exists: the greater the total number of pores, the more detrimental it is to the flexural and compressive strength of cementitious materials. However, a slight increase in pore count, corresponding to 5% and 15% latex doping, enhanced flexural strength. When the increase in total pores became too significant, there was a noticeable deterioration in flexural strength.

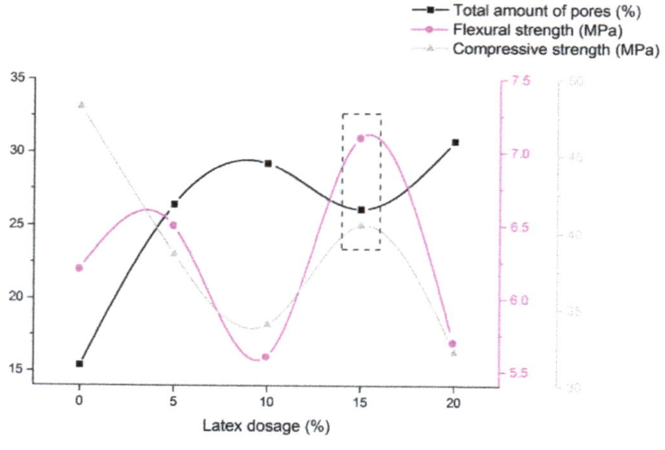

Figure 10. Effect of total porosity on the flexural and compressive strength of the cementitious materials.

The impact of pore count on compressive strength consistently indicated a weakening effect, suggesting that latex inclusion hinders the development of strength in cementitious materials. Interestingly, at 15% latex doping, a reduction in the total pore count aligns with an improvement of nearly 1 MPa in flexural strength and a lesser reduction in compressive strength. This implies that a reduction in pores could enhance the mechanical properties of the material. Conversely, at 10% latex doping, the presence of excessively large pores (>1000 µm) and a higher total pore count substantially impaired the mechanical properties of the material.

Figure 11 delves into the impact of various pore sizes on the flexural and compressive strength of the material. The data revealed a limited correlation between changes in flexural and compressive strength and the content of tiny pores (5–200 μm). Notably, at 15% latex doping, an increased concentration of these tiny pores corresponded to a favorable increase in flexural strength and a gradual reversal of the decreasing trend in compressive strength.

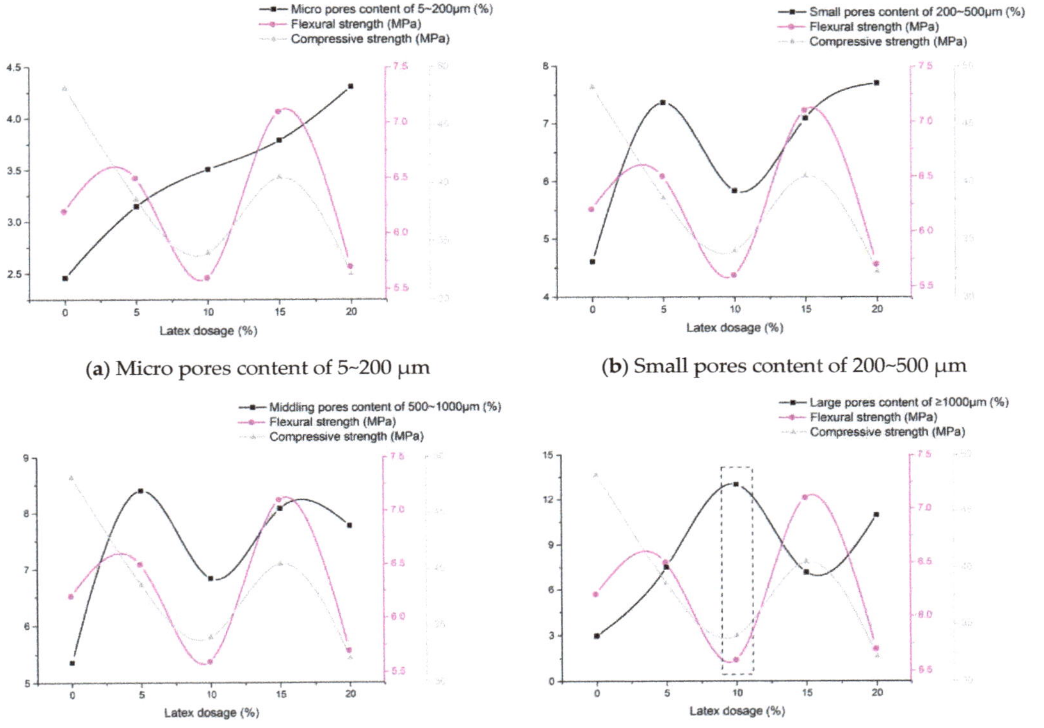

(**a**) Micro pores content of 5~200 μm (**b**) Small pores content of 200~500 μm

(**c**) Middling pores content of 500~1000 μm (**d**) Large pores content of ≥1000 μm

Figure 11. Effect of pore content with different size pores on the flexural and compression strength of the cementitious materials.

Figure 11b,c illustrates that the curves representing changes in the content of small pores (200–500 μm) and medium pores (500–1000 μm) roughly aligned with alterations in the flexural and compressive strengths of cementitious materials. This implies that an increase in the content of small and medium pores was beneficial for enhancing the mechanical properties of these materials. At 15% latex doping, both the decline in compressive strength and the improvement in flexural strength were moderated.

Figure 11d reveals that the curves reflecting changes in macropore content (≥1000 μm) were inversely related to the curves for both flexural and compressive strength in cementitious materials. This indicates that an increase in macropore content had a negative impact on the material's mechanical properties. Specifically, at 10% latex doping, the macropore content reached its peak, resulting in a significant decrease in both flexural and compressive strength. Conversely, at 15% latex doping, both the macropore content and total porosity were relatively low. This led to improved flexural strength and a less pronounced decrease in compressive strength, creating a more balanced relationship between these two mechanical properties. The effects on both flexural strength and compressive strength were relatively balanced.

In summary, the test results indicated that at low dosages, latex was unevenly dispersed throughout cementitious materials, resulting in relatively isolated pores that occupied limited space. During this phase, the viscoelastic properties of the latex improved the material's elasticity, enabling greater flexural deformation and mitigating stress concentration, thereby enhancing the material's flexural strength. However, as the latex dosage increased, it began to agglomerate within the cementitious matrix, introducing a significant number of air bubbles. While the viscoelastic nature of the latex initially contributed to filling and improving the matrix, the increased number of air bubbles began to compromise the material's structural integrity. This lack of compaction was exacerbated by the presence of a considerable number of large pores, severely undermining the mechanical properties of the material, particularly its compressive strength.

In essence, the mechanical properties of cementitious materials were influenced by both the total pore volume and the size distribution of these pores. When using admixtures, it was crucial to consider not only the overall pore count but also the specific distribution of pore sizes, especially the concentration of large pores (>1000 μm). This comprehensive approach aided in the accurate evaluation of the material's mechanical properties and helped establish effective feedback mechanisms for optimizing admixture applications. Pore content analysis provided a microscopic lens through which to explore changes in the material's properties, and our tests validated the feasibility and precision of this method for assessing external influences. Unlike methods requiring expensive instrumentation, complex sample preparation, or intricate operational steps, our experimental study presented a cost-effective and straightforward approach for analyzing pore evolution in cementitious materials. This offered both a practical methodology and a theoretical foundation for enhancing the material's performance, optimizing admixture types, and determining appropriate dosages.

4. Conclusions

(1) A characterization method was introduced for analyzing the pore distribution within this two-dimensional context, which involved enhancing image grayscale processing and binarization to achieve greater precision in distinguishing between pore-dense and pore-sparse regions. The distribution characteristics of pores of varying sizes in the two-dimensional structure of the materials were examined, and a pore distribution model was developed. Based on this model, the overall pore volume in the material, the distribution of pores across different size ranges, and the resulting impact on the material's mechanical properties were investigated. The key conclusions from this study included the following:

(2) Analysis of the total pore volume within the complex solid–liquid–gas system of cementitious materials revealed that the addition of latex increased the overall porosity. Remarkably, when the latex dosage reached 15%, there was an inflection point in the increase in the total pore volume, after which it showed a modest rebound.

(3) The pore content in all four categories tended to increase with the addition of latex. However, it was crucial to note that increasing the latex dosage did not linearly result in more significant porosity. This suggested that pore formation was influenced not just by the admixture itself but also by the interaction between the latex and the cementitious material, among other factors.

(4) The study revealed that tiny pores marginally increased with higher latex dosages. In contrast to other pore sizes, the absolute content of tiny pores remained low, suggesting that latex additives had a limited impact. Small and medium pores behaved similarly with latex addition; both types showed a general increase but experienced a decline at a latex dosage of 10%.

(5) In the case of large pores, the highest content was observed at 10% latex dosage, reaffirming the trend of smaller pores evolving into larger ones. These findings confirmed that smaller pores tended to develop into larger pores over time.

(6) Higher porosity and micropore content generally weakened mechanical performance. However, at a small latex dosage, there was an improvement in flexural strength. When the latex dosage reached 15%, the total pore and micropore content declined, resulting in a balanced increase in flexural strength and a mitigated decline in compressive strength.
(7) In the binarization process of planar images of cementitious materials using digital imaging techniques, a combination of automated threshold determination and manual comparison was employed to maximize accuracy. Despite these efforts, the method still has room for improvement, as it is susceptible to human error. Further research is necessary to enhance the quantitative accuracy of binarization thresholds.

Author Contributions: Methodology, Y.Y. and S.Z.; software, Z.H. and G.Y.; formal analysis, K.N.; investigation, B.T.; data curation, W.B. and Y.Y.; writing—original draft, Y.Y.; writing—review and editing, S.Z. and G.Y.; visualization, Z.H. and G.Y.; supervision, K.N., B.T. and L.C.; project administration, Y.Y. and S.Z. All authors have read and agreed to the published version of the manuscript.

Funding: This research was funded by 2021 Key Technology Projects of the Transportation Industry (grant number 2021-MS1-036), the Transportation Technology Project of Hainan province (grant number hizw20200903007; Qiong Jiao Ke Xin [2020] No. 293), the Technology Innovation and Demonstration Project of Yunnan Provincial Department of Transportation (2022-81), and the Yunnan Transportation Investment and Construction Group Co., Ltd. Technology Innovation Project (YCIC-YF-2021-09), and Science and Technology Plan Project of Yunnan Provincial Department of Science and Technology (202103AA080013). The funder had the following involvement with the study: the study design, collection, analysis, interpretation of data, the writing of this article, and the decision to submit it for publication.

Institutional Review Board Statement: Not applicable.

Informed Consent Statement: Not applicable.

Data Availability Statement: Data are contained within the article.

Conflicts of Interest: Author Y.Y. was employed by the company Chongqing Jiaotong University and Guangxi Nanbai Expressway Co., Ltd. The remaining authors declare that the research was conducted in the absence of any commercial or financial relationships that could be construed as a potential conflict of interest.

References

1. Han, J. *Research on the Performance of High-Performance Cement Concrete Pavement in Seasonal Frozen Areas*; Chang'an University: Xi'an, China, 2009; pp. 124–125. (In Chinese)
2. Czarnecki, L. Concrete-polymer composites: Trends shaping the future. *Int. J. Soc. Mater. Eng. Resour.* **2007**, *15*, 1–5. [CrossRef]
3. Ollitrault-Fichet, R.; Gauthier, C.; Clamen, G.; Boch, P. Microstructural aspects in a polymer-modified cement. *Cem. Concr. Res.* **1998**, *28*, 1687–1693. [CrossRef]
4. Li, Z.; Liang, N.; Wu, D.; Jing, W. Mechanism analysis of polymer cement-based materials. *J. Highw. Trans. Res. Dev.* **2005**, *2*, 63–66+86. (In Chinese)
5. Zhang, G.; Wang, P. Effect of hydroxyethyl methylcellulose on the formation of cement hydration products. *J. Build. Mat.* **2010**, *13*, 573–577. (In Chinese)
6. Yang, Z. *Research on the Control and Durability of Concrete Pore Structure in High Altitude Low Pressure Environment*; Chang'an University: Xi'an, China, 2023. (In Chinese)
7. Sakai, E.; Kasuga, T.; Sugiyama, T.; Asaga, K.; Daimon, M. Influence of superplasticizers on the hydration of cement and the pore structure of hardened cement. *Cem. Concr. Res.* **2006**, *36*, 2049–2053. [CrossRef]
8. Kondraivendhan, B.; Bhattacharjee, B. Effect of age and water-cement ratio on size and dispersion of pores in ordinary portland cement paste. *J. ACI Mat.* **2010**, *107*, 147–154.
9. Zhang, L.; Wang, W. Research on the influence of polymer on the pore structure and performance of foam concrete. *J. Concr Cem. Prod.* **2011**, *185*, 9–12. (In Chinese) [CrossRef]
10. Wu, Z. Reflection on concrete science and technology. *J. Concr. Cem. Prod.* **1988**, *6*, 4–5. (In Chinese)
11. Wu, Z.; Lian, H. *High Performance Cement Concrete*; China Railway Press: Beijing, China, 1999. (In Chinese)
12. Wang, T.; Dra, Y.A.S.S.; Cai, X.; Cheng, Z.; Zhang, D.; Lin, Y.; Yu, H. Advanced cold patching materials (CPMs) for asphalt pavement pothole rehabilitation: State of the art. *J. Clean. Prod.* **2022**, *366*, 133001. [CrossRef]
13. Shen, A. *Cement and Cement Concrete*; Beijing Communications Press: Beijing, China, 2000; pp. 93–103. (In Chinese)

14. CEB European Committee on Concrete. *CEB Design Guidelines for Durable Concrete Structures*, 2nd ed.; Zhou, Y.; Di, X.; Han, J.; Cai, L., Translators; China Academy of Building Sciences: Beijing, China, 1989.
15. He, J. *Research on the Microstructure and Properties of Road Cement Concrete*; Chang'an University: Xi'an, China, 2009. (In Chinese)
16. Wang, T.; Li, M.; Cai, X.; Cheng, Z.; Zhang, D.; Sun, G. Multi-objective design optimization of composite polymerized asphalt emulsions for cold patching of pavement potholes. *Mater. Today Commun.* **2023**, *35*, 105751. [CrossRef]
17. Wang, T.; Weng, Y.; Cai, X.; Li, J.; Xiao, F.; Sun, G.; Zhang, F. Statistical modeling of low-temperature properties and FTIR spectra of crumb rubber modified asphalts considering SARA fractions. *J. Clean. Prod.* **2022**, *374*, 134016. [CrossRef]
18. Yan, X. *Study on the Influence of Pore Structure on the Salt Freezing Resistance of Cement-Based Materials*; Southeast University: Nanjing, China, 2015. (In Chinese)
19. Yang, W. *Effect of Inorganic Salts on Pore Structure and Frost Resistance of Concrete*; Harbin Institute of Technology: Harbin, China, 2009. (In Chinese)
20. Jo, B.W.; Park, S.K.; Kim, D.K. Mechanical properties of nano-MMT reinforced polymer composite and polymer concrete. *Constr. Build. Mater.* **2008**, *22*, 14–20. [CrossRef]
21. Pascal, S.; Alliche, A.; Pilvin, P. Mechanical behavior of polymer modified mortars. *J. Mater. Sci. Eng. A* **2004**, *380*, 1–8. [CrossRef]
22. Beeldens, A.; Gemert, D.V.; Ohama, Y. Integrated model of structure formation in polymer modified concrete. In Proceedings of the 11th International Congress on the Chemistry of Cement 2003, Durban, South Africa, 11–16 May 2003; pp. 206–216.
23. *JTG 3420-2020*; Test Regulations for Cement and Cement Concrete for Highway Engineering. People's Communications Press: Beijing, China, 2020.
24. Beeldens, A. Evolution in modeling microstructure formation in polymer-cement concrete: Restoration of buildings and monuments. *J. Restor. Build Monum.* **2013**, *19*, 97–108.
25. Abdulateef, Y.M.; Fahad, B.M.; Eweed, K.M. Repairing of concrete by using polymer-mortar composites. *Int. J. Mod. Eng. Res.* **2014**, *1*, 63–72.
26. Ukrainczyk, N.; Rogina, A. Styrene–butadiene latex modified calcium aluminate cement mortar. *Cem. Concr. Compos.* **2013**, *41*, 16–23. [CrossRef]
27. Petit, J.Y.; Wirquin, E. Evaluation of various cellulose ethers performance in ceramic tile adhesive mortars. *Int. J. Adhes. Adhes.* **2013**, *40*, 202–209. [CrossRef]
28. Bühler, T.; Zurbriggen, R.; Pieles, U.; Huwiler, L.; Raso, R. Dynamics of early skin formation of tiling mortars investigated by microscopy and diffuse reflectance infrared Fourier transformed spectroscopy. *Cem. Concr. Compos.* **2013**, *37*, 161–170. [CrossRef]
29. Ohama, Y. Principle of latex modification and some typical properties of latex-modified mortars and concretes. *J. ACI Mat.* **1987**, *84*, 511–518.
30. Pourchez, J.; Ruot, B.; Debayle, J.; Pourchez, E.; Grosseau, P. Some aspects of cellulose ethers influence on water transport and porous structure of cement-based materials. *Cem. Concr. Compos.* **2010**, *40*, 242–252. [CrossRef]
31. Bülichen, D.; Kainz, J.; Plank, J. Working mechanism of methyl hydroxyethyl cellulose (MHEC) as water retention agent. *Cem. Concr. Compos.* **2012**, *42*, 953–959. [CrossRef]
32. Ahmed, S.F.U. Mechanical and durability properties of mortars modified with combined polymer and supplementary cementitious materials. *J. Mater. Civil. Eng.* **2011**, *23*, 1311–1319. [CrossRef]
33. Tabassi, A.A.; Ramli, M. Effects of different curing regimes on engineering properties of polymer-modified mortar: An experimental study. *J. Mater. Civil. Eng.* **2012**, *24*, 468–478.
34. Linares-Unamunzaga, A.; Pérez-Acebo, H.; Rojo, M.; Gonzalo-Orden, H. Flexural strength prediction models for soil–cement from unconfined compressive strength at seven days. *J. Mater.* **2019**, *12*, 387. [CrossRef]
35. Tran, V.Q. Hybrid gradient boosting with meta-heuristic algorithms prediction of unconfirmed compressive strength of stabilized soil based on initial soil properties, mix design and effective compaction. *J. Clean. Prod.* **2022**, *355*, 131683. [CrossRef]

coatings

MDPI

Article

Study on the Performance of Nano-Zinc Oxide/Basalt Fiber Composite Modified Asphalt and Mixture

Chaojie Li [1,2,3], Zhenxia Li [3], Tengteng Guo [3], Yuanzhao Chen [3,4,*], Junying Ma [3], Jing Wang [3] and Lihui Jin [3]

1 Henan Transportation Research Institute Co., Ltd., Zhengzhou 450045, China; lcj13676936229@163.com
2 Xi'an Changda Highway Engineering Inspection Center Co., Ltd., Xi'an 710064, China
3 School of Civil Engineering and Communication, North China University of Water Resources and Electric Power, Zhengzhou 450045, China; zhenxiali2009@ncwu.edu.cn (Z.L.); guotth@ncwu.edu.cn (T.G.); junying6568@163.com (J.M.); wj@ncwu.edu.cn (J.W.); lihuijin2023@163.com (L.J.)
4 Henan Province Engineering Technology Research Center of Environment Friendly and High-Performance Pavement Materials, Zhengzhou 450006, China
* Correspondence: cyz740513@ncwu.edu.cn

Abstract: In order to improve the service quality of roads and resolve the problem of defects in the conventional asphalt pavement in service, this paper uses a 5.3% aluminate coupling agent to modify the surface of nano-ZnO and prepares a composite-modified asphalt with nano-ZnO and basalt fiber (BF) as modifiers. First, the basic performance of different types of asphalt was investigated by means of a rotary film oven experiment. Then, a dynamic shear rheology experiment was carried out to analyze the high-temperature anti-rutting performance of the composite-modified asphalt at different temperatures and frequencies. Then, using a bending creep stiffness test, the low-temperature properties of the composite-modified asphalt were investigated. Finally, the microstructure and modification mechanisms of the composite-modified asphalt were analyzed with scanning electron microscopy and infrared spectroscopy. The results indicate that the anti-aging performance of the nano-ZnO/BF composite-modified asphalt is significantly improved after adding fibers to the modified asphalt. The average mass loss ratio is only 0.192%. At 46 °C, the rutting coefficient of the composite-modified asphalt was increased by 62.3%. The frequency master curve is always at the highest position and continues to rise, indicating a significant improvement in the high-temperature anti-rutting performance of the composite-modified asphalt. At 24 °C, the creep stiffness modulus S value of the composite-modified asphalt increased by 24.9%; moreover, there is no obvious effect of improving low temperature, but the variation range of creep tangent slope m of the modified asphalt after aging is decreased, which further shows that the addition of a modifier can decrease the influence of aging on asphalt. Nanoparticles are uniformly dispersed in the asphalt and form a three-dimensional interconnected structure with BF, which effectively improves the overall performance of the asphalt. Nano-ZnO and fibers have weak chemical reactions in matrix asphalt, but they are physically dispersed and compatible.

Keywords: road engineering; modification mechanism; experimental study; composite-modified asphalt; microscopic morphology

Citation: Li, C.; Li, Z.; Guo, T.; Chen, Y.; Ma, J.; Wang, J.; Jin, L. Study on the Performance of Nano-Zinc Oxide/Basalt Fiber Composite Modified Asphalt and Mixture. *Coatings* **2024**, *14*, 23. https://doi.org/10.3390/coatings14010023

Academic Editor: Valeria Vignali

Received: 19 November 2023
Revised: 18 December 2023
Accepted: 22 December 2023
Published: 25 December 2023

1. Introduction

With the fast development of the highway transportation system, the total mileage of the national expressway is increasing year by year. During the 14th Five-Year Plan, China will achieve the goal of 25,000 km of expressway reconstruction and realize the main line of the "71,118" national highway network [1]. Asphalt pavement has the advantages of simple construction and comfortable driving, and more than 90% of expressways in China are paved with asphalt. Due to complex environmental changes and the increasing proportion of heavy-duty channelized traffic, conventional asphalt pavement is liable to be damaged during its service period, which makes the service quality of the road decline rapidly [2,3]. This is one of the main difficulties in the construction of high-quality modern highway systems in the new era. Therefore, improving the original asphalt to improve its road performance has become an important research direction in the field of transportation [4,5]. The unique and outstanding efficacy of nano-ZnO makes the performance of nano-ZnO-modified asphalt incomparable with other modified asphalt. The nano-ZnO-modified asphalt has a lower complex module and a higher phase angle, which shows that the asphalt has excellent heat-oxidation and photo-oxidation resistance, and the nano-ZnO modifier can significantly improve the high-temperature properties of matrix asphalt after PAV aging [6,7]. The increase in nano-ZnO content causes the Marshall stability, the indirect tensile strength, the pull-off adhesion, and the fracture energy of asphalt mixture to obviously increase, and the coating degree of asphalt on aggregate increases [8,9]. Nano-ZnO has a cross-linking reaction with asphalt, and nano-ZnO has physical and chemical modifications on matrix asphalt [10–12]. Compared to other modified asphalt, nano-ZnO-modified asphalt has outstanding road performance, but the addition of only nano-ZnO modifiers to improve the low-temperature performance of the asphalt mixture is not significant [13,14]. Studying nano-ZnO in asphalt and how to improve its low-temperature properties and deepen its high-temperature properties is an urgent problem.

Because of the high strength and green environmental protection of BF, the related research on BF-modified asphalt has entered people's field of vision [15,16]. When the BF content is 2% and the length is 9 mm, BF-modified asphalt enhances the stress dissipation capacity at low temperatures to a certain extent, showing excellent rheological properties, and the BF is spatially distributed in the asphalt mortar, which has better stability and enhancement [17,18]. Compared to normal asphalt, adding BF greatly reduces the freeze–thaw splitting strength of aged specimens, effectively prevents the asphalt from aging, and improves the resistance to low temperatures and water damage. BF can effectively improve the combination of mineral aggregate and asphalt. Adding a small amount of fiber can significantly increase the viscosity of the asphalt mortar, hinder the effective flow of the asphalt, and, to some extent, lower the cracking probability of asphalt mortar, thus improving its low-temperature crack resistance. This provides the theoretical foundation for the popularization and application of BF in asphalt pavement [19–22].

In summary, as opposed to conventional asphalt blends, the addition of nanoparticles can significantly improve the performance of the asphalt mixture at high temperatures and water stability. The BF is added to the asphalt to strengthen the anti-crack and fatigue resistance of the asphalt mixture, thus increasing the durability of the pavement materials. However, in the results of recent research, there are few literature studies on the addition of multi-dimensional, multi-scale fibers and nanomaterials to the matrix bitumen. Therefore, nano-ZnO and BF are used as composite admixtures to modify the matrix asphalt in this paper. Nano-ZnO and BF are adopted as compound admixtures for modification of base asphalt. The performance of nano-ZnO/BF compound-modified asphalt at high and low temperatures is evaluated, and the micro-modification mechanism and road performance of composite-modified bitumen are also analyzed. This study may provide new optimization and improvement ideas for the development of China's transportation field.

2. Raw Material

2.1. Asphalt

This article uses grade 70# asphalt from the Jintan Municipal Construction Co., Ltd. (Jintan, China). The basic performance index is determined according to the test procedure [23]. The results of the test are given in Table 1. In accordance with the requirements of the Technical Specification [24], the 70# asphalt specification meets the specification conditions.

Table 1. Basic performance index of 70# asphalt.

Test Items		Unit	Test Results	Technical Requirements	Test Method
Penetration (25 °C, 100 g, 5 s)		0.1 mm	64.8	60~80	T0640
Ductility (5 cm/min, 5 °C)		cm	13.4	≥0	T0605
Ductility (5 cm/min, 15 °C)		cm	127.6	≥100	T0650
Softening point (Ring and ball method)		°C	47.5	≥46	T0606
Flash point		°C	280	≥260	T0611
Density (25 °C)		g/cm^3	1.205	Measured value	T0603
After RTFOT	Quality change	%	−0.264	−0.8~+0.8	T0610
	Penetration ratio	%	73.5	≥61	T0604
	Ductility (5 cm/min, 5 °C)	cm	11.2	≥6	T0605
	Ductility (5 cm/min, 15 °C)	cm	130.4	≥15	T0605

2.2. Nano-Zinc Oxide

The nano-Zinc Oxide (nano-ZnO) used in the test is provided by Pinyu New Material Co., Ltd. (Shanghai, China). The shape of the nano-ZnO particles is approximately spherical, and the particle diameter is about 30 nm. It has the morphological characteristics of nanomaterials and belongs to the standard nanomaterials. Its technical index is shown in Table 2.

Table 2. Technical Indicators of Nano-ZnO.

Performance	Appearance	Purity (%)	Specific Surface Area (m^2/g)	Loose Density (g/cm^3)
Nano-ZnO	White powder	99.6	58	0.94

2.3. Basalt Fiber

The basalt fiber (BF) used is a high-quality fiber provided by Hunan Basalt Stone Co., Ltd. (Changsha, China). The BF is linearly distributed, the surface is smooth, and it is not easy to deform and distort. The conventional performance index parameters are shown in Table 3.

Table 3. Performance index of BF.

Performance	Length (mm)	Diameter (μm)	Density (g/cm^3)	Fracture Elongation (%)	Elastic Modulus (GPa)	Tensile Strength (MPa)
BF	6	12	2.94	2.958	95	3500

3. Test Scheme

3.1. Optimization of Surface Properties of Nano-ZnO

In this paper, an aluminate coupling agent was selected to modify the surface of nano-ZnO [25]. The experimental equipment was a heat-collecting constant temperature heating magnetic stirrer. The concrete steps are as follows:

(1) Different doses of aluminate coupling agent were added to the prepared ethanol solution (anhydrous ethanol:water = 9:1). The mixture was stirred with a glass rod evenly and then added to a conical flask. By stirring the mixture for 5 min at room temperature with a magnetic stirrer, the nano-ZnO was brought into full contact with the coupling agent.

(2) A certain amount of nano-ZnO was weighed and dried in the electric blast drying box and slowly added to a conical bottle. The motor temperature was adjusted to 70 °C, the speed to 300 rpm, and stirring was continued for 40 min.

(3) The activated nano-ZnO was placed in an electric blast drying furnace, and the temperature was controlled at about 100 °C. Finally, the dried nano-ZnO was placed in a mortar for grinding.

Each type of test includes three samples, including parallel experiments. The lipophilicity of the nano-ZnO was tested after modification, and the dose modification effect corresponding to the maximum lipophilicity value was the best. Finally, a 5.3% aluminate coupling agent was selected to activate the nano-ZnO.

3.2. Preparation of Composite-Modified Asphalt

The matrix asphalt was placed in an electric hot blast drying box and heated to melting at a temperature of 180 °C for dehydration. Then, 500 g of the matrix asphalt was quickly weighed into the heating furnace of the high-speed shear machine to keep it flowing. The temperature was set to about 150 °C for heating. In the course of heating, continuous mixing of the base asphalt with the glass rod can avoid local heating and aging of the asphalt. First, 4% of the surface-modified dry nano-ZnO was slowly added in batches to the matrix asphalt, and then the high-speed shear machine was used to slowly shear for about 15 min. When the white powder in the asphalt was evenly dispersed, 2% of the dry BF was slowly added in batches, and it was necessary to continuously carry out manual stirring to ensure it was evenly dispersed during the addition process. The shear rate was kept at 5000 r/min. The shear rate was cut for 60 min, then, it was placed in an electrothermal blast drying oven at 180 °C to swell and develop until the asphalt surface was smooth and free of bubbles. In this way, the nano-ZnO/BF compound-modified asphalt was prepared.

3.3. Rolling Thin Film Oven Test

The aging test of 50 ± 0.5 g matrix asphalt, nano-ZnO-modified asphalt and nano-ZnO/BF composite-modified asphalt was carried out according to the requirements of T0610 in the specification [23] using an asphalt rotating film oven. The oven temperature was kept at 163 ± 0.5 °C for a total duration of 85 min. All test procedures should be completed within 72 h. The basic performance of three kinds of asphalt after aging was analyzed and the anti-aging properties of the compound-modified asphalt were compared and evaluated.

3.4. Rheological Test of High-Temperature Dynamic Shear

The rheological behavior of asphalt was investigated by means of a DHP-1 dynamic shear rheometer according to T0628 in the regulation [23]. A sample of 25 mm in diameter and 1 mm thick was used in the test. The strain level selected for temperature scanning was 10%, and the shear frequency was 10 rad/s. The trend of complex shear modulus G^*, the phase angle δ, and the rutting factor $G^*/\sin\delta$ at different temperatures was analyzed. The high-temperature performance of the matrix asphalt, the nano-ZnO-modified asphalt, and the nano-ZnO/BF composite-modified asphalt before and after aging was studied. Frequency scanning was used to select the strain level of 1%, and the frequency range was within 0.1 to 100 rad/s. The trend of G^* at different angular frequencies was analyzed, and the anti-deformation ability of the three types of asphalt before and after aging was further studied. Based on the theory of the time–temperature equivalence principle, the change rule of the three types of asphalt G^* was compared with that of the main curve of viscoelasticity.

3.5. Rheological Test of Low-Temperature Bending Beam

An ATS low-temperature bending rheometer made in the United States was adopted to investigate the low-temperature properties with reference [26]. Preparation of the asphalt was performed for 127 mm; it was 12.7 mm wide and 6.35 mm thick. The specimens were subjected to continuous stress loading for 4 min in a cold environment. Then, the creep stiffness modulus S value and the creep tangent slope m value of the three kinds of trabecular bending specimens were recorded, respectively. To ensure the data were accurate, the creep stiffness, $S \leq 300$ Mpa, and the creep rate, $m \geq 0.3$ at 60 s, were required in the specification. Comparison and analysis of the experimental results were performed to evaluate the low-temperature properties of matrix asphalt, nano-ZnO-modified asphalt, and nano-ZnO/BF composite asphalt before and after aging.

3.6. *Scanning Electron Microscope Experiments*

The morphology of matrix asphalt, nano-ZnO-modified asphalt, and nano-ZnO/BF composite-modified asphalt were studied using a field emission scanning electron microscope. The magnification of the matrix asphalt and nano-ZnO-modified asphalt was 2000 times, and the magnification of the nano-ZnO/BF composite-modified asphalt was 500 times. The sample size was in the size range of 10 mm × 10 mm × 5 mm, in order to more clearly study the combination of admixture and asphalt and the specific distribution characteristics, so as to better distinguish the differences in microstructure.

3.7. *Fourier Infrared Spectroscopy Test*

A VERTEX70 Fourier transform infrared spectrometer was used. The test scanning range was 4000 cm^{-1}~500 cm^{-1}, 64 times of scan, and 4 cm^{-1} resolution. The matrix asphalt, nano-ZnO-modified asphalt, and nano-ZnO/BF composite-modified asphalt were tested using infrared spectroscopy. Based on the position and strength of the characteristic peaks, the modification mechanism of the two modifiers on the base asphalt was studied.

4. Experiment Results and Analysis

4.1. *Aging Performance Analysis*

The matrix asphalt, nano-ZnO-modified asphalt, and nano-ZnO/BF composite-modified asphalt were subjected to RTFOT short-term aging, respectively. The comparison results are shown in Table 4.

Table 4. Short-term aging test results of modified asphalt.

Asphalt Type	Before RTFOT				After RTFOT			
	25 °C Penetration (0.1 mm)	5 °C Ductility (cm)	Softening Point (°C)	Quality (g)	Residual Penetration Ratio (%)	Residual Ductility Ratio (%)	Softening Point Increment (°C)	Quality Loss (%)
Matrix asphalt	64.8	13.4	47.5	49.862	73.5	83.6	6.2	0.264
Nano-ZnO-modified asphalt	60.5	24.6	53.4	50.047	75.2	87.4	5.7	0.237
Nano-ZnO/BF composite-modified asphalt	52.3	21.9	57.8	50.236	76.9	85.3	5.4	0.192

As shown in Table 4, the residual penetration ratio of the modified asphalt increased by 1.7% and 3.4%, respectively, the residual ductility ratio increased by 3.8% and 1.7%, respectively, the softening point increment decreased by 0.5 °C and 0.8 °C, respectively, and the mass change decreased by 0.027% and 0.072%, respectively. The greater the residual penetration ratio of asphalt after aging, the smaller the increment of softening point increment and the change in mass, the better the anti-aging performance of asphalt. This shows that the addition of two additives promoted the aging of asphalt and effectively reduced the volatilization of light components in asphalt. Both modified asphalts have better heat-aging resistance than matrix asphalt. It also proves that the nano-ZnO/BF composite-modified asphalt has the strongest anti-aging ability.

4.2. *Analysis of Dynamic Shear Rheological Test*

4.2.1. Temperature Scanning

The rutting factor $G^*/\sin\delta$ is used to predict the viscoelastic properties of asphalt. The temperature scanning test of the three types of asphalt before and after aging was carried out at 46~82 °C. The temperature interval was 6 °C. The results are shown in Figures 1 and 2.

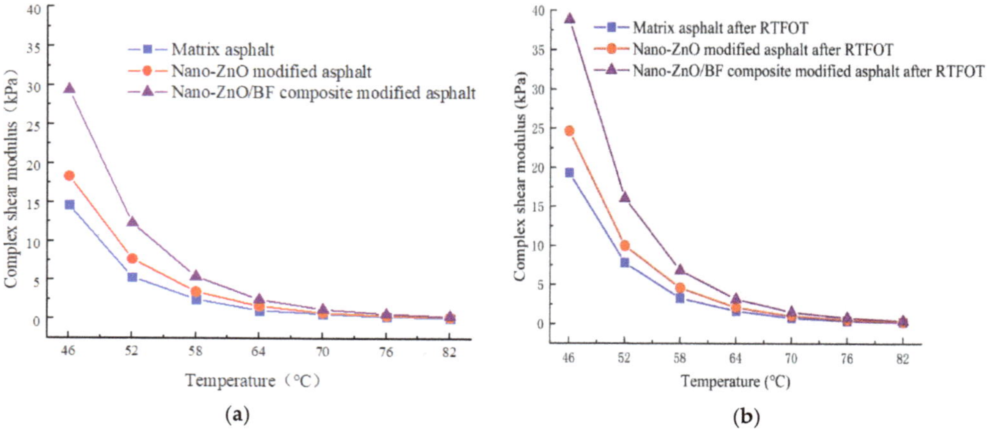

Figure 1. Complex shear modulus–temperature variation diagram of three kinds of asphalt. (**a**) Unaged asphalt. (**b**) Aged asphalt.

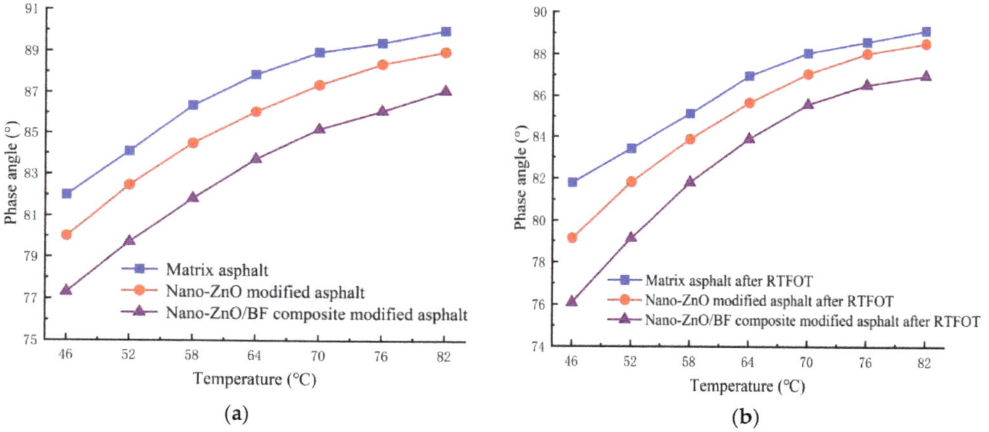

Figure 2. Phase angle–temperature variation diagram of three kinds of asphalt. (**a**) Unaged asphalt. (**b**) Aged asphalt.

Analysis of Figures 1 and 2 shows:

① When the temperature condition is relatively low, the complex shear modulus of the three kinds of asphalt is larger, but the phase angle value at the same temperature is smaller, indicating that the asphalt is more elastic at low temperatures, and has a stronger capacity to withstand deformation. As the temperature rises, the complex shear modulus decreases, and the phase angle increases gradually, which demonstrates that the viscosity of the asphalt is stronger under the condition of a high-temperature environment, and the anti-deformation capability of the asphalt is reduced to a certain extent. After the nano-ZnO and BF were added to the matrix asphalt, under the same temperature conditions, the complex shear modulus G^* of the nano-ZnO-modified asphalt and nano-ZnO/BF composite-modified asphalt showed an upward trend compared with the matrix asphalt, and the phase angle δ showed a corresponding decreasing trend, but the increase in G^* and the decrease in δ of the composite-modified asphalt were obviously greater than that of nano-ZnO-modified asphalt. Therefore, the addition of BF to nano-ZnO-modified asphalt can effectively promote resistance to permanent deformation.

② After aging, the G* and δ of the three types of asphalt are almost the same as those before aging. From Figure 1b, it can be seen that the G* values of the three kinds of asphalt will increase when aging occurs. However, compared with the G* value changes of nano-ZnO-modified asphalt and matrix asphalt, it can be seen that after adding BF, the G* value increases greatly. From Figure 2b, it can be seen that the δ value of asphalt decreases during aging, but the decreasing trend also increases with the addition of BF. Through the above analysis, it can be seen that in the process of asphalt RTFOT aging, adding BF to modified asphalt can play a helpful role in resisting RTFOT aging.

According to the result data, three diverse classifications of asphalt rutting factor–temperature change trend diagrams are further analyzed. As shown in Figure 3, the temperature scanning test results before and after aging can be studied more clearly and intuitively.

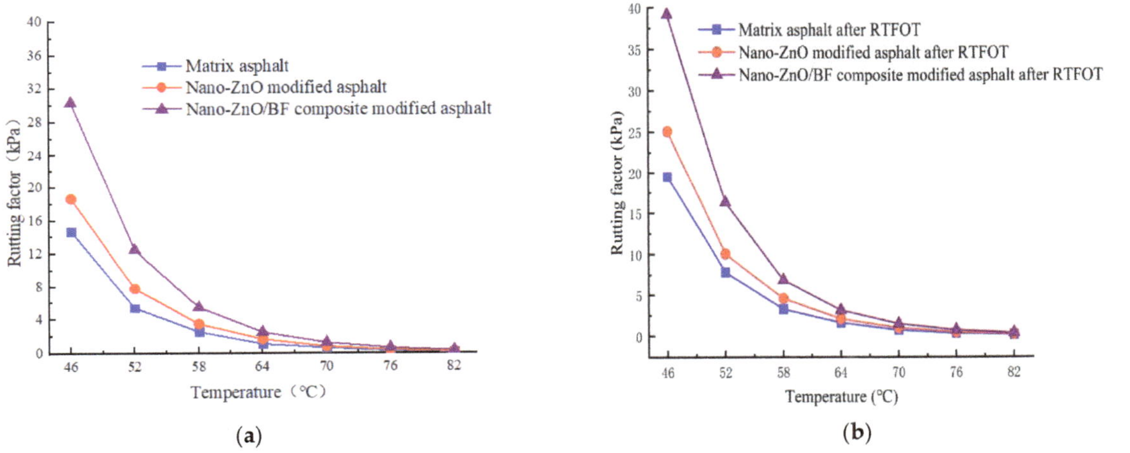

Figure 3. Rutting factor–temperature variation diagram of three kinds of asphalt. (**a**) Unaged asphalt. (**b**) Aged asphalt.

Analysis of Figure 3 shows:

① As the test temperature rises, the rutting factors of the original asphalt and the aged asphalt have the same decreasing tendency. Rutting factor in the environmental conditions of 46 °C gradually warming process, the overall trend of rutting factor suddenly decreased, but the three kinds of asphalt curve after 64 °C numerical change slowly and gradually overlap. The reason is that the relative proportion of viscous behavior is increased with the increase in test temperature, and the viscous components increase while the elastic components gradually decrease. Under the same temperature conditions, the rutting factor G*/sinδ from large to small is nano-ZnO/BF-modified asphalt > nano-ZnO-modified asphalt > matrix asphalt. If the asphalt shows a large G*/sinδ value, it shows that asphalt has strong high-temperature deformation resistance, which can more effectively resist the traffic gravity load so that the rutting deformation phenomenon appears slowly or decreases significantly.

② RTFOT aging makes the rutting factor of asphalt larger. Analysis of the reasons shows that aging changes asphalt's composition, and the gradual transformation of light components to heavy components causes the nano-ZnO/BF composite-modified asphalt to be harder. When the asphalt pavement undergoes heavy traffic under high-temperature conditions, its good viscoelastic behavior makes it less prone to rutting deformation, indicating that aging causes the modified asphalt to have better thermal stability. The increase in the rutting factor of modified asphalt after aging also shows that the high strength characteristics of BF itself are effectively combined with nanoparticles to form a unified system, and the integral performance of modified asphalt is effectively reinforced. The incorporation of nano-ZnO and BF into the matrix asphalt significantly reduces the probability of rutting deformation of the asphalt, which is of great importance for the anti-damage ability of the asphalt under high-temperature conditions.

4.2.2. Frequency Scanning

(1) Frequency scanning test

The frequency scanning experiment can accurately and comprehensively analyze the dynamic high-temperature characteristics of different asphalts at different shear angle frequencies. In this section, frequency scanning tests in the three types of asphalt were carried out at temperatures of 40~88 °C, and the temperature interval was 12 °C. The results of the test are shown in Figures 4–6.

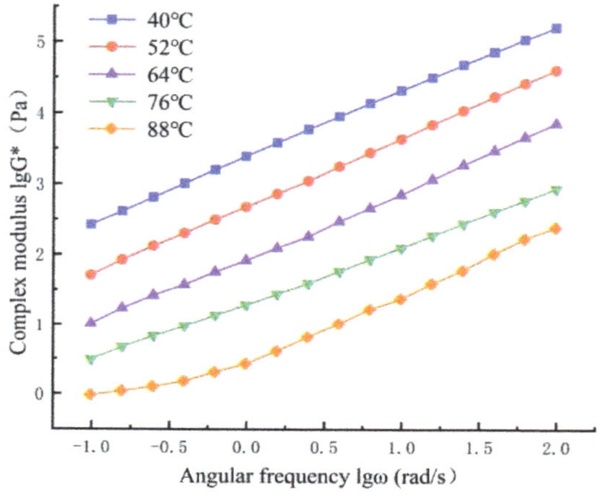

Figure 4. Complex modulus–angular frequency change diagram of base asphalt.

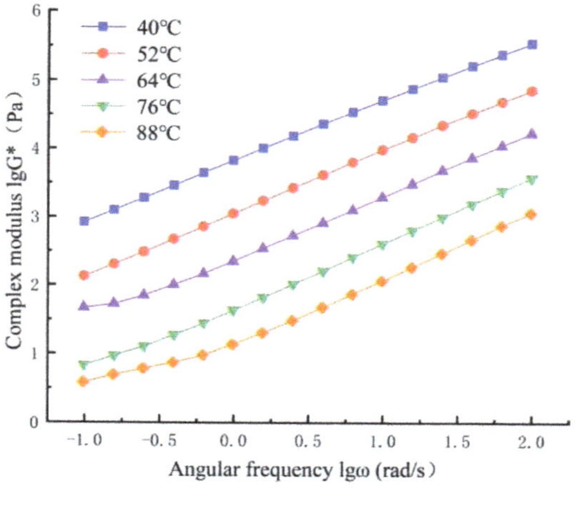

Figure 5. Complex modulus–angular frequency change diagram of nano-ZnO-modified asphalt.

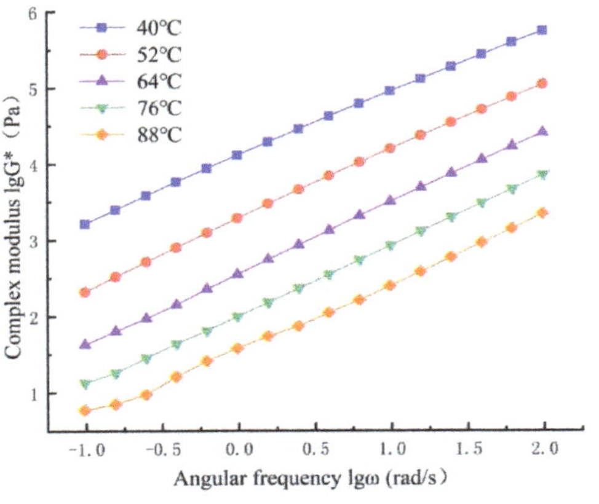

Figure 6. Complex modulus–angular frequency change diagram of nano-ZnO/BF composite-modified asphalt.

Analysis of Figures 4–6 shows:

① Under the same temperature conditions, as the angular frequency gradually increases, the complex modulus values of the three asphalts also increase and are almost linear. In practical road applications, the increase in the frequency of pavement loading indicates that the higher the vibration frequency of the road in unit time, that is, the total shear strain generated in unit time decreases, and the complex modulus of asphalt increases. Under the state of relatively low temperature, the increase in load frequency has little effect on asphalt road pavement, and the deformation of asphalt road pavement is small. However, under the state of relatively high temperatures, when the pavement is subjected to heavy traffic vehicles, its G* is small, and the asphalt pavement is prone to deformation, which may result in the destruction of the road surface.

② The change rate of G* of matrix asphalt, nano-ZnO-modified asphalt, and nano-ZnO/BF composite-modified asphalt is basically the same and roughly parallel under five different test temperature conditions. Analysis of experimental data shows that with the change in environmental conditions, the G* of asphalt is negatively correlated with temperature. The reason is that the asphalt softens in a high-temperature environment, which leads to the weakening of the interaction force between the interior of the asphalt, the weakening of the elastic properties of the modified asphalt, and a change in the direction of viscoelasticity. The final result is that the G* decreases, and the viscous component of asphalt increases. The transformation of asphalt to viscous characteristics reduces its ability to resist high-temperature deformation. It shows that in a high-temperature environment, the probability of road rutting will increase significantly, and the performance of asphalt pavement will be weakened.

(2) Main curve analysis of frequency scanning results

First, the data of the complex modulus and angular frequency of the three original asphalts are fitted to the curve equation. Then, by selecting the G* value, the displacement factors corresponding to the matrix asphalt, nano-ZnO-modified asphalt, and nano-ZnO/BF composite-modified asphalt are calculated in turn. Finally, with the help of the time–temperature equivalence principle, the main curve diagram of $lgG^* - lg\omega$ is constructed to increase the research interval. The time–temperature equivalence principle is based on the fact that there is a certain distance between the data curves of other temperatures and the data curves of this temperature at a certain temperature. This distance is called the displacement factor.

① Determination of displacement factors of the three original asphalts.

Through the analysis of the frequency scanning data of matrix asphalt at different test temperatures, and then the double logarithm curve equation is fitted. The results are presented in Table 5.

Table 5. Matrix asphalt double logarithmic fitting curve summary table.

Test Temperature (°C)	Fitting Curve Equation	R^2
40	$lgG^* = 0.9270lg\omega + 3.3825$	0.9996
52	$lgG^* = 0.9608lg\omega + 2.6847$	0.9999
64	$lgG^* = 0.9386lg\omega + 1.9455$	0.9987
76	$lgG^* = 0.9024lg\omega + 1.3011$	0.9997
88	$lgG^* = 0.8793lg\omega + 0.7813$	0.9811

In this section, by selecting G^* = 1000 Pa, it is substituted into lgG^* to obtain $lgG^* = 3$, and then the lgG^* value is substituted into the fitting curve equation of Table 5, respectively. The corresponding $lg\omega$ values at various temperatures can be obtained. On the basis of the displacement factor of the test temperature of 40 °C, the displacement factor values corresponding to the angular frequency $lg\omega$ under the remaining temperatures are calculated in turn. The results are in Table 6 below.

Table 6. Matrix asphalt displacement factor table.

Test Temperature (°C)	$Lg\omega$ (Rad/s)	Displacement Factor
40	−0.4126	0
52	0.3282	−0.7408
64	1.1235	−1.5361
76	1.8826	−2.2952
88	2.5233	−2.9359

Taking the same analysis method as matrix asphalt, the fitting curve equations and corresponding displacement factor values of nano-ZnO-modified asphalt and nano-ZnO/BF composite-modified asphalt were calculated in turn, as given in Tables 7 and 8, respectively.

Table 7. Two modified asphalt double logarithmic fitting curve equation summary table.

Types of Modified Asphalt	Test Temperature (°C)	Fitting Curve Equation	R^2
Nano-ZnO-modified asphalt	40	$lgG^* = 0.8777lg\omega + 3.8214$	0.9997
	52	$lgG^* = 0.9193lg\omega + 3.0539$	0.9998
	64	$lgG^* = 0.8956lg\omega + 2.4102$	0.9962
	76	$lgG^* = 0.9389lg\omega + 1.6799$	0.9984
	88	$lgG^* = 0.8419lg\omega + 1.2662$	0.9807
Nano-ZnO/BF composite-modified asphalt	40	$lgG^* = 0.8456lg\omega + 4.1064$	0.9993
	52	$lgG^* = 0.9108lg\omega + 3.2771$	0.9992
	64	$lgG^* = 0.9404lg\omega + 2.5598$	0.9998
	76	$lgG^* = 0.9220lg\omega + 2.0145$	0.9997
	88	$lgG^* = 0.8827lg\omega + 1.5499$	0.9966

② Main curve analysis of the three original asphalts.

When the time–temperature equivalence principle is used for graphic drawing, the curves of different temperatures will move horizontally on the basis of the displacement factor. Then, by studying the generated $lgG^* - lg\omega$ curve of a wider area, the temperature susceptibility of different asphalts under the action of the external environment can be evaluated and analyzed. The diagrammatic drawing of the displacement factor principle is illustrated in Figure 7.

Table 8. Two modified asphalt displacement factor tables.

Types of Modified Asphalt	Test Temperature (°C)	Lgω (Rad/s)	Displacement Factor
Nano-ZnO-modified asphalt	40	−0.9359	0
	52	−0.0586	−0.8773
	64	0.6586	−1.5945
	76	1.4060	−2.3419
	88	2.0594	−2.9953
Nano-ZnO/BF composite-modified asphalt	40	−1.3084	0
	52	−0.3042	−1.0042
	64	0.4681	−1.7765
	76	1.0689	−2.3773
	88	1.6428	−2.9512

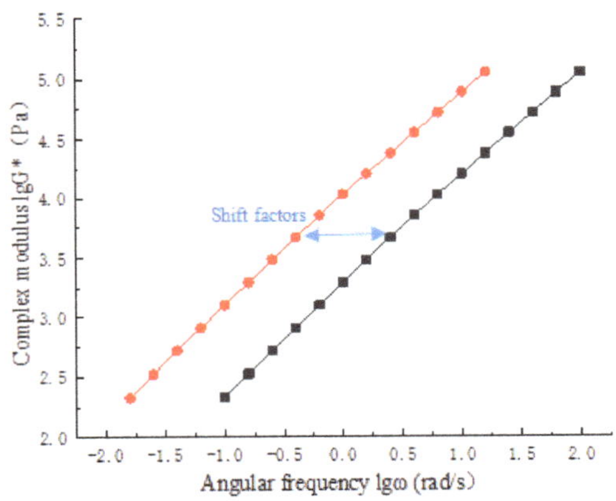

Figure 7. Schematic diagram of displacement factor.

Based on the principle of time–temperature equivalence, the temperature of 40 °C is used as the benchmark. According to the displacement scale factor values of the three original asphalts calculated in Tables 6 and 8, the remaining complex modulus change diagrams are moved horizontally to the left, and the main curve of $lgG^* - lg\omega$ is obtained, as shown in Figure 8.

According to the analysis of Figure 8, under the same loading conditions of angular frequency, especially in the high-temperature and low-frequency section, the G^* value of nano-ZnO/BF composite-modified asphalt is larger than that of nano-ZnO-modified asphalt, and the complex modulus value of matrix asphalt is the smallest, indicating that matrix asphalt has poor rutting resistance. BF can improve the high-temperature properties of asphalt, but the improvement effect is not as obvious as that of composite materials. It can be seen intuitively from Figure 8 that in the low-temperature and high-frequency section, as the angular frequency increases, the G^* of the three types of asphalt also increases. The main curve of matrix asphalt and nano-ZnO-modified asphalt gradually approaches, and the complex modulus values are not much different, indicating that the nanoparticles as modifiers in the low-temperature environment have no significant effect in improving the temperature sensitivity of the asphalt. However, the complex modulus of the nano-ZnO/BF compound-modified asphalt is still at the highest position and is in a continuous rising stage; this shows that the two admixtures can effectively upgrade the high-temperature anti-rutting properties of asphalt, and the promotional effects are still remarkable in the high-frequency section.

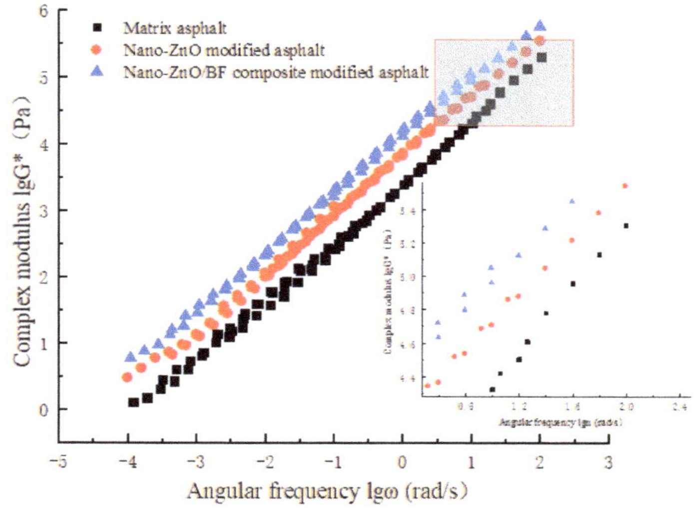

Figure 8. The complex modulus–angular frequency main curves of the three original asphalts.

4.3. Analysis of Bending Beam Rheological Test

Three temperature conditions were selected for the three kinds of asphalt: −12 °C, −18 °C, and −24 °C. The loading time was determined to be 240 s. The stiffness modulus S value and tangent slope m value of the bending creep stiffness test were automatically collected and recorded by the sensing system, as given in Figures 9 and 10.

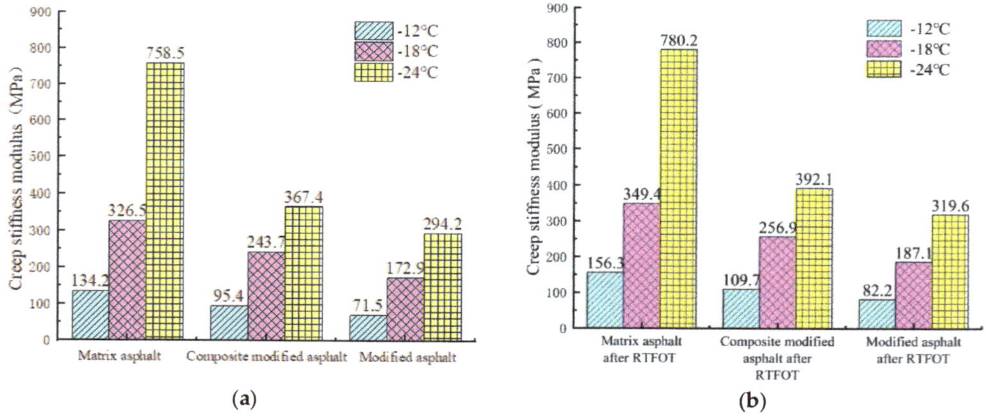

(a) (b)

Figure 9. Creep stiffness modulus of asphalt at different temperatures. (**a**) Unaged asphalt. (**b**) Aged asphalt.

From the analysis of Figures 9a and 10a, it can be seen that the creep stiffness modulus of the three different types of asphalt increases, and the creep tangent slope decreases as the temperature decreases. Among them, the increase (decrease) of matrix asphalt is the most significant, which reflects that matrix asphalt shows more significant brittleness under low-temperature conditions and has poor deformation resistance. After the nanoparticle modifier is added, the S value decreases and the m value increases. After the addition of BF to the nano-modified asphalt, the S value increases and the m value decreases. The reason may be that the BF absorbs a part of the light component content in the asphalt so that the content of heavy components is relatively high. Nanoparticles and BF form a dense three-dimensional mesh system, which increases its synergy, thus promoting the creep stiffness modulus of nano-ZnO/BF composite-modified asphalt to be higher than that of nano-ZnO-modified asphalt.

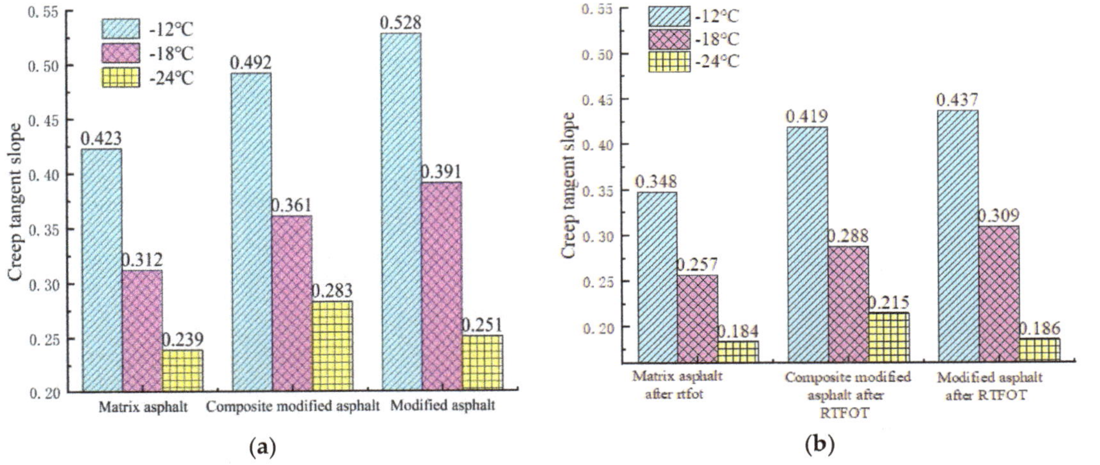

Figure 10. Creep tangent slope of asphalt at different temperatures. (**a**) Unaged asphalt. (**b**) Aged asphalt.

It can be seen from Figures 9b and 10b that the parameter S of matrix asphalt, nano-ZnO-modified asphalt, and nano-ZnO/BF composite-modified asphalt increased to a certain extent after RTFOT aging. Analysis at −12 °C shows that the S value of matrix asphalt increased by 16.5% after aging, the S value of nano-ZnO-modified asphalt increased by 14.9% after aging, and the S value of composite-modified asphalt increased by 15.0% after aging. The creep tangent slope of the three test asphalts decreased after short-term aging. The m values of matrix asphalt, nano-ZnO-modified asphalt, and nano-ZnO/BF composite-modified asphalt at −12 °C decreased by 21.6%, 14.8%, and 13.4%, respectively. In summary, nanoparticles can improve the low-temperature creep property of asphalt to some degree, while the effect of adding fiber is not obvious. After short-term aging, the modified asphalt with admixture weakens the change range of creep rate, which further verifies that the addition of nanoparticles and fibers can enhance the anti-aging properties of asphalt.

4.4. Micromorphology Analysis

4.4.1. Study on the Microstructure of Raw Materials

The scanning electron microscope morphology of nano-ZnO and basalt fiber was studied, as shown in Figures 11 and 12.

It can be seen from Figure 11 that the shape of the surface-modified nano-zinc oxide particles is approximately spherical, and the particle diameter is about 30 nm. The aluminate coupling agent reacts with nano-zinc oxide so that the surface of the particles is coated with a layer of film, which improves its dispersion.

It can be seen from Figure 12 that the basalt fiber is linearly distributed, the surface is smooth, and it is not easy to deform and distort, indicating that the fiber has high strength and can have a good reinforcement effect on the asphalt mixture.

(**a**) (**b**)

Figure 11. Scanning electron microscopy of two multiples of nano-ZnO. (**a**) The magnification is 20,000 times. (**b**) The magnification is 50,000 times.

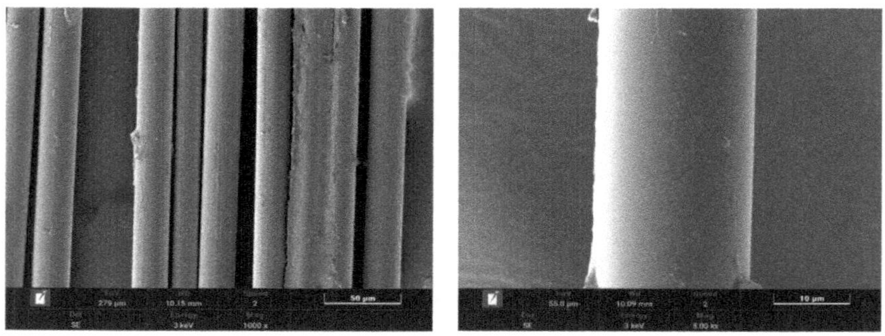

(**a**) (**b**)

Figure 12. Scanning electron microscopy of two kinds of basalt fiber. (**a**) The magnification is 1000 times. (**b**) The magnification is 5000 times.

4.4.2. Morphology Characterization Analysis of Composite-Modified Asphalt

The morphology of matrix asphalt, nano-ZnO-modified asphalt, and nano-ZnO/BF composite-modified asphalt was studied using scanning electron microscopy (SEM), as shown in Figures 13–15.

Figure 13. Ase asphalt scanning electron microscope.

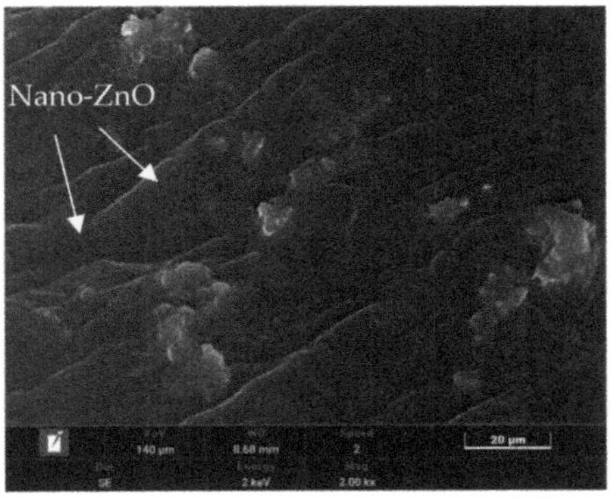

Figure 14. Scanning electron microscopy of nano-ZnO-modified asphalt.

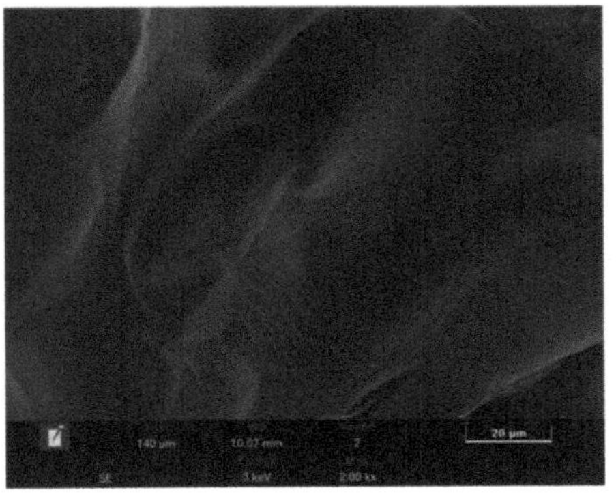

Figure 15. Scanning electron microscopy of nano-ZnO/BF composite-modified asphalt.

It can be seen from Figure 13 that after the matrix asphalt is magnified by 2000 times, no other impurities or particles appear in the scanning electron microscope image, the surface is very smooth, and the micro-morphology is evenly distributed. This also proves that the matrix asphalt selected in this paper is relatively clean and pure, which is convenient for later comparison with the images of nano-ZnO-modified asphalt and nano-ZnO/BF composite-modified asphalt.

It can be seen from Figure 14 that, compared with the matrix asphalt, the surface of the modified asphalt added with nano-ZnO has some protrusions, and a small amount of agglomeration occurs, but on the whole, it is still evenly dispersed in the asphalt, and a relatively stable network structure is formed with the asphalt. This may be due to the damage to the structure of the nanoparticles themselves during the high-speed stirring process. The chemical structures, such as functional groups and covalent bonds, change, and the increase in their surface activity makes it easier to integrate with the asphalt, increasing the viscosity and toughness of the overall structure, thereby improving the asphalt's performance.

It can be seen from Figure 15 that the microstructure of the composite-modified asphalt becomes more complex, and the surface presents an intricate irregular wrinkle phase. This is because the fiber itself has high oil absorption. After the adsorption of the surrounding asphalt reaches saturation, the fiber is completely coated by the asphalt, forming a wrinkle phenomenon around the vacancy, indicating that the fiber and the asphalt have good compatibility. There is no reaction between the two admixtures. The mutual extension of the surface of the asphalt phase and the fiber phase forms a denser structure, which helps to improve the stability of the asphalt.

4.5. Infrared Spectrum Test Analysis of Composite-Modified Asphalt

The data obtained from the wavenumber-transmittance test of the three kinds of asphalt are summarized into an infrared spectrum, and then the characteristic peaks are compared and analyzed. The test results are shown in Figure 16.

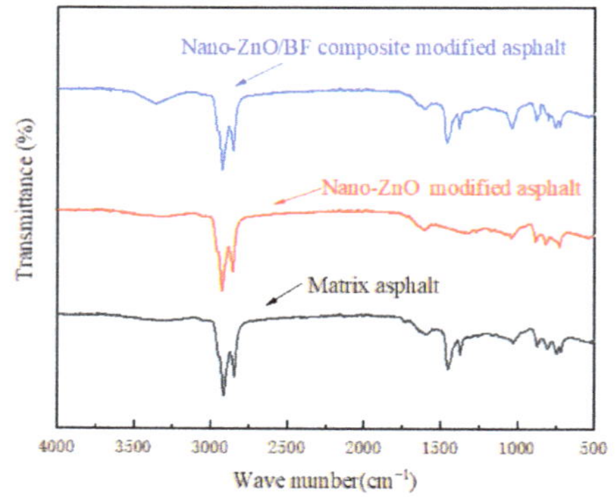

Figure 16. Infrared spectra of the three kinds of modified asphalt.

The infrared spectra of the three kinds of asphalt in Figure 14 are studied and analyzed. The main conclusions are as follows:

(1) The infrared spectra of matrix asphalt are analyzed as follows: There are two significant characteristic absorption peaks at 2905 cm^{-1} and 2817 cm^{-1}, which are caused by the stretching vibration of the C–H bond in asymmetric methylene and symmetric methylene (-CH_2-), respectively. The stretching vibration peaks at 1486 cm^{-1} and 1378 cm^{-1} were observed, which may be caused by the bending vibration of the C–H bond in the asymmetric group and the symmetrical methyl (-CH_3-). A weak absorption peak was found at a wave number of about 1035 cm^{-1}, corresponding to the stretching vibration of the functional group S=O in the sulfoxide (R^1-SO-R^2). The absorption peaks found at 815 cm^{-1} and 738 cm^{-1} at the end are due to the out-of-plane bending of the =C–H group in the olefin. It can be seen that the matrix asphalt contains aromatic hydrocarbon compounds.

(2) After adding nanoparticles into the matrix asphalt, the peak position of the infrared spectrum of the asphalt changed significantly. The absorption peak of nano-ZnO-modified asphalt gradually weakens and disappears in the range of 1300 cm^{-1} to 1530 cm^{-1}, indicating that there is a certain amount of strong oxidizing hydroxyl (-OH) on the surface of the surface-treated nano-materials. Under the action of high-speed shear, a certain chemical reaction occurs between the matrix asphalt. The intensity and position of the absorption peaks found at 726 cm^{-1} and 613 cm^{-1} at the end changed slightly. The reason may be that nano-ZnO has a certain influence on the out-of-plane swing vibration of CH_2 olefins and the in-plane swing of long-chain alkanes CH_2 groups. In short, nano-ZnO and matrix asphalt have a certain chemical reaction, but mainly a physical reaction.

(3) On the whole, nano-ZnO/BF composite-modified asphalt has a certain wave number absorption peak band in the range of 3150~3460 cm^{-1}. On the one hand, the CH functional group in the asphalt may have a weak chemical reaction with the composite modifier during the preparation process. On the other hand, it is caused by the stretching vibration of O–H and N–H bonds in the phenolic hydroxyl group. At the wave numbers 815 cm^{-1} and 726 cm^{-1}, there are moderate stretching vibration

peaks, which are mainly caused by the bending vibration of the crystalline long chain (-$(CH_2)n$-, ($n \geq 4$)). At the wave number 1035 cm^{-1}, the absorption peak with obvious strength is found. The reason may be the degradation reaction of the polymer chain segment, which leads to the change in the content of the related group C=C, and the C=C group is a conjugated double bond. The value can characterize the mechanical properties of asphalt, which further indicates that nano-ZnO/BF composite-modified asphalt has strong mechanical properties.

In summary, when nano-ZnO and BF are added to matrix asphalt, the types of functional groups remain almost unchanged, and only the content of functional groups changes. It shows that after adding the composite modifier, there is basically no chemical reaction inside the asphalt, mainly physical dispersion and compatibility.

5. Conclusions

(1) After adding nano-ZnO and BF to matrix asphalt, the three performance indicators before and after RTFOT aging improved to varying degrees. Compared with nano-ZnO-modified asphalt, the residual ductility ratio, softening point increment, and mass change in nano-ZnO/BF composite-modified asphalt decreased by 1.7%, 0.3 °C and 0.045%, respectively, and the residual penetration ratio increased by 1.7%, indicating that the fiber can reduce the effect of aging on asphalt and further improve its anti-aging performance.

(2) The rutting factor of the three kinds of original asphalt and thermal aging asphalt decreases with the increase in test temperature, and at the same temperature, the $G^*/\sin\delta$ of the three kinds of asphalt from large to small is nano-ZnO/BF composite-modified asphalt > nano-ZnO-modified asphalt > matrix asphalt, indicating that the composite-modified asphalt has the strongest high-temperature deformation resistance. RTFOT aging makes the rutting factor of asphalt larger, which is of great significance to its damage resistance in a high-temperature environment.

(3) The complex modulus of three kinds of original asphalt and aged asphalt increased gradually with the increase in angular frequency, almost linear relationship; nano-ZnO/BF composite-modified asphalt has good deformation resistance in both high-frequency and low-frequency regions, which improves the pavement's performance from a macro perspective. After short-term aging, the complex modulus of the three kinds of asphalt showed a significant growth trend; that is, aging improved the high-temperature stability of the asphalt.

(4) After adding BF to nano-ZnO-modified asphalt, the S value increases and the m value decreases, that is, nanoparticles can improve the low-temperature creep performance of asphalt to a certain extent, while the low-temperature improvement effect of nano-ZnO/BF composite-modified asphalt is not obvious. The S value of nano-ZnO-modified asphalt and composite-modified asphalt increased by 14.9% and 15.0%, respectively, while the m value decreased by 14.8% and 13.4%, respectively, and the change range of creep rate of modified asphalt decreased at −12 °C. After RTFOT aging, the creep rate of modified asphalt decreased, indicating that the incorporation of modifiers can improve the anti-aging performance of asphalt.

(5) The modified asphalt mixed with nanoparticles and BF increases the viscosity and toughness of the whole structure and forms a three-dimensional network structure, which can effectively improve the performance of the asphalt. It can be seen from the infrared spectrum that nano-ZnO/BF composite-modified asphalt has strong mechanical properties. Nano-ZnO and BF have weak chemical reactions in matrix asphalt, but they are mainly physically dispersed and compatible.

Author Contributions: Writing—original draft preparation, C.L., Z.L. and J.M.; Writing—review and editing, T.G., Y.C. and J.M.; Investigation, J.W. and L.J.; Discussion, T.G., Y.C. and J.M. All authors have read and agreed to the published version of the manuscript.

Funding: This work was supported by Key R&D and Promotion of Special Scientific and Technological Research Projects of Henan Province: [Grant Number 182102210061, 212102310089]; Key Scientific Research Projects of Colleges and Universities in Henan Province in 2021: Study on pavement performance of cotton straw cellulose modified asphalt [Grant Number 21A580004].

Institutional Review Board Statement: Not applicable.

Informed Consent Statement: Not applicable.

Data Availability Statement: Some or all data, models, or code that support the findings of this study are available from the corresponding author upon reasonable request.

Conflicts of Interest: Author Chaojie Li was employed by the company Henan Transportation Research Institute Co., Ltd., Zhengzhou 450045, Henan, China. The remaining authors declare that the research was conducted in the absence of any commercial or financial relationships that could be construed as a potential conflict of interest.

References

1. Rao, Z.; Wang, Y.; Cui, Z. Key Interpretation of '14th Five-Year' Expressway Construction. *China Highw.* **2022**, *623*, 28–32.
2. Shen, L.; Ma, Q. Comparative Analysis of Long-life Asphalt Pavement and Traditional Asphalt Pavement. *Traffic Stand.* **2014**, *42*, 60–62.
3. Wang, H. *Comparative Study on Construction Equipment of MOH Semi-Flexible Pavement and Traditional Asphalt Pavement*; Chang'an University: Xi'an, China, 2019.
4. Wang, F. Analysis of the influence of different kinds of modifiers on the performance of matrix asphalt. *Appl. Chem.* **2021**, *50*, 2132–2135+2139.
5. Wang, Y. Application research status of nano-modified asphalt materials in pavement engineering. *Aging Appl. Synth. Mater.* **2022**, *51*, 159–161.
6. Zhu, C.; Zhang, H.; Shi, C.; Li, S. Effect of nano-zinc oxide and organic expanded vermiculite on rheological properties of different bitumens before and after aging. *Constr. Build. Mater.* **2017**, *146*, 30–37. [CrossRef]
7. Xu, X.; Guo, H.; Wang, X.; Zhang, M.; Wang, Z.; Yang, B. Physical properties and anti-aging characteristics of asphalt modified with nano-zinc oxide powder. *Constr. Build. Mater.* **2019**, *224*, 732–742. [CrossRef]
8. Hamedi, G.H.; Nejad, F.M.; Oveisi, K. Estimating the moisture damage of asphalt mixture modified with nano zinc oxide. *Mater. Struct.* **2016**, *49*, 1165–1174. [CrossRef]
9. Mansour, F.; Ehsan, S. The effects of nano zinc oxide (ZnO) and nano reduced graphene oxide (RGO) on moisture susceptibility property of stone mastic asphalt (SMA). *Case Stud. Constr. Mater.* **2021**, *15*, e00655.
10. Zhang, H.; Zhu, C.; Wu, C. Effects of multi-scale nanomaterials on the rheological and aging properties of asphalt. *J. Build. Mater.* **2019**, *22*, 238–244.
11. Wang, J.; Li, Y. Preparation of nano modified asphalt and its mixture road performance. *Road Constr. Mach. Constr. Mech.* **2020**, *37*, 22–28.
12. Bao, M.; Xie, X.; Li, G. Phase analysis of nano zinc oxide modified asphalt under ultraviolet irradiation. *Highway* **2022**, *67*, 228–236.
13. Dong, T. *Study on the Properties of Nano-ZnO/SBS/SBR Composite Modified Materials*; Chongqing Jiaotong University: Chongqing, China, 2020.
14. Tao, H.; Liu, H.; Xie, X.; Sun, T.; Dong, R.; Lu, X. Preparation and Properties of Nano-ZnO Combined with Biomass Heavy Oil Composite-Modified Asphalt. *Adv. Mater. Sci. Eng.* **2022**, *2022*, 5179787. [CrossRef]
15. Vamsikrishna, D.; Manikanta, K.V. Tyre Rubber Modified Bitumen for Ashpalt Mixture. *J. Trend Sci. Res. Dev.* **2019**, *3*, 42–46.
16. Abdelsalam, M.; Yue, Y.; Khater, A.; Luo, D.; Musanyufu, J.; Qin, X. Laboratory Study on the Performance of Asphalt Mixes Modified with a Novel Composite of Diatomite Powder and Lignin Fiber. *Appl. Sci.* **2020**, *10*, 5517. [CrossRef]
17. Gu, Q.; Kang, A.; Li, B.; Xiao, P.; Ding, H. Effect of fiber characteristic parameters on the high and low temperature rheological properties of basalt fiber modified asphalt mortar. *Case Stud. Constr. Mater.* **2022**, *17*, e01247. [CrossRef]
18. Celauro, C.; Praticò, F. Asphalt mixtures modified with basalt fibres for surface courses. *Constr. Build. Mater.* **2018**, *170*, 245–253. [CrossRef]
19. Zhao, Y. Experimental study on crack resistance of basalt fiber reinforced asphalt concrete. *Highw. Eng.* **2014**, *39*, 48–51.
20. Fu, Z.; Huang, Z.; Ma, F. Effect of basalt fiber on road performance of aged asphalt mixture. *Mater. Bull.* **2016**, *30*, 118–122.
21. Yan, J.; Zheng, J.; Li, N. Study on the crack resistance of basalt fiber asphalt mortar. *J. Build. Mater.* **2019**, *22*, 800–804.
22. Wang, G.; Li, B.; Xiao, P. Analysis of crack resistance of basalt fiber recycled asphalt mixture. *J. Yangzhou Univ. (Nat. Sci. Ed.)* **2021**, *24*, 69–73.
23. *JTG E20-2011*; Standard Test Methods of Bitumen and Bituminous Mixtures for Highway Engineering. People's Transportation Press: Beijing, China, 2011.
24. *JTG F40-2004*; Standard Specification for Construction and Acceptance of Highway Asphalt Pavement. People's Transportation Press: Beijing, China, 2004.

25. Zhou, L.; Zang, S.; Hu, X. Study on Surface Modification of Nanometer Zinc Oxide. *J. Petrochem. Univ.* **2009**, *22*, 5–8.
26. *SH/T 0775-2005*; Standard Test Methods for Determining the Flexural Creep Stiffness of Asphalt Binder Using the Bending Beam Rheometer (BBR). National Development and Reform Commission of the People's Republic of China: Beijing, China, 2005.

coatings

MDPI

Article

Study on the Performance of SBS/Polyphosphoric Acid Composite Modified Asphalt

Chaojie Li [1,2,3], Zhenxia Li [3,4,5], Tengteng Guo [3,4,5], Yuanzhao Chen [3,4,5,*], Qi Liu [3], Jing Wang [3] and Lihui Jin [3]

1 Henan Transportation Research Institute Co., Ltd., Zhengzhou 450006, China; lcj13676936229@163.com
2 Xi'an Changda Highway Engineering Inspection Center Co., Ltd., Xi'an 710064, China
3 School of Civil Engineering and Communication, North China University of Water Resources and Electric Power, Zhengzhou 450045, China; zhenxiali2009@ncwu.edu.cn (Z.L.); guotth@ncwu.edu.cn (T.G.); lq_9898@163.com (Q.L.); wj@ncwu.edu.cn (J.W.); lihuijin2023@163.com (L.J.)
4 Technology Innovation Center of Henan Transport Industry of Utilization of Solid Waste Resources in Traffic Engineering, North China University of Water Resources and Electric Power, Zhengzhou 450045, China
5 Henan Province Engineering Technology Research Center of Environment Friendly and High-Performance Pavement Materials, Zhengzhou 450045, China
* Correspondence: cyz740513@ncwu.edu.cn

Abstract: To address the issue of bad compatibility between a single polymer modifier and asphalt and high preparation cost, the composite modified asphalt with polyphosphoric acid (PPA) and SBS as a modifier was prepared. Basic properties, viscosity characteristics, high-temperature and low-temperature rheological performance, and aging-resistant performance of SBS/PPA composite modified asphalt were comprehensively evaluated, and the best content of PPA was obtained through the experimental results below. By performing an infrared spectrum test and a scanning electron microscope test, structural changes and modifying mechanisms of composite modified asphalt were analyzed. The results indicate the optimal PPA content is 0.75%. After adding PPA, the penetration and ductility of composite modified asphalt were reduced, while the softening point increased. At 135 °C, the viscosity was 1.2 times that of SBS modified asphalt. The average weight loss ratio was 0.163%. When the content of PPA was 0.75% and 1%, the rutting factor increased significantly. Therefore, PPA can not only improve the thermal oxidation aging resistance of asphalt, it can also improve the shear resistance, high-temperature performance, and temperature sensitivity of asphalt. Between 1027 and 1150 cm^{-1}, the composite modified asphalt forms a new absorption peak, and from 1610 cm^{-1}, the absorption peak presents an upward trend, suggesting that PPA reacts chemically with asphalt, produces the new substance, and also increases a large number of hydrocarbon components with chain structure. The surface appearance of the compound modified asphalt gradually presents a smooth wrinkle state due to the increase of PPA, so the issue of easy segregation of SBS in asphalt is improved.

Keywords: environmental protection material; composite modified asphalt; rheological properties; polyphosphoric acid; microscopic analysis

Citation: Li, C.; Li, Z.; Guo, T.; Chen, Y.; Liu, Q.; Wang, J.; Jin, L. Study on the Performance of SBS/Polyphosphoric Acid Composite Modified Asphalt. *Coatings* **2024**, *14*, 72. https://doi.org/10.3390/coatings14010072

Academic Editors: Giorgos Skordaris and Valeria Vignali

Received: 2 December 2023
Revised: 30 December 2023
Accepted: 2 January 2024
Published: 4 January 2024

1. Introduction

Hole cracks, pit collapse, and other diseases occur in asphalt pavement all year round. In addition to construction reasons, pavement aging is also an important cause. The principal cause for pavement aging is the decrease of cohesion and even movement of aggregate and asphalt in asphalt mixture subjected to repeated heavy loads and high temperatures [1]. Conventional polymer modified asphalt has the problems of poor compatibility with asphalt and easy segregation during transportation, which ultimately results in a decrease in service life and a rise in the cost of building and maintaining [2]. Therefore, it is of great significance for practical road applications to combine modifiers with different modification mechanisms into base asphalt to achieve complementary advantages among modifiers to improve asphalt performance.

As the third generation of synthetic rubber, an SBS modifier is widely used because of its excellent thermal stability and high-temperature performance, but it still cannot avoid the problem of transportation segregation caused by poor blending compatibility, and the performance deterioration is also common [3–6]. Cortizo et al. [7] indicated that the penetration of SBS modified asphalt was reduced and the softening point was raised after RTFOT. The infrared spectrum showed that changes in performance were mainly due to changes in microstructure, as well as the molecular weight of SBS modified asphalt under thermal oxidative degradation. The modification of asphalt with a different modifying agent was investigated. It was found that although both SBS and rubber powder were physically modified, its performance was obviously superior to single mixing [8–12]. The physical index, storage stability, and aging resistance of SBS modified asphalt can be improved by applying surface-active nanomaterials to produce composite modified asphalt [13,14]. When the chemical modifier gradually entered the academic field of vision, the composite modified asphalt co-doped with the reactive modification agent based on polyurethane precursor (PRM) and SBS appeared, which increased the compatibility of the polymer modifier and asphalt by chemical crosslinking and formed a more stable network structure [15,16].

As an acidic modifier, polyphosphoric acid was first popularized abroad, accounting for more than 16% of the modified asphalt market in the United States [17]. This is enough to show that polyphosphoric acid can play an active role in modified asphalt, and its cost is only half or even lower than that of polymer. Therefore, for large-volume asphalt pavement construction projects, the use of polyphosphoric acid is a choice to save economic costs and has economic benefits for practical road applications. Jafari et al. [18,19] found that adding PPA could improve the properties of high-temperature and aging resistance through various rheological tests. Amirkhanian et al. [20,21] found that the addition of phosphorus-containing acidic modifiers represented by PPA could facilitate the conversion of asphalt colloids to asphaltenes, so that asphalt can be transformed into a solution-gel type asphalt colloid structure; the effect of improving the high-temperature property of asphalt is realized by a stable homogeneous system. Cao et al. [22] used a standard penetration test (SD) method and a semi-circular bending (SCB) test, pointed out that PPA can change the asphalt structure and that the high-temperature stabilization is enhanced, and revealed that the reason for the increase of asphalt brittleness is that PPA increases the viscosity. Adding PPA into asphalt mixture makes low-temperature crack resistance worse, and the negative impact is greater. Wang [23] put forward that the phosphorylation reaction caused the macromolecular long-chain hydrocarbon components to appear in the asphalt, so the asphalt became hard and the penetration decreased. Hou [24], through multiple stress creep recovery (MSCR) tests, found that when the content of PPA was 1~1.25%, high-temperature stability can be enhanced regardless of whether the polymer is added into asphalt through the MSCR test.

Nowadays, there are some problems in combining SBS with base asphalt, such as unstable compatibility, easy segregation, and insufficient anti-aging performance, which will have a non-negligible impact on the asphalt structure. The content of PPA as a modifier in composite modified asphalt still needs to be carefully studied. It is generally believed that the content of PPA in 0.5~2% is more appropriate. In addition, the modification mechanism of PPA is still controversial, so it is important to systematically study its modification mechanism from the micro level. For this reason, SBS/PPA compound modified asphalt was prepared, and its properties and modifying mechanism were compared and explored. The changes in conventional properties and rheological properties of SBS modified asphalt with different PPA contents were revealed; then, further analysis about microstructure changes in modified asphalt was carried out.

2. Raw Materials

2.1. Asphalt

A-grade 70# road petroleum asphalt, produced by Zhengzhou Municipal Engineering Corporation, was adopted in this study. Basic properties of base asphalt were tested in accordance with the test program [25], and its technical indicators were assessed against the technical specifications [26]. The experimental results are shown in Table 1, which suggests the asphalt specification is up to standard.

Table 1. Technical indicators of 70# base asphalt.

Test Items		Unit	Test Results	Technical Requirements	Test Method
Penetration (100 g, 5 s)		0.1 mm	67.6	60~80	T0604
Ductility (5 cm/min, 10 °C)		cm	48	≥20	T0605
Softening point		°C	46.5	≥46	T0606
Penetration index PI		—	−0.768	−1.5~+1.0	T0604
After RTFOT	Quality change	%	−0.272	−0.8~+0.8	T0610
	Penetration ratio (25 °C)	%	62.4	≥61	T0604
	Ductility (5 cm/min, 10 °C)	cm	6.8	≥6	T0605

2.2. Polyphosphoric Acid

The polyphosphoric acid (PPA) with a purity of 117% (H3PO4 content), provided by Lanjue Chemical Co., Ltd. (Zhongshan, China), is used in this study. It is a transparent viscous liquid. The results presented in Table 2 show that the polyphosphate modifier meets the specification.

Table 2. Technical indicators of polyphosphoric acid.

Test Items	Unit	Test Results
P_2O_5 concentration	%	82.6
25 °C vapor pressure	Pa	2.61×10^{-6}
Boiling point	°C	558
Chloride (Cl) content	%	0.0002
Iron (Fe) content	%	0.0010
Arsenic (As) content	%	0.0066
Heavy metal (Pb) content	%	0.0014

2.3. Styrene-Butadiene-Styrene Block Copolymer

In this study, SBS-YH791 SBS1301 modified material, produced in Yueyang, Baling Petrochemical (Yueyang, China), was used, and its state is white solid particles. It can be concluded from the test results presented in Table 3 that the SBS modifier satisfies the specification.

Table 3. Technical indicators of SBS.

Test Items	Unit	Test Results	Technical Requirements
Molecular structure	—	Linear	Linear
Ash	%	0.07	≤0.20
300% tensile stress	MPa	4.8	≥3.5
Tensile strength	MPa	28.4	≥24.0
Elongation at break	%	742	≥730
Volatile matter	%	0.54	≤0.70
Styrene/Butadiene mass ratio	—	20/80	20/80

3. Experimental

3.1. Formulate Sample Content and Modified Asphalt Preparation Method

3.1.1. Preliminary Formulation of Sample Content

According to the actual construction situation of the current pavement and the comprehensive analysis of fuzzy mathematics, it was finally determined that the comprehensive evaluation of various physical properties is the best when the SBS content is 5% [27]. As a result, SBS content in SBS modified asphalt was 5% [28–30]. Considering the advantages of PPA in high-temperature performance and its negative influence on low-temperature performance [31], a large number of studies [32,33] believe that the content of PPA should not be higher than 1%. Because a small amount of PPA can achieve a high amount of polymer modification effect, it is appropriate to reduce the amount of SBS when preparing composite modified asphalt. In this study, the SBS content was selected to be 3.5% of the asphalt quality, and the contents of PPA were 0.5%, 0.75%, 1.0%, 1.25%, and 1.5% of asphalt quality, respectively.

3.1.2. Preparation of SBS Modified Asphalt

First, the asphalt was heated to its melting flow in an electrothermal blast drying oven at 160 °C, then taken out and placed on a constant temperature heating table. The weighed 5% SBS was slowly and uniformly added at 3000 r/min. After that, shear 45~50 min at 4500 r/min. During the shearing period, glass rods were continuously stirred to prevent uneven shearing. After the stirring was completed, the rotational speed was set to 1500 r/min for 30 min, so that SBS was further dissolved in asphalt in a finer state. Finally, the sheared sample was put into an electrothermal blast drying oven at 180 °C to swell and develop for 1 h to achieve the purpose of full compatibility between the modifier and the asphalt.

3.1.3. Preparation of SBS/PPA Composite Modified Asphalt

First, the asphalt was put into the electric hot blast drying box, and the asphalt was heated to melt and dehydrate. It continued to heat to 165 °C, then was taken out and placed on a constant temperature heating table. A total of 3.5% SBS was added at a slow and uniform speed of 3000 r/min; then, the asphalt was sheared at a speed of 4500 r/min for 35~40 min and finally maintained a constant temperature. It was mixed at a low speed for about 25 min to fully swell SBS modified asphalt. After the sample was fully prepared and swelled, the constant temperature heating platform was heated to 170 °C, and the speed was increased to 4500 r/min. A different dosage of PPA modifiers, which were weighed by the electronic balance, were added at a constant speed, and the speed was set to 5000 r/min for 30 min after complete addition. Lastly, the sheared sample was put into an electrothermal blast drying oven at 180 °C to swell and develop for 1 h.

3.2. Basic Performance Test

According to the specification [25], the short-term aging of asphalt and the above prepared samples were carried out by an SYD-3061 rotating film oven. A total of 14 groups of samples prepared above were tested for three major indicators, and the basic property changes of samples with or without aging were compared.

3.3. Brookfield Viscosity Test

In accordance with the T 0625-2011 test method in the specification [25], the NDJ series Brookfield viscometer of Changji Geological Instrument Co., Ltd. in Shanghai, China was used to carry out 135 °C and 175 °C rotational viscosity tests on above prepared samples. The effect of the modifier on viscosity-temperature characteristics was studied.

3.4. DSR

In accordance with the T 0628-2011 test method in specification [25], the temperature scanning experiments were conducted on the prepared samples in the original sample group and the aging group by using the instrument. A large sample with a diameter of

25 mm and a thickness of 1 mm was used. The angular frequency was adjusted to 10 rad/s, the experiment temperature was set to start from 46 °C, and the interval was 6 °C until 82 °C. The phase angle δ and rutting factor G*/sinδ of all samples were compared. In addition, frequency scanning tests were performed on 6 groups of modified asphalt in the original and aging groups. The experiment temperature was set to start from 40 °C, and the interval was 12 °C until 88 °C. The frequency was set from 0.1 to 100 rad/s and the strain level was 1%. The change trend of logG*/sinδ, with frequency at 52 °C and 76 °C, was analyzed to study the high-temperature property and aging resistance of composite modified asphalt.

3.5. BBR

According to the T 0627-2011 in the specification [25], the trabecular bending specimens of all samples were prepared, as shown in Figure 1. The specimen size is 127 × 12.7 × 6.35 mm. Experiment temperature was set at −12 °C, −18 °C, and −24 °C, and all specimens were loaded, unloaded, and dead loaded. The stiffness modulus and creep curve slope of each asphalt group before and after aging were compared to study its low-temperature crack resistance.

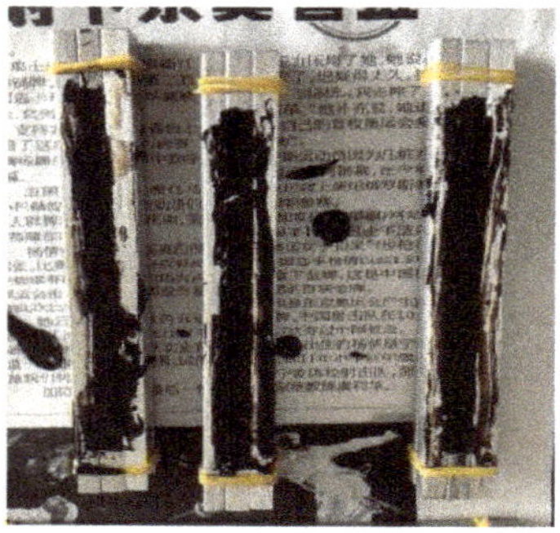

Figure 1. Low-temperature bending beam specimens.

3.6. FTIR

With the help of the Nicolet iS10 FTIR spectrometer produced by Thermo Fisher Scientific (Waltham, MA, USA), as shown in Figure 2, the infrared spectrum test of 7 kinds of asphalt in the original sample group was carried out, the change rule of absorbance in the spectrum after the addition of a modifier was analyzed, and the reaction type between modifier and asphalt was judged. The test parameters mainly include a scanning range of 500 cm^{-1}~4000 cm^{-1}, scanning time of 64 times, and a minimum resolution of 0.019 cm^{-1}.

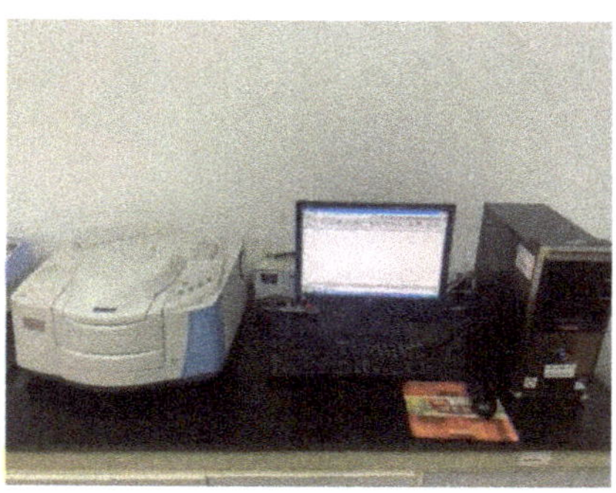

Figure 2. Infrared spectrometer.

3.7. SEM

The test of SEM was carried out by SU8010 field-emission scanning electron microscope produced by HITACHI in Tokyo, Japan, the combination of the modifier and asphalt was observed, and its microstructure change was analyzed. The preparation process for the specimens was as follows: first, the prepared sample was put into the oven and heated to melt. Then, the sample was dipped quickly with a glass rod, and the sample with a volume of 10 mm × 10 mm × 1 mm (or 2 mm) was formed on the clean glass slide. After cooling, the specimens were sprayed with gold and then the experiments were conducted. The test specimens are illustrated in Figure 3.

Figure 3. SEM test specimens.

4. Test Results and Analysis

4.1. Analysis of Basic Performance Results

4.1.1. Analysis of Performance Results of Original Group Modified Asphalt

The tests of penetration, softening point, and ductility performance were conducted on asphalt and six groups of modified asphalt in the original group. The test results are presented in Table 4.

Table 4. Three index results of composite modified asphalt.

Test Number and Schemes		Penetration (25 °C, 0.1 mm)	Softening Point (K)	Ductility (5 °C, cm)
1	70# Base asphalt	67.6	319.7	10.3
2	5% SBS	51.9	339.5	29.7
3	0.5% PPA + 3.5% SBS	54.7	322.9	26.6
4	0.75% PPA + 3.5% SBS	51.8	331.0	20.3
5	1.0% PPA + 3.5% SBS	49.1	336.4	16.8
6	1.25% PPA + 3.5% SBS	44.7	342.6	14.4
7	1.5% PPA + 3.5% SBS	42.4	73.2	12.9

After adding SBS, the penetration is reduced by 23.2%, the softening point increased from 319.7 K to 339.5 K—an increase of 6.2%—and the ductility is 2.9 times that of asphalt, suggesting that SBS can remarkably improve the high- and low-temperature properties of asphalt. The penetration of each modified group is smaller than that of asphalt, and it will be reduced gradually as the dosage is increased. The ductility of each modified group exceeds that of asphalt, but when the content exceeds a certain limit, the ductility will show a downward trend. When PPA content is 1.5%, it tends to be closer to basic asphalt. The softening point of each modified group exceeds that of the asphalt, and when PPA content is 1.5, it is 1.5 times that of the asphalt, indicating that even a little amount of PPA can increase the viscosity of asphalt, which makes a positive function on the property of asphalt at both high and low temperatures.

The penetration of each composite modified asphalt group decreases as the increment of the dosage. When PPA dosage is 0.75%, the penetration is similar to that of SBS modified asphalt. When PPA dosage is less than 1%, although the softening point is smaller than that of SBS modified asphalt, the softening point is increased as the dosage is increased, indicating that adding PPA will increase the viscosity of SBS modified asphalt and enhance its high-temperature performance [34]. The 5 °C ductility of each composite modified asphalt group reduces as the increment of the dosage; even if PPA dosage is 0.5%, it is not as good as 5% SBS modified asphalt, illustrating that adding PPA makes it gradually harden so that its ability to extend and deform at a low temperature decreases. Therefore, adding PPA is not conducive to the low-temperature property of SBS modified asphalt.

4.1.2. Analysis of Performance Results of Aging Group Modified Asphalt

Experimentation results of the mass change, 25 °C penetration, 5 °C ductility, and softening point of seven groups of asphalt after aging are presented in Table 5.

Table 5. Aging test results of composite modified asphalt.

Test Number and Schemes		Mass Change (%)	Penetration (25 °C, 0.1 mm)	Softening Point (°C)	Ductility (5 °C, cm)
1	70# Base asphalt	0.272	42.2	54.2	7.8
2	5% SBS	0.218	34.2	74.7	21.4
3	0.5% PPA + 3.5% SBS	0.204	34.4	56.4	20.7
4	0.75% PPA + 3.5% SBS	0.192	33.2	63.9	14.9
5	1.0% PPA + 3.5% SBS	0.167	32.9	67.1	12.6
6	1.25% PPA + 3.5% SBS	0.133	32.1	72.7	11.2
7	1.5% PPA + 3.5% SBS	0.119	31.3	75.1	10.3

As shown in Table 5, as the increment of PPA dosage increases from 0.5% to 1.5%, the average mass loss rate of composite modified asphalt is 0.163%, which is less than 0.264% of base asphalt and 0.218% of 5% SBS modified asphalt. The results suggest that adding PPA could enhance the thermal oxidation resistance of asphalt, and the more the content, the less the weight loss. The penetration of seven groups of aged asphalt decreases after aging, and the higher the PPA dosage, the smaller the penetration, and the base asphalt decreases the most. When the PPA content increases from 0.5% to 1.5%, the residual penetration ratio increases from 62.9% to 73.8%, and the change range is more than 10%. The residual penetration ratio of SBS modified asphalt is between the composite modified asphalt group with PPA content of 0.75~1.0%, indicating that adding PPA could decrease the thermal oxidation of SBS/PPA compound modified asphalt and improve its aging resistance. After aging, the softening point of seven groups of asphalt increases, and the increment of PPA content causes the increment of the softening point of each compound modified asphalt to reduce, indicating that adding PPA is beneficial to greatly improving the aging resistance of SBS modified asphalt. After aging, the ductility of the seven groups of asphalt decreases, and the greater the PPA content, the lower the difference in ductility pre- and post-aging, and the residual ductility ratio increases from 70% to nearly 80%. The ductility of composite modified asphalt and its ductility difference is less than 5% SBS modified asphalt, but still better than matrix asphalt, indicating that thermal oxygen aging will weaken the ductility and low-temperature property of compound modified asphalt.

4.2. Analysis of Viscosity-Temperature Characteristics Results

The viscosity experiment results of seven groups of samples at 135 °C and 175 °C are summarized in Figure 4. Through analysis, adding SBS to the base asphalt at 135 °C can increase the viscosity by 1.4 times, and the viscosity is increased by 3.8 times at 175 °C. At 135 °C, the viscosity of each compound modified asphalt group increased by 34%, 73.7%, 134.2%, 166.7%, and 179.8%, respectively, compared with SBS modified asphalt. However, when PPA dosage exceeds 1% to 1.25%, the viscosity exceeds 3 Pa·s, which does not meet the construction requirements. At 175 °C, the viscosity of the seven groups of asphalt has decreased by no less than 50%, but the viscosity after adding PPA is always greater than that without PPA. This is because the viscosity is mainly determined by the asphaltene content; adding PPA will increase the asphaltene content—which is mainly converted from the resin—and further enhance the high-temperature property of asphalt. It proves that adding PPA can increase the shear deformation resistance of asphalt mixture.

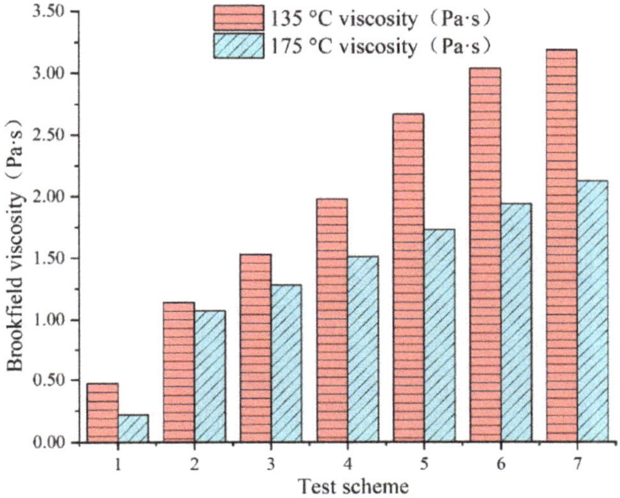

Figure 4. Brookfield viscosity test results.

4.3. *Analysis of DSR Test Results*

4.3.1. Analysis of Temperature Sweep

Figure 5 illustrates the change curves of G*/sinδ and δ concerning temperature in the original group, and Figure 6 presents the change curves of G*/sinδ and δ concerning temperature in the aging group. The analysis shows that:

(1) As the experimental temperature gets higher and higher, the experimental results of all samples in the original group and the aging group are generally that G*/sinδ decreases and δ increases. The main cause of this phenomenon is that the higher the temperature is, the weaker the close force between asphalt molecules is. The asphalt binder presents a viscous fluid at high temperature, which leads to a decrease in its recovery deformation ability and a poor anti-deformation ability.

(2) Regardless of aging or not, the G*/sinδ of each composite modified asphalt group increases with increasing PPA content, suggesting the viscosity is further strengthened by adding PPA. At the same test conditions, when the content of PPA is 0.75% and 1%, the G*/sinδ value increases significantly, and the absolute value of slope reaches more than 0.5, indicating that the best dosage range of PPA in composite modified asphalt is 0.75~1%.

(3) Regardless of aging or not, the δ of each group has a significant downward trend with increasing PPA dosage. When the PPA dosage is 0.75%, the phase angle is significantly smaller than that of others, suggesting that under high temperature load, when the PPA dosage is 0.75%, the improvement degree of elastic element in the composite modified asphalt is the largest, closest to the elastomer, and is easier to recover after load deformation.

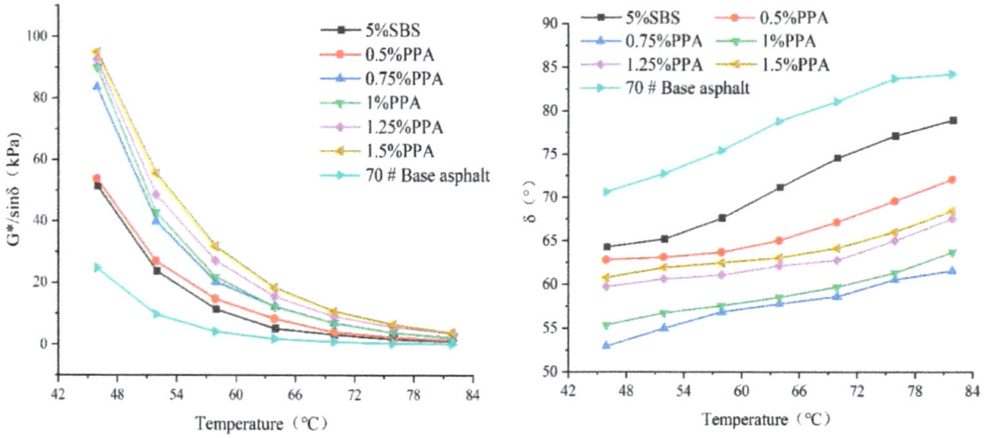

Figure 5. Rutting factor and phase angle-temperature change curve of the original group.

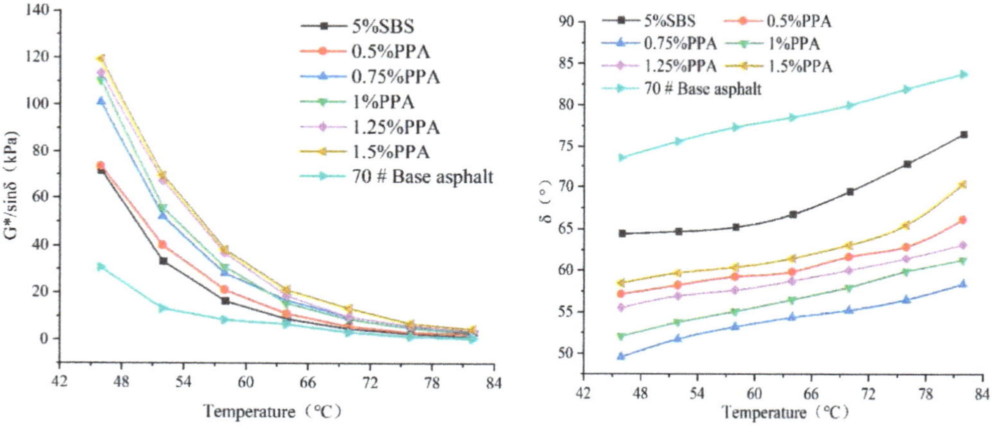

Figure 6. Rutting factor and phase angle-temperature change curve of the aging group.

4.3.2. Analysis of Frequency Scanning

Figure 7 illustrates the logarithmic curve of the change about G*/sinδ in the six groups of modified asphalt in the original group and the aging group at 52 °C and 76 °C. The analysis shows that:

(1) The G*/sinδ values in the original sample group and the aging group generally increase in parallel with the change curve of frequency, and the logarithm of the rutting factor has a good linear correlation with the logarithm of frequency, illustrating that the influence of each modified asphalt group on frequency is basically the same. The increase of G*/sinδ value in the aging group is larger than that in the original group, and the logarithmic value of G*/sinδ in each group gradually approaches with increasing frequency, suggesting the high-temperature performance has changed significantly as a result of aging. The reason why G*/sinδ in the aging group is higher than that in the original sample group is that under the action of aging, asphalt lightweight component volatilizes, asphaltene increases, elasticity of the asphalt gets stronger, and the load resistance and deformation resistance are strengthened.

(2) With regard to the G*/sinδ value, the composite modified asphalt of the original group and the aging group is always greater than the SBS modified asphalt in all

frequency ranges. Especially for 0.75% SBS/PPA composite modified asphalt, no matter how much the temperature is, the G*/sinδ value corresponding to this dosage is the highest. This is mainly because adding PPA can accelerate the change of asphalt structure to gel type more quickly, and the elastic behavior is more obvious when the aging effect is applied to the same degree.

(3) At 76 °C, the G*/sinδ value of the compound modified asphalt with 0.75% content increases by 40.4% after aging, while the measured G*/sinδ values of the SBS modified asphalt original group and the aging group are relatively low, and the G*/sinδ value increases by 27.7% after aging. The effect of aging reduces the volatilization of the light components, which results in a lower viscosity and a hardening of the asphalt binder. However, aging will also destroy the original stable spatial structure that was initially formed by SBS decomposition, so its high-temperature rutting resistance is worse than that of composite modified asphalt.

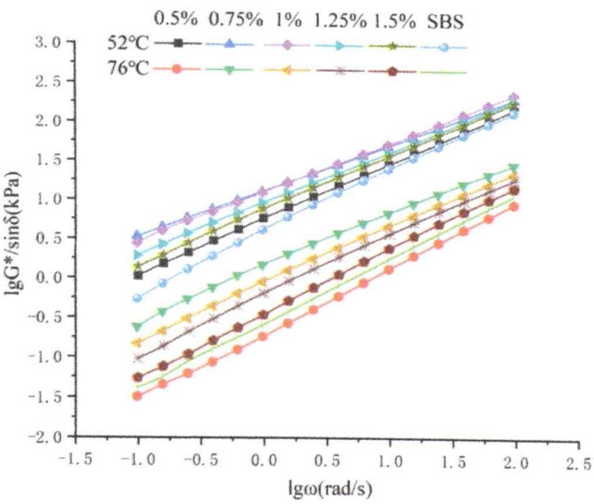

(**a**) The original group.

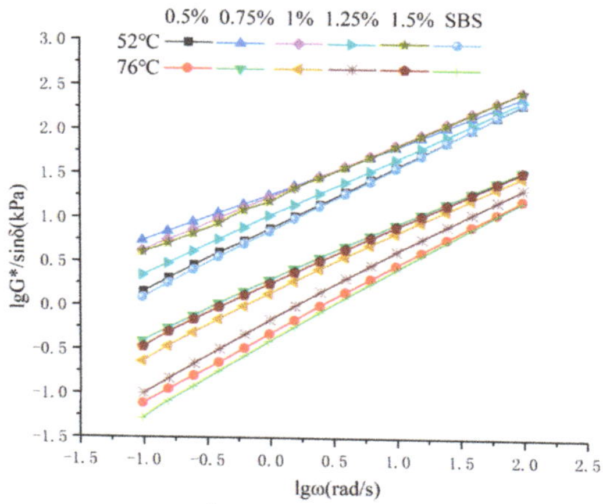

(**b**) The aging group.

Figure 7. Logarithmic curves of the rutting factor and frequency of modified asphalt.

4.4. Analysis of BBR Test Results

Figures 8 and 9 show the variation curves of stiffness modulus S and creep rate m with respect to temperature in the original group and the aging group, respectively. The analysis shows that:

(1) Before aging, the lower the temperature of the original sample group is, the larger the S is, and the smaller the m is. Comparing the S value and m value at the same temperature, the base asphalt has the highest S value and the minimum m value. The S value and m value of each composite modified asphalt group are all below the SBS modified asphalt, indicating that after adding PPA, its deformation resistance and stress relaxation property have different degrees of decline, but compared with the matrix asphalt, the low-temperature creep ability of the modified group is still better. By comparing the different PPA content of compound modified asphalt, the S value is increased and the m value is reduced as the PPA content is increased, indicating that PPA will weaken the low temperature creep property of SBS modified asphalt, and that the degree of low temperature variation is positively correlated with PPA content. Therefore, the low-temperature property of SBS modified asphalt in the original sample group is the best.

(2) After aging, the variation tendency in asphalt with temperature is consistent with that in the original group. The S value of the compound modified asphalt in the interval of −18~−24 °C is the same as that in the original sample group. The slope is significantly larger, and the S value of all asphalts in this temperature range increases the fastest, indicating that adding PPA will reduce its low-temperature performance regardless

of aging or not. The aging group asphalt is still hard in this temperature range, and the load acting on it makes it more easily damaged by load deformation. The increment of S value becomes smaller after aging, indicating that adding PPA could weaken the aging effect and further improve the aging resistance, but it will have a reverse effect on the low-temperature creep property.

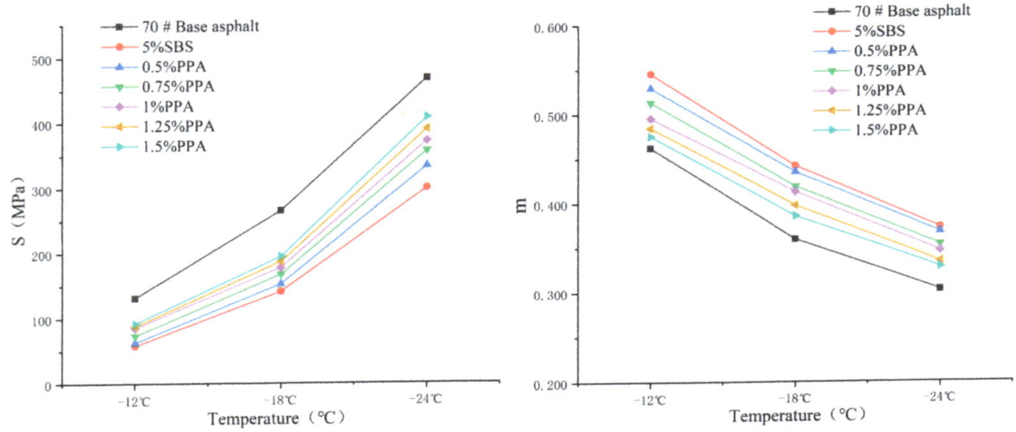

Figure 8. Stiffness modulus and creep rate of the original group.

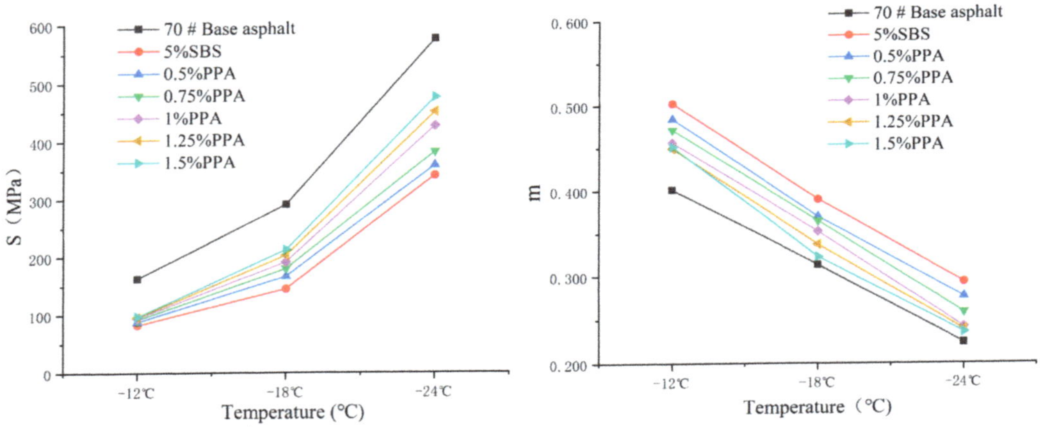

Figure 9. Stiffness modulus and creep rate of aging asphalt.

4.5. Determination of Optimum Dosage Combination

Based on the basic property test, it is found that adding PPA can decrease the penetration and ductility of SBS modified asphalt, increase the softening point, and reduce mass loss after aging. It shows that the more PPA content there is, the more obvious the improvement effect on high-temperature properties, thermal oxidation resistance, and aging resistance of SBS modified asphalt, while it is negatively correlated with the low-temperature anti-cracking performance. The Brookfield viscosity test indicates that the more PPA content there is, the higher the viscosity increases, but it should not exceed 1%. The high-temperature rheological test shows that the improvement effect is the best when 0.75% PPA is added. The low-temperature creep test shows that the higher the PPA content is, the more unfavorable the low-temperature ductility is. Therefore, taking into account the properties of the high and low temperature, viscosity temperature characteristics as well as aging resistance of composite modified asphalt, the optimum amount of PPA is recommended to be 0.75%.

4.6. Analysis of FTIR Test

Figure 10 is the infrared spectrum of 70# base asphalt. The small functional group absorption peak within 1000 cm^{-1} is due to the presence of aromatics in the asphalt, and the bending vibration of the benzene ring C-H bond is generated. At 1028 cm^{-1}, the energy fluctuation is because of the extensional vibration of the sulfoxide group S=O in the asphalt molecule. At 1375 cm^{-1}, it is owing to the symmetric extensional vibration of the C-H bond in methyl -CH3-, and at 1457 cm^{-1}, it is for the sake of the antisymmetric extensional vibration, in which the antisymmetric stretching vibration is stronger. The absorption peak transmittance at 1616 cm^{-1} is because the vibration of the toluene C=C double bond skeleton is low. Two absorption peaks with strong transmittance are generated at 2917 cm^{-1} and 2847 cm^{-1}. The reason is that the methylene group has antisymmetric and symmetric stretching vibrations, and the vibration amplitude generated by antisymmetric is larger than the symmetry. In summary, the constituent of asphalt is complicated and is a hydrocarbon containing a variety of hydrocarbons.

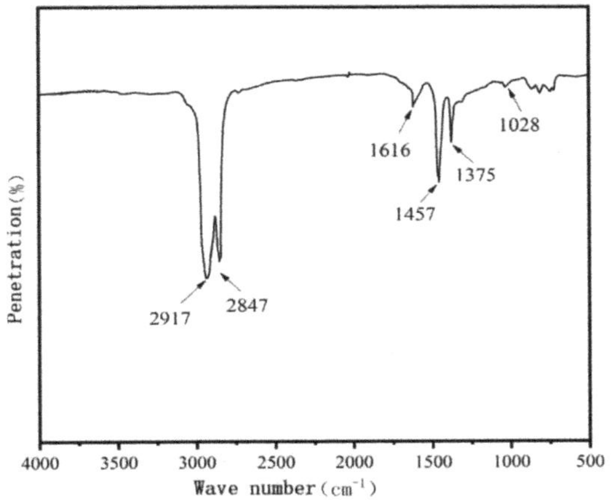

Figure 10. Infrared spectrogram of base asphalt.

By analyzing the infrared spectra of each group of modified asphalt shown in Figure 11, it can be seen that:

(1) After the addition of SBS, the vibration of the C-H bond in the benzene ring of the polystyrene block in SBS causes an absorption peak at 697 cm^{-1}, and the -C=C- double bond bending vibration of the butadiene block causes an absorption peak at 964 cm^{-1}. In the region of 1000~2000 cm^{-1}, the absorption peaks at 1027 cm^{-1}, 1457 cm^{-1}, and 1610 cm^{-1} are simply superimposed compared with the base asphalt, so SBS and base asphalt only produce physical changes.

(2) Compared with modified asphalt under different PPA dosages, the absorption peaks of SBS modified asphalt after adding PPA are mainly concentrated in the above five listed: 697 cm^{-1}, 964 cm^{-1}, 1027 cm^{-1}, 1457 cm^{-1}, and 1610 cm^{-1}. Among them, the absorption peak intensity at 964 cm^{-1} is weakened because the addition of PPA will undergo a reduction reaction to open the -C=C- double bond of SBS, further promoting the copolymerization reaction between asphalt and SBS. The greater the content, the better the compatibility, and the stronger the high-temperature stability.

(3) Because SBS is an immiscible system, the infrared spectrum of the SBS is only a simple superposition of the infrared spectrum of polystyrene and polybutadiene. The position and intensity of the absorption peak are basically unchanged, and no new

absorption peak appears. The C=C double bond bending vibration of the butadiene block in the SBS modifier causes an absorption peak at 964 cm^{-1}, and the C-H bond out-of-plane bending vibration in the benzene ring of the polystyrene block in the SBS modifier causes an absorption peak at 697 cm^{-1}.

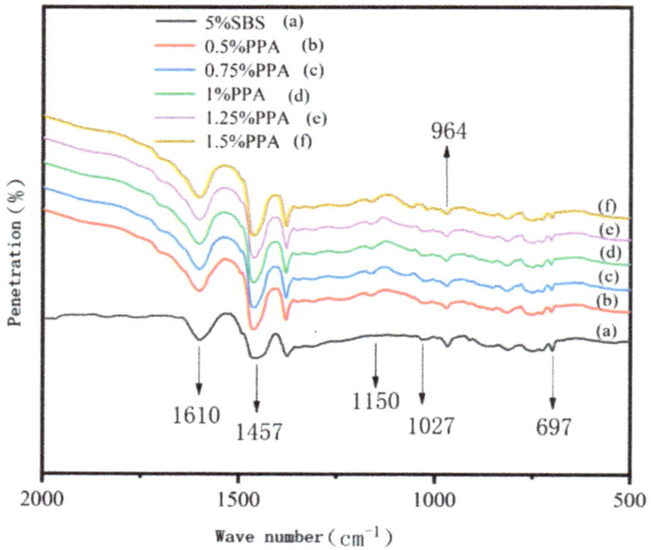

Figure 11. Infrared spectrogram of composite modified asphalt.

4.7. Analysis of SEM Test

4.7.1. Microstructure of Asphalt

The base asphalt is magnified 500 times to obtain Figure 12. It can be seen that there are some textures on the surface of the base asphalt, but no other substances or particles appear, indicating that the clean base asphalt is selected in this study.

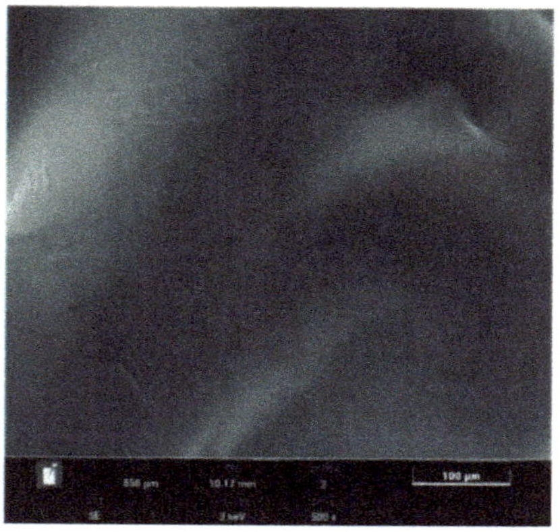

Figure 12. Base asphalt scanning electron micrograph.

4.7.2. Microstructure of SBS Modified Asphalt

As shown in Figure 13, the microstructure of SBS modified asphalt has changed greatly after being enlarged by 2000 times. The SBS modifier is basically dissolved in asphalt due to shear action, forming a more uniform distribution of wavy mesh space structure. When the magnification is 5000 times, it is observed that there are still stacked SBS particles in the lower right corner area and the particle size is different, indicating that there is an issue of poor compatibility between SBS and asphalt. The SBS modifier is crosslinked with asphalt in the state of elastic particles, so the surface is uneven [35].

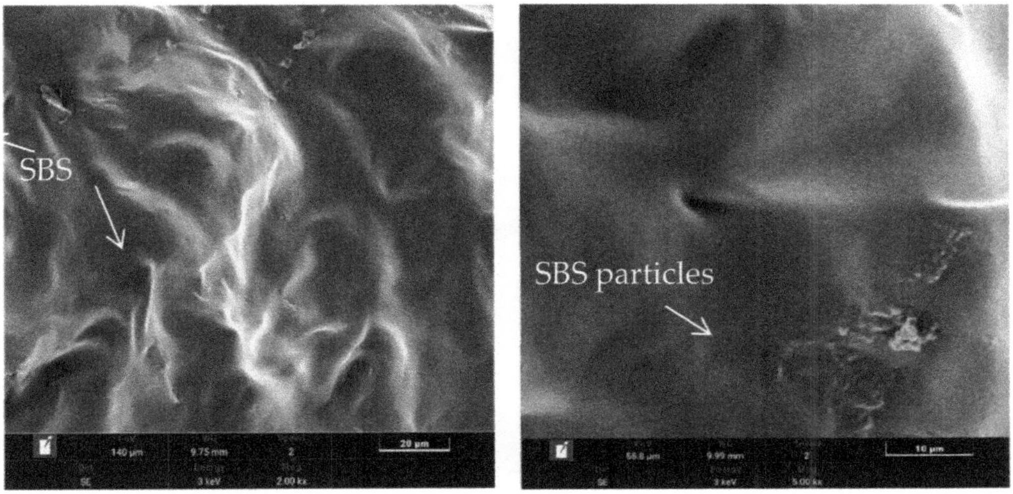

(a) (b)

Figure 13. SBS modified asphalt scanning electron micrographs. (**a**) Amplified 2000 times; (**b**) Amplified 5000 times.

4.7.3. Microstructure of SBS/PPA Composite Modified Asphalt

Figure 14a–j are the SEM images of compound modified asphalt amplified by 5000 times and 10,000 times under different dosages. Figure 14a,b show the compound modified asphalt with 0.5% PPA. It can be seen that the number of elastic particle protrusions formed by the SBS modifier is significantly reduced, forming uneven surface morphology. After 10,000 times magnification, it can still be seen that there are dispersed SBS particles on the surface of compound modified asphalt. Figure 14c,d are composite modified asphalt with 0.75% PPA content. When the magnification is 5000 times, there are still some SBS particles in the upper left corner of the picture. When the region is further enlarged, a smooth plane can be seen in the region, indicating that the increase of PPA dosage makes SBS particles more compatible with asphalt. The PPA content in Figure 14e,f is 1%. The graininess in the figure is obviously weakened, and the distribution is more scattered. After 10,000 times magnification, the smooth area increases significantly, and occasionally SBS particles exist. The PPA content in Figure 14g,h is 1.25%. The microstructure of composite modified asphalt with this content basically shows a slight wrinkle mechanism. The compatibility between SBS particles and asphalt is good, and there is almost no obvious particle bulge. The PPA content in Figure 14i,j is 1.5%. At this time, the microscopic image is smoother, there are many clear folds, and the SBS particles in the composite modified asphalt have completely disappeared.

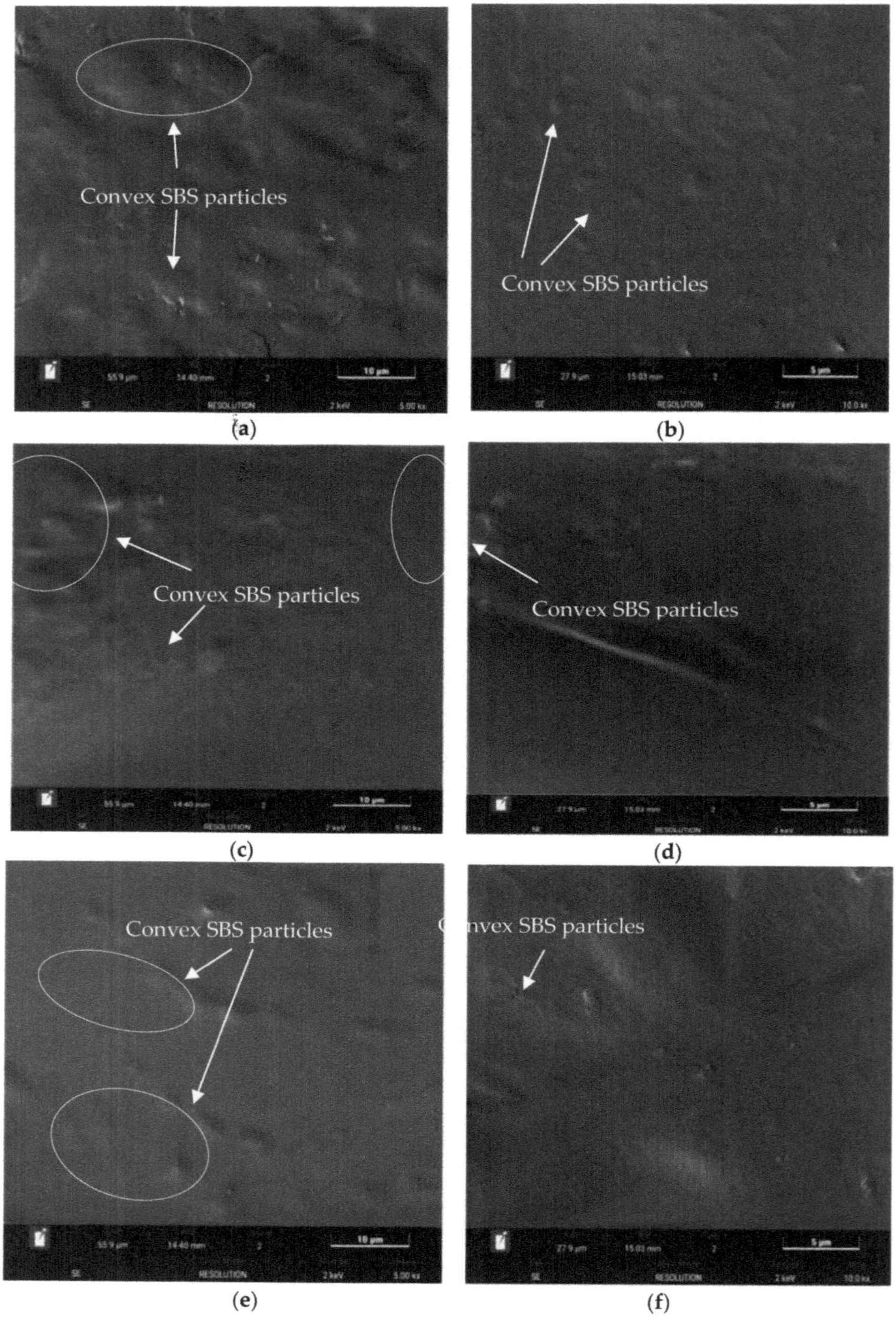

Figure 14. *Cont.*

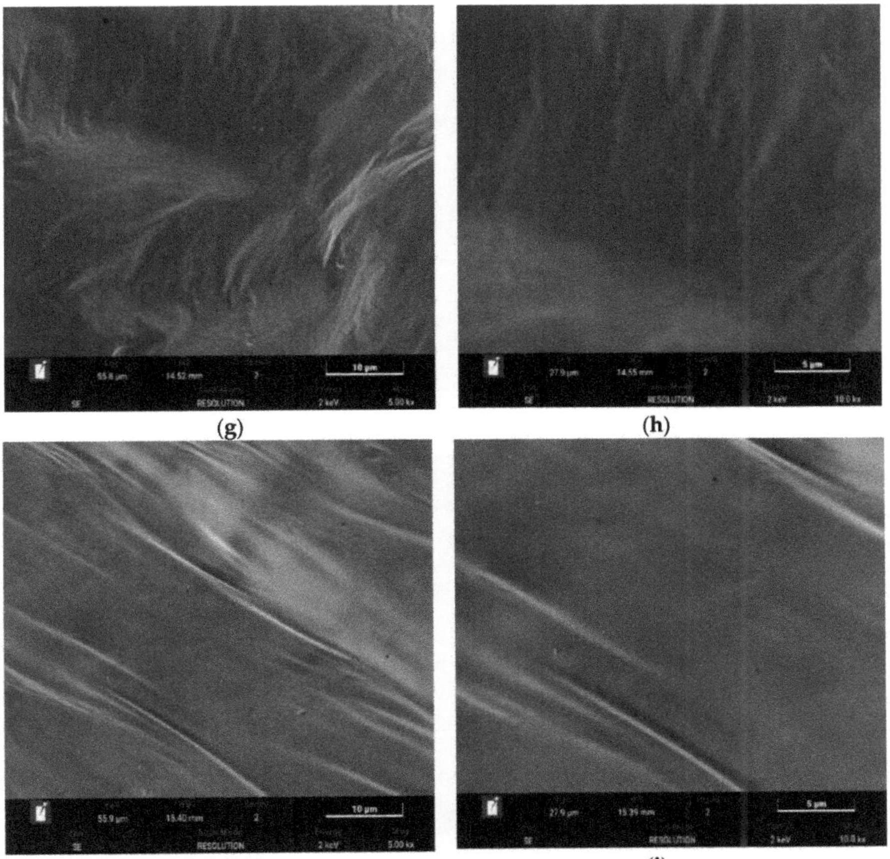

(g) (h) (i) (j)

Figure 14. SBS/PPA composite modified asphalt scanning electron micrographs. (**a**) 0.5% PPA magnified 5000 times; (**b**) 0.5% PPA magnified 10,000 times; (**c**) 0.75% PPA magnified 5000 times; (**d**) 0.75% PPA magnified 10,000 times; (**e**) 1% PPA magnified 5000 times; (**f**) 1% PPA magnified 10,000 times; (**g**) 1.25% PPA magnified 5000 times; (**h**) 1.25% PPA magnified 10,000 times; (**i**) 1.5% PPA magnified 5000 times; (**j**) 1.5% PPA magnified 10,000 times.

As the increment of PPA dosage increases from 0.5% to 1.5%, the microstructure is basically smooth from the initial uniform bulge to the final smooth state, and the SBS particles are completely dissolved in the matrix asphalt. From the microscopic mechanism analysis, the surface of the cementing material is smooth because adding PPA will reduce the size and number of SBS particles. Therefore, the uniformity distribution of asphalt is improved, and a more stable phase structure is formed. The swelling adsorption between the light component (mainly oil) and SBS is enhanced by the rise of the dosage. Meanwhile, the addition of PPA has a chemical reaction, and the asphaltenes are increased; therefore, when the content is 1.25% and 1.5%, a large number of uniform folds will be distributed in the image, which is also the reason why adding PPA in SBS modified asphalt further improves the high-temperature stability.

5. Conclusions

(1) The softening point and viscosity of SBS/PPA composite modified asphalt increase with the increase of content, and the penetration decreases. The change range of resid-

ual penetration ratio, mass change rate, and softening point increment of aged composite modified asphalt are smaller than that of SBS modified asphalt, indicating that the addition of PPA can improve the thermal oxidation resistance, high-temperature performance, and anti-aging performance of SBS modified asphalt. After adding PPA, the ductility of composite modified asphalt decreases, and the residual ductility ratio increases after aging, indicating that PPA will weaken the low-temperature crack resistance. At 135 °C, the average viscosity of composite modified asphalt after PPA addition is 2.2 times that of SBS modified asphalt, indicating that PPA can improve the viscosity-temperature characteristics, but once the PPA content exceeds 1%, it will not meet the construction requirements due to high viscosity.

(2) Under the same test temperature, when the PPA dosage is between 0.75% and 1%, the anti-rutting factor increases significantly, indicating the composite modified asphalt in this range is less sensitive to temperature, resulting in better plastic deformation resistance than other test groups. When the PPA content is 0.75%, the phase angle is the lowest. In summary, adding PPA will increase the elastic element in SBS modified asphalt and exert better high-temperature rheological properties. The S value is increased while the m value is reduced as the PPA content is increased. At −12 °C, as PPA dosage increases, the S value of aged composite modified asphalt is reduced from 40.2% to 5.3% compared with that before aging, and the creep rate m is reduced from 8.5% to 5.1%, suggesting that the aging resistance of the composite modified asphalt is improved, but adding PPA will make the low-temperature creep worse. Therefore, considering the above indexes of composite modified asphalt, the optimal PPA dosage in composite modified asphalt is 0.75%.

(3) The physical effect of SBS is mainly manifested in the bending vibration of the C-H bond of the benzene ring and the bending vibration of the butadiene block C=C, resulting in new absorption peaks at 697 cm^{-1} and 964 cm^{-1}, while the remaining positions are simply superimposed with asphalt. The incorporation of PPA has undergone chemical modification. The reduction reaction weakens the absorption peak intensity at 964 cm^{-1}, new absorption peaks are produced from 1027 to 1150 cm^{-1}, and the absorption peak after 1610 cm^{-1} continues to rise. The microstructure of SBS modified asphalt shows a more uniform distribution of wavy mesh space structure, and there are still stacked SBS particles with different particle sizes, indicating that the combination of SBS and asphalt is difficult. The microstructure of composite modified asphalt gradually forms a smooth wrinkle state with an increasing PPA dosage, and SBS particles basically disappear in composite modified asphalt. PPA could improve the segregation state of SBS in asphalt, and the composite modified asphalt is existed in a more stable phase structure. The results indicate that adding PPA into low-content SBS modified asphalt could shape a more stable network structure, and the macro properties are the high-temperature properties, the viscosity-temperature characteristics, and the aging resistance of the compound modified asphalt.

Author Contributions: Writing—original draft preparation, C.L., Z.L. and Q.L.; writing—review and editing, C.L., T.G., Y.C. and Q.L.; investigation, C.L., J.W. and L.J.; discussion, T.G., Y.C. and Q.L. All authors have read and agreed to the published version of the manuscript.

Funding: This work was supported by Key R&D and Promotion of Special Scientific and Technological Research Projects of Henan Province (Grant Number: 182102210061).

Institutional Review Board Statement: Not applicable.

Informed Consent Statement: Not applicable.

Data Availability Statement: Some or all data, models, or code that support the findings of this study are available from the corresponding author upon reasonable request.

Conflicts of Interest: Author Chaojie Li was employed by the Henan Transportation Research Institute Co., Ltd. and Xi'an Changda Highway Engineering Inspection Center Co., Ltd. The remaining authors declare that the research was conducted in the absence of any commercial or financial relationships that could be construed as a potential conflict of interest.

References

1. Zhou, L.; Huang, W.; Xiao, F.; Lv, Q. Shear adhesion evaluation of various modified asphalt binders by an innovative testing method. *Constr. Build. Mater.* **2018**, *183*, 253–263. [CrossRef]
2. Lu, Y.Y.; Yu, L.M.; Fang, J.; Yun, Q.; Lu, J. Research development of polymer modified asphalt. *New Chem. Mater.* **2020**, *48*, 222–225+230. [CrossRef]
3. Zhang, J.B. Application of SBS modified asphalt in highway pavement. *TranspoWorld* **2021**, *12*, 62–63.
4. Min, X.B.; Ni, Y.L.; Zhang, J. Analysis on the stability of on-site storage SBS modified asphalt. *Pet. Asph.* **2020**, *34*, 34–40.
5. Jiang, Z.X. Research summary of SBS modified asphalt index system. *Highway* **2019**, *64*, 272–277.
6. Ma, Z.M.; Liang, J.L.; Wang, X.R.; Liu, X.; Lu, J.Y. Simulation and Optimization of Three-Step lon Polymerization Process for SBS. *J. Petrochem. Univ.* **2021**, *34*, 22–28.
7. Cortizo, M.S.; Larsen, D.O.; Bianchetto, H.; Alessandrini, J.L. Effect of the thermal degradation of SBS copolymers during the ageing of modified asphalts. *Polym. Degrad. Stab.* **2004**, *86*, 275–282. [CrossRef]
8. Yu, L.M.; Chen, Z.G. Stability analysis of waste rubber powder and SBS composite modified asphalt. *Highway* **2019**, *64*, 267–270.
9. Dong, Z.J.; Zhou, T.; Luan, H.; Yang, C.; Wang, P.; Leng, Z. Composite Modification Technology and Mechanism of SH Blended Bio-asphalt by Combining SBS with Crumb Rubber. *China J. Highw. Transp.* **2019**, *32*, 215–225.
10. Chen, Q.; Li, Y.; Wang, C.; Feng, L.; Zhang, Z. Degradation of mechanical properties of polyurethane elastomer coatings under different pavement pollution conditions. *Constr. Build. Mater.* **2023**, *409*, 134181. [CrossRef]
11. Yang, X.; Liu, G.; Rong, H.; Meng, Y.; Peng, C.; Pan, M.; Wang, G. Investigation on mechanism and rheological properties of Bio-asphalt/PPA/SBS modified asphalt. *Constr. Build. Mater.* **2022**, *347*, 128599. [CrossRef]
12. Wei, J.; Shi, S.; Zhou, Y.; Chen, Z.; Yu, F.; Peng, Z.; Duan, X. Research on performance of SBS-PPA and SBR-PPA compound modified asphalts. *Materials* **2022**, *15*, 2112. [CrossRef] [PubMed]
13. Fu, Z.; Tang, Y.J.; Sun, Q.Y.; Wang, Y.J.; Liu, S.R. Pavement performance and compatibility of carbon nanotube/SBS composite modified asphalt. *Appl. Chem. Ind.* **2021**, *50*, 2031–2036.
14. Shafabakhsh, G.; Rajabi, M.; Sahaf, A. The fatigue behavior of SBS/nanosilica composite modified asphalt binder and mixture. *Constr. Build. Mater.* **2019**, *229*, 116796. [CrossRef]
15. Li, P.F.; Hu, G.F.; Wang, D.W.; Tao, J.X.; Wang, D.X.; Hong, B. Polyurethane-precursor/Styrene-Butadiene-Styrene Composite Modified Asphalt and Its Modification Mechanism. *J. Beijing Univ. Technol.* **2022**, *48*, 655–666.
16. Liu, G.; Zhang, W.; Yang, X.; Ning, Z. Study on the conventional performance and microscopic properties of PPA/SBS-modified bio-mixed asphalt. *Materials* **2022**, *15*, 4101. [CrossRef] [PubMed]
17. Liu, C.; Li, Z.J.; Liu, S.H.; Fu, L. Analysis and prospect of polyphosphoric acid modified asphalt technology. *China Adhes.* **2020**, *29*, 63–67.
18. Jafari, M.; Babazadeh, A. Evaluation of polyphosphoric acid-modified binders using multiple stress creep and recovery and linear amplitude sweep tests. *Road Mater. Pavement Des.* **2016**, *17*, 859–876. [CrossRef]
19. Cuadri, A.A.; Navarro, F.J.; Partal, P. Synergistic ethylcellulose/polyphosphoric acid modification of bitumen for paving applications. *Mater. Struct.* **2020**, *53*, 42–82. [CrossRef]
20. Xiao, F.; Amirkhanian, S.; Wang, H.; Hao, P. Rheological property investigations for polymer and polyphosphoric acid modified asphalt binders at high temperatures. *Constr. Build. Mater.* **2014**, *64*, 316–323. [CrossRef]
21. Yadollahi, G.; Mollahosseini, H.S. Improving the performance of Crumb Rubber bitumen by means of Poly Phosphoric Acid (PPA) and Vestenamer additives. *Constr. Build. Mater.* **2010**, *25*, 3108–3116. [CrossRef]
22. Cao, X.J.; Zhang, Z.; Hao, P.W.; Dong, L.P. Effect of Polyphosphoric Acid on the High-and-low Temperature Property of Matrix Asphalt Mixture. *J. Wuhan Univ. Technol.* **2014**, *36*, 47–53.
23. Wang, Z.X. Research on the Warm Asphalt Binders Modified by Polyphosphoric Acid. Master's Thesis, Hunan University, Changsha, China, 2015.
24. Hou, T.J. Study on preparation and properties of polyphosphoric acid composited polymel. *New Build. Mater.* **2019**, *46*, 8–22.
25. *JTG E20-2011*; Standard Test Methods of Bitumen and Bituminous Mixtures for Highway Engineering. China Communications Press: Beijing, China, 2011.
26. *JTG F40-2004*; Technical Specifications for Construction of Highway Asphalt Pavements. China Communications Press: Beijing, China, 2004.
27. Ma, F.; Li, C.; Fu, Z.; Huang, Y.; Dai, J.; Feng, Q. Evaluation of high temperature rheological performance of polyphosphoric acid-SBS and polyphosphoric acid-crumb rubber modified asphalt. *Constr. Build. Mater.* **2021**, *306*, 124926. [CrossRef]
28. Zhu, W. The Influence of SBS Dosage to the Pavement Performance of Modified Asphalt Mixture. Master's Thesis, Chang'an University, Xi' an, China, 2012.
29. Ma, F.; Wen, Y.L.; Fu, Z.; Feng, Q.; Jin, Y.X.; Guo, X.L. Pavement performance of PPA composite modified asphalt mixture. *Appl. Chem. Ind.* **2021**, *50*, 887–891.

30. Wang, G.Z. Study on Properties of Polyphosphoric Acid Composited Polymer Modified Asphalt and Its Mixture. *Highw. Eng.* **2019**, *44*, 225–231.
31. Zhang, L.; Huang, W.; Wei, M.; Ouyang, X.; Wang, Y. Analysis of Rheological Properties of Polyphosphoric Acid Modified Asphalt. *J. Mater. Sci. Eng.* **2020**, *38*, 638–642.
32. Dou, Q.P.; Shi, Y. Effect of PPA Content on the Performance of Polyphosphate Modified Asphalt. *J. Lanzhou Inst. Technol.* **2020**, *27*, 24–27.
33. Huang, G.; Wang, P.; Liu, B.S. Research on Polyphosphoric Acid Modified Asphalt Performance. *West. China Commun. Sci. Technol.* **2018**, 5–10.
34. Ju, Z.; Ge, D.; Wu, Z.; Xue, Y.; Lv, S.; Li, Y.; Fan, X. The performance evaluation of high content bio-asphalt modified with polyphosphoric acid. *Constr. Build. Mater.* **2022**, *361*, 129593. [CrossRef]
35. Wang, F. Effect of aging time on functional groups of SBS modifier and asphalt binder. *Sci. Technol. Eng.* **2023**, *23*, 13765–13773.

MDPI AG
Grosspeteranlage 5
4052 Basel
Switzerland
Tel.: +41 61 683 77 34

Coatings Editorial Office
E-mail: coatings@mdpi.com
www.mdpi.com/journal/coatings

www.ingramcontent.com/pod-product-compliance
Lightning Source LLC
Chambersburg PA
CBHW061504210926
51850CB00008B/1821
* 9 7 8 3 7 2 5 8 1 7 5 9 7 *